From Darwin to Hitler

From Darwin to Hitler ᔓ

Evolutionary Ethics, Eugenics, and Racism in Germany

Richard Weikart

First published 2004 by
PALGRAVE MACMILLAN™
175 Fifth Avenue, New York, N.Y. 10010 and
Houndmills, Basingstoke, Hampshire, England RG21 6XS
Companies and representatives throughout the world

PALGRAVE MACMILLAN is the global academic imprint of the Palgrave Macmillan division of St. Martin's Press, LLC and of Palgrave Macmillan Ltd. Macmillan® is a registered trademark in the United States, United Kingdom and other countries. Palgrave is a registered trademark in the European Union and other countries.

ISBN 1–4039–6502–1

Library of Congress Cataloging-in-Publication Data
Weikart, Richard, 1958–
    From Darwin to Hitler : evolutionary ethics, eugenics, and racism in Germany / by Richard Weikart.
        p. cm.
    Includes bibliographical references and index.
    ISBN 1–4039–6502–1
        1. Eugenics—Germany—History. 2. Ethics, Evolutionary. 3. Racism—Germany. 4. Germany—Race relations. I. Title.
HQ755.5.G3W435 2004
305.8′00943—dc22   2003065613

A catalogue record for this book is available from the British Library.

Design by Newgen Imaging Systems (P) Ltd, Chennai, India.

First edition: May 2004
10   9   8   7   6   5   4

Printed in the United States of America.

Contents ✍

Illustrations   vii

Preface   ix

Introduction   1

1. Laying New Foundations for Ethics   **19**

1. The Origin of Ethics and the Rise of Moral Relativism   21
2. Evolutionary Progress as the Highest Good   43
3. Organizing Evolutionary Ethics   59

2. Devaluing Human Life   **71**

4. The Value of Life and the Value of Death   73
5. The Specter of Inferiority: Devaluing the Disabled
   and "Unproductive"   89
6. The Science of Racial Inequality   103

3. Eliminating the "Inferior Ones"   **127**

7. Controlling Reproduction: Overturning Traditional
   Sexual Morality   129
8. Killing the "Unfit"   145
9. War and Peace   163
10. Racial Struggle and Extermination   183

4. Impacts   **207**

11. Hitler's Ethic   209

Conclusion   229

Notes   235

Bibliography   279

Index   305

Illustrations ᕲ

1.1   "Ape-like" Italian criminal   39
6.1   Frontispiece of Haeckel's *Natürliche Schöpfungsgeschichte*   107
6.2   *Pithecanthropus alalus*   109
6.3   Cover of *Der Brummer* magazine   125
10.1 Spaniards' dogs mauling Indians   190

Preface ✍

I became fascinated with the topic of evolutionary ethics while doing research for my dissertation, *Socialist Darwinism: Evolution in German Socialist Thought from Marx to Bernstein* (published in 1999). Little did I suspect the course my study would take. While examining Darwinian discourse in Germany, I found that many Darwinists believed that Darwinism had revolutionary implications for ethics and morality, providing a new foundation for ethics and overturning traditional moral codes. Intrigued by these ideas, I intended at first merely to describe and analyze the development of evolutionary ethics in Germany and elsewhere. However, as I began to read the writings of Ernst Haeckel and other early Darwinists, my focus shifted to a specific field of ethics—what is today called biomedical ethics.

One cause for this shift was my study of Ernst Haeckel's works, in which—to my surprise—he advocated infanticide for babies having certain kinds of disabilities. Second, I discovered that many German eugenicists wrote essays and passages in their books on how to apply Darwinism to ethics. I had not initially intended eugenics to play an important part in this study, but I could not avoid it—leaders of the eugenics movement were some of the most prominent promoters of evolutionary ethics.

Last, but certainly not least, James Rachel's book, *Created from Animals: The Moral Implications of Darwinism* (Oxford, 1990), stimulated my thinking. Rachel's argument that Darwinism undermines the sanctity of human life and his support for euthanasia seemed remarkably similar to some of the ideas I encountered in late nineteenth and early twentieth-century Germany. Rachel's book—together with what I already knew about the views of Haeckel, some social Darwinists, and eugenicists—suggested to me a new question to explore: Did German Darwinists use Darwinian theory to undermine the traditionally held doctrine of the sanctity of human life? To pose the question a different way, what did Darwinism—or at least influential proponents of Darwinism—have to say about the value of human life? As I framed the question in this way, other issues relating to life and death emerged naturally, especially war and racial conflict.

As I reformulated my study on evolutionary ethics to include discussions on the value of human life, another topic became inescapable: the influence of this discourse on Hitler. Hitler was not even on my radar screen when I began my research, and Daniel Gasman's one-sided attempt to link Haeckel and Hitler made me wary. However, the more books, articles, and documents I read by Darwinists and eugenicists in the late nineteenth and early twentieth centuries, and the more I read by and about Hitler, the more I became convinced that there were significant historical connections between Darwinism and Hitler's ideology. I will leave it to the reader to decide how straight or twisted the path is from Darwinism to Hitler after reading my account. It is my task now to trace this path from Darwin's time to approximately World War I (in the case of Hitler, I extend the discussion a little later chronologically, since all of Hitler's speeches and writings came after World War I).

Some of the material presented in this book has appeared previously in journal articles: "The Origins of Social Darwinism in Germany, 1859–1895," *Journal of the History of Ideas* 54 (1993): 469–88; "Darwinism and Death: Devaluing Human Life in Germany, 1860–1920," *Journal of the History of Ideas* 63 (2002): 323–44; and "Progress through Racial Extermination: Social Darwinism, Eugenics, and Pacifism in Germany, 1860–1918," *German Studies Review* 26 (2003): 273–94. Thanks to the anonymous readers and to Diethelm Prowe, editor of *German Studies Review*, for their helpful comments on these essays.

I would also like to thank the many other people who made this book possible. First and foremost, California State University, Stanislaus, provided many resources, including a sabbatical and research grants. Crucial for this project was the CSU, Stanislaus, Inter-Library Loan department (thanks, Julie Reuben), without which this study would have been extremely difficult or impossible. My colleagues in the History Department have given constant encouragement and inspiration. Many thanks also to the Center for Science and Culture (especially Jay Richards and Steve Meyer), which provided crucial funding and much encouragement, without which this project would have taken much longer to complete. I also want to thank the Templeton Foundation for their funding of a Faculty Summer Seminar in 2001 on "Biology and Purpose: Altruism, Morality, and Human Nature in Evolutionary Theory," which helped stimulate my thinking.

I also thank the many libraries and archives who allowed me to gain access to the information I needed for my research: University of California, Berkeley Library, Stanford University Library and Archives, Hoover Institution, Staatsbibliothek Preussischer Kulturbesitz (Berlin),

Akademie der Künste Archives (Berlin), Humboldt University Archives (Berlin), Berlin-Brandenburgische Akademie der Wissenschaft Archives (Berlin), Bundesarchiv Koblenz, Bayerische Staatsbibliothek (Munich), Ernst-Haeckel-Haus Archives (Jena), University of Freiburg Archives, University of Zurich Archives, Wiener Land- und Stadtarchiv (Vienna), Österreichische Nationalbibliothek (Vienna), Forschungsstelle und Dokumentationszentrum für österreichische Philosophie (Graz), League of Nations Archives (Geneva), University of Geneva Archives, and the University of Wroclaw Archives. Special thanks to Wilfried Ploetz, who allowed me to examine the papers of his father, Alfred Ploetz, and who was wonderfully hospitable.

I have also benefited from my interaction with many colleagues, who have contributed immensely to my intellectual development and without whom this project would have been impossible. I would especially like to thank Mitch Ash and Allan Megill for grounding me in German intellectual history and the history of science. I'm especially grateful to Edward Ross Dickinson for reading part of the manuscript and making suggestions for improvement, as well as his input at conferences and via e-mail exchanges. Many others—too numerous to name—provided input at conferences, through e-mail, or especially through their books and articles. Whatever faults remain are mine alone, but any merit in this work owes much to these and other scholars, most of whose names may be found in my bibliography.

My editor, Brendan O'Malley, did a terrific job. He kept me well-informed at each stage of the review process and answered all my concerns in a timely fashion. Many thanks to him for all this.

Finally, I would like to thank my parents, Ray and Lois, for their support and encouragement in ways far too numerous to list. I dedicate this book to my wife, Lisa, and my six precious children: Joy, John, Joseph, Miriam, Christine, and Hannah. They brought me great joy while I was working on this project and were a constant reminder of the immense value of human life.

Introduction ↬

Controversy began raging immediately after Darwin's *Origin of Species* appeared in 1859, and the dispute was not only about whether organisms arose through supernatural or natural means. Many of Darwin's contemporaries found the moral implications of his theory unsettling, despite the fact that he did not publicly discuss human evolution or its implications for morality until 1871 in *The Descent of Man*. A good deal of the initial resistance to Darwinism sprang from a perceived threat to the moral order. Adam Sedgwick, Darwin's former mentor in natural science at the University of Cambridge, expressed this fear poignantly in a letter to Darwin in 1859, shortly after reading *The Origin of Species*. He stated, "Passages in your book . . . greatly shocked my moral taste." He further explained,

> There is a moral or metaphysical part of nature as well as a physical. A man who denies this is deep in the mire of folly. Tis the crown and glory of organic science that it does, thro' final cause, link material to moral; . . . You have ignored this link; and, if I do not mistake your meaning, you have done your best in one or two pregnant cases to break it. Were it possible (which, thank God, it is not) to break it, humanity, in my mind, would suffer a damage that might brutalize it, and sink the human race into a lower grade of degradation than any into which it has fallen since its written records tell us of its history.[1]

Sedgwick was not the only one to accuse Darwin of undermining morality. William Jennings Bryan's campaign against Darwinism in early twentieth-century America was largely motivated by concern over the moral implications of Darwinism. As a pacifist, Bryan was outraged by the Darwinian rhetoric of German militarists, whom he held responsible for the outbreak of World War I. Horrified by the wanton slaughter of supposedly civilized nations, he agreed with Sedgwick that Darwinism had a brutalizing and degrading effect on people. Germans also expressed concern about the moral implications of Darwinism. A Protestant pastor,

Rudolf Schmid, reported in his 1876 book that many critics of Darwinism view it as "only an unproven hypothesis that threatens to become a torch, which could reduce the most noble and highest cultural achievements of the past century to a heap of ashes."[2]

Creationists still routinely accuse Darwinism of undermining morality, while, on the opposite side of the spectrum, some Darwinists today exult in the moral liberation of Darwinism. Daniel Dennett, a leading materialist philosopher, extols *Darwin's Dangerous Idea*, which he calls a "universal acid," dissolving traditional ideas about religion and morality. The famous bioethicist Peter Singer and his compatriot James Rachels argue that because Darwinism effectively discredits the Judeo-Christian conception of the sanctity of human life, therefore abortion, euthanasia, and infanticide can be morally justified. While Singer and Rachels may see this as morally liberating, I suspect that Sedgwick would have considered their views a dramatic confirmation of his prediction about the brutalizing tendencies of Darwinism.

Indeed, Sedgwick did not have to wait very long to have his fears confirmed. Many Darwinists in the late nineteenth century began applying Darwinism to ethical issues, including questions about the value of human life. Robby Kossmann, a German zoologist who later became a medical professor, was perhaps more forthright than most when in an 1880 essay, "The Importance of the Life of an Individual in the Darwinian World View," he declared

> that the Darwinian world view must look upon the present sentimental conception of the value of the life of a human individual as an overestimate completely hindering the progress of humanity. The human state also, like every animal community of individuals, must reach an even higher level of perfection, if the possibility exists in it, *through the destruction of the less well-endowed individual,* for the more excellently endowed to win space for the expansion of its progeny. . . . The state only has an interest in preserving the more excellent life *at the expense of the less excellent.*[3]

Kossmann's views on life and death were shocking and provocative for his time, but, as we shall see, many of his fellow Darwinists expressed similar ideas.

By the early twentieth century, ideas like Kossmann's had spread widely, especially as the Darwinian-inspired eugenics movement—defining itself as the science of improving human heredity—blossomed. Not all eugenicists agreed, however, on where to focus their efforts. Who fit into Kossmann's categories of "less well-endowed" or "less excellent"? Whose life was less valuable or—to use the term constantly bandied about by eugenicists—"inferior"? The question itself and the presuppositions behind it are, in my view, pernicious, but social Darwinists and eugenicists were fearful that various aspects of modern civilization contributed to biological degeneration. Their campaign to combat the dreaded degeneration focussed on two groups allegedly threatening the health and vitality of the human species: the disabled and those of non-European races. Though sometimes disagreeing on which group posed the greater danger, many—probably most—eugenicists regarded both the disabled and non-European races (sometimes even non-Germanic Europeans) as inferior and favored measures to eliminate them in some way, either now or in the future.

Among those embracing social Darwinism and a racist version of eugenics was an Austrian-born German politician, whose name—Hitler—immediately conjures up images of evil and death. Since Hitler is the epitome of wickedness, while Darwin is generally held in high esteem, any link between them immediately arouses incredulity, and with good reason. Obviously, Darwin was no Hitler. The contrast between the personal lives and dispositions of these two men could hardly be greater. Darwin eschewed politics, retreating to his country home in Down for solitude to conduct biological research and to write. Hitler as a demagogue lived and breathed politics, stirring the passions of crowds through frenzied speeches. Politically Darwin was a typical English liberal, supporting laissez-faire economics and opposing slavery. Like most of his contemporaries, Darwin considered non-European races inferior to Europeans, but he never embraced Aryan racism or rabid anti-Semitism, central features of Hitler's political philosophy.

So, what are the connections between Darwinism and Hitler and are they really all that significant? Perhaps we should pose the question this way: Did Hitler hijack Darwinism and hold it hostage to his own malevolent political philosophy, or did he merely climb on board and follow it to its destination? The latter view might be oversimplified as follows: First, Darwinism undermined traditional morality and the value of human life. Then, evolutionary progress became the new moral imperative. This aided the advance of eugenics, which was overtly founded on Darwinian principles. Some eugenicists began advocating euthanasia and infanticide for the disabled. On a parallel track, some prominent Darwinists argued that human racial competition and war is part of the Darwinian struggle for existence. Hitler imbibed these social Darwinist ideas, blended in virulent anti-Semitism, and—there you have it: Holocaust.

Many scholars have, in fact, argued for the importance of Darwinism—or at least social Darwinism—in preparing the ground for Nazi ideology and the Holocaust.[4] In his study on the influence of social Darwinism on the formulation of Nazi ideology, Hans-Günther Zmarzlik wrote that "an analysis of Social Darwinism reveals a process of declining standards, accompanied by a tendency to sacrifice the individual to the species, to devaluate the humanitarian idea of equality from the standpoint of a 'natural' inequality, to subordinate ethical norms to biological needs."[5] Richard J. Evans has recently defended Zmarzlik's position against revisionists who downplay the significance of social Darwinism in helping produce Nazi ideology.[6]

The opposing view—that Hitler hijacked Darwinism—has significant supporting arguments, for many scholars have pointed out that Darwinism did not lead to any one particular political philosophy or practice. Social Democrats with impeccable Marxist credentials were enthusiastic about Darwinism and even considered it a corroboration of their own worldview. After reading Darwin's *Origin of Species*, Karl Marx wrote to Friedrich Engels, "Although developed in a coarse English manner, this is the book that contains the foundation in natural history for our view."[7] Furthermore, many pacifists, feminists, birth control advocates, and homosexual rights activists—some of whom were persecuted and even killed by the Nazis—were enthusiastic Darwinists and used Darwinian arguments to support their political and social agendas. Eugenics discourse was commonplace all across the political spectrum, causing the historian Atina Grossmann to convincingly argue that the path from eugenics and sex reform to Nazism was "a convoluted and highly contested route."[8] Nazism was not predetermined in Darwinism or eugenics, not even in racist forms of eugenics.

The multivalence of Darwinism and eugenics ideology, especially when applied to ethical, political, and social thought, together with the multiple roots of Nazi ideology, should make us suspicious of monocausal arguments about the origins of the Nazi worldview. The Jewish historian Steven Aschheim, however, has rightly warned that, despite the daunting complexity of the task, we should never cease trying to trace the intellectual influences on the Nazis. Just because Darwinism does not lead inevitably to Nazism does not mean that we can strike Darwinism off the list of influences that helped produce Hitler's worldview and thus paved the way to the Holocaust. Aschheim states,

> The path from Darwinism, Wagnerism, Nietzscheanism, and even racism and anti-Semitism to Nazis, it is clear, was never simple or direct. Different roads did, of course, lead in different directions. Nevertheless, twisted though it may have been, one did, in point of fact, lead to Auschwitz.

However great the perils of teleology, they should not blunt our determination to understand the processes and impulses that, at least in one instance, led to this destination. The fear of complexity is a poor reason, I believe, to abandon cultural history.[9]

So, while remaining ever cognizant of the multiple potentialities of Darwinian, eugenic, and racist discourse in the pre-Nazi period, we should not close our eyes to the many similarities and parallels with later Nazi thinking, either. (I would also like to make clear from the outset that, while stressing intellectual history in this work, I recognize the influence of political, social, economic, and other factors in the development of ideologies in general and of Nazism in particular—but these topics are outside the scope of this study.)

Kevin Repp maintains precisely this balance in his fascinating work on German social reformers active around the turn of the twentieth century. Repp's thesis is that the social reform milieu in Wilhelmine Germany—which included many moral reformers and eugenicists (some of whom are prominent in this study)—contained many different possibilities, some benign, some malevolent. There were multiple paths to modernity, and most did not lead in the direction of Nazism. Nonetheless, he acknowledges that some aspects of German social reform in the early twentieth century did contribute to the development of Nazi ideology, and Darwinism played a key role in this: "Confused, distorted, repulsive though it was, however, the Nazi synthesis of Darwinism and national community bore more than a surface resemblance to the discursive terrain Wilhelmine reformers were attempting to reclaim in the vastly altered landscape of Weimar politics."[10] While continually reminding us that the Wilhelmine social reformers were not proto-Nazis, Repp nonetheless does not exonerate them completely, insisting,

> Yet the enthusiasms of Wilhelmine feminists and other reformers for racial hygiene and population policy are also linked to the inhuman brutality of the holocaust, as Greven-Aschoff and Gerhard suggest, since they cast an aura of legitimacy around such concepts, a legitimacy they would otherwise never have enjoyed in many circles, and that far outlived the fleeting historical circumstances that initially evoked those enthusiasms.[11]

Not only eugenics and social reform movements, but German anthropology—a crucial element in our study because of its legitimization of scientific racism on Darwinian grounds—also contained political and moral ambiguities. In his study of German anthropology in the late nineteenth and early twentieth centuries, Andrew Zimmerman points out the "multivalent and contradictory potentials" inherent in that discipline. Nonetheless, he acknowledges that German anthropology "provided a series of practices, theories, and ideologies for some of the greatest evils of human history: colonialist and Nazi genocide."[12] No matter how crooked the road was from Darwin to Hitler, clearly Darwinism and eugenics smoothed the path for Nazi ideology, especially for the Nazi stress on expansion, war, racial struggle, and racial extermination.

One's perspective on how straight or how crooked the path was from Darwinism to Nazism also depends on which aspects of Nazism one is considering, since social Darwinism was only one component—albeit a central one—in Nazi ideology. If one concentrates on anti-Semitism, surely an important part of Hitler's worldview, then there does not seem to be any direct connection between Darwinism and Nazism.[13] Some Jews were avid Darwinists, some were eugenicists, and a few will figure prominently in the pages of this work.[14] Many other aspects of Hitler's worldview and political practice, too—dictatorship, for instance—seem to have little to do with Darwinism. However, if we focus more narrowly on the question of ethics, the value of human life, and racism, as I will do in the succeeding pages, the historical connections appear more significant. Sheila Faith Weiss, after adequately demonstrating the Darwinian roots of eugenics, is probably right when she contends,

> Finally, one might add, to categorize people as "valuable" and "valueless," to view people as little more than variables amenable to manipulation for some "higher end," as Schallmayer and all German eugenicists did, was to embrace an outlook that led, after many twists and turns, to the slave-labor and death camps of Auschwitz.[15]

Let us briefly explore these connections between Darwinism and Hitler before moving on.

First, it is important to understand that on the whole Hitler's view of ethics and morality was by no means conservative or reactionary, despite the fact that some of his specific positions were. For example, some scholars focussing on Hitler's rejection of feminism and abortion have branded Hitler's ideology as reactionary.[16] However, as Michael Burleigh and Wolfgang Wippermann have shown in *The Racial State: Germany, 1933–1945*, Nazi ideology cannot be pigeon-holed so easily, because Hitler opposed feminism and abortion on totally different grounds than did traditional conservatives. He believed that feminism and abortion were biologically deleterious and thus antiscientific (and he only opposed abortion for "Aryans"). Though some of his policies lined up with conservatism,

Hitler saw himself as a revolutionary who would bring advancement and progress to Germany and the world.[17] Burleigh is right to argue that Nazism was "a dystopian attempt to fabricate 'new' men and women by erasing or transforming their 'inherited' ethical values in favor of others derived from a modernized and scientized version of pre–Judeo-Christian conduct. In other words, it was a case of ancient or primitive civilizations put through the refracting mirrors of Darwin and Nietzsche."[18]

Hitler's view of ethics can probably be summed up in the following quotation: "The ethical ideal demands of us, that we place our entire life in its service; and the racial ideal is such that we really can live according to it. With every deed and with every inaction, we have to ask ourselves: does it benefit our race? And then make our decision accordingly." Neither Hitler nor any of his entourage penned this. Rather a leading Darwinian biologist, the geneticist Fritz Lenz, who in 1923 became professor of eugenics at the University of Munich, made this assertion in his 1917 article, "Race as a Principle of Value: Toward Renovating Ethics." In 1933, Lenz boasted that this article "contained all the basic characteristics of the National Socialist world view."[19]

One fundamental component of this Nazi worldview was human inequality, the notion that humans have differing values depending on their biological characteristics. Hitler in *Mein Kampf* expressed this view repeatedly. He explained that his worldview

> by no means believes in the equality of races, but recognizes along with their differences their higher or lower value, and through this knowledge feels obliged, according to the eternal will that rules this universe, to promote the victory of the better, the stronger, and to demand the submission of the worse and weaker. It embraces thereby in principle the aristocratic law of nature and believes in the validity of this law down to the last individual being. It recognizes not only the different value of races, but also the different value of individuals. . . . But by no means can it approve of the right of an ethical idea existing, if this idea is a danger for the racial life of the bearer of a higher ethics.[20]

I will explain this statement in greater detail later (in chapter 11 on Hitler), but for now suffice it to say that Hitler believed that these biologically unequal humans, just as all other organisms, were locked in an eternal, inescapable Darwinian struggle for existence. The strong triumph and the weak perish.

Darwinian terminology and rhetoric pervaded Hitler's writings and speeches, and no one to my knowledge has ever even questioned the common assertion by scholars that Hitler was a social Darwinist. It is too obvious to deny.[21] In order to rescue Darwinian science from the taint of Nazism, however, some historians assert that Hitler's views were pseudo-scientific or eccentric, or they refer to his views of Darwinism as crude or vulgar. In her wonderful book on *Hitler's Vienna*, for example, Brigitte Hamann stated, "Almost all of the theories Hitler preferred have in common that they were not in agreement with academic science but were the products of the idiosyncratic thought processes of private scholars who were full of contempt for established scientists, who hardly ever accepted them either, and for good reason."[22] On the contrary, however, many recent studies about Nazi science, especially those relating to biology, medical science, and eugenics, demonstrate that many mainstream scientists, professors, and physicians—including those identifying with the political Left—upheld views about Darwinism and eugenics quite similar to Hitler's.[23] Fritz Lenz was not the only scientist to recognize the affinity of his views with Nazism. My study will demonstrate that many of Hitler's ideas derived ultimately from respectable scientists and scholars who were grappling with the implications of Darwinism for ethics and society (though Hitler probably imbibed them mostly second or third hand). These included not only prominent scientists and physicians, but also professors of philosophy, economics, and geography.

Even Hitler's social Darwinist views on racial extermination were upheld by prominent Darwinian scientists and social thinkers, though often without reference to Jews. Many Darwinian biologists and social theorists explained that racial extinction was inescapable and even beneficial, for it brought about evolutionary progress for the species as a whole.[24] A prominent Darwinian ethnologist, Oscar Peschel, editor of the journal *Das Ausland*, explained already in 1870—before Darwin published *The Descent of Man* and long before Hitler's birth—that ethics could not oppose the natural process of racial annihilation:

> Everything that we acknowledge as the right of the individual will have to yield to the urgent demands of human society, if it is not in accord with the latter. The decline of the Tasmanians therefore should be viewed as a geological or paleontological fate: the stronger variety supplants the weaker. This extinction is sad in itself, but sadder still is the knowledge, that in this world the physical order treads down the moral order with every confrontation.[25]

Peschel would thus have us believe that nature trumps ethics every time, with science teaching us to resign ourselves to the fact that there are no universal human rights, not even the right to life. No wonder Sedgwick was worried.

Hitler will be discussed at length in chapter 11, where I explain how Hitler drew upon a bountiful fund of social Darwinist thought to construct his own racist philosophy. What really concerns me in this work, though, is not so much Hitler, but Darwinism, especially the implications of Darwinism for ethics and for our understanding of human life and death. When I use the term Darwinism in this study, I mean the theory of evolution through natural selection as advanced by Darwin in *The Origin of Species*. In the late nineteenth century, however, the term Darwinism was often used loosely. Sometimes it meant the idea of biological evolution in general, other times it referred to Darwin's particular theory of natural selection (as I am using it in this work), and elsewhere it meant an entire naturalistic worldview with biological evolution as its centerpiece. Among those who accepted the validity of biological evolution in the late nineteenth century, disagreements arose concerning the mechanism. Many biologists adopted Lamarckism, the now-discredited view that organisms pass on acquired characteristics to their offspring. However, Lamarckian explanations were not contradictory to Darwin's idea of natural selection (even Darwin accepted a measure of Lamarckism), and most German biologists in the nineteenth century followed Ernst Haeckel's lead in synthesizing Darwinism and Lamarckism.

I also need to make clear from the start that this is a historical study. When I draw connections between Darwin, German Darwinists, eugenicists, racial theorists, or militarists, I am not thereby endorsing their logic—I leave it to the reader to decide the logic of their case. Nor am I making the absurd claim that Darwinism of logical necessity leads (directly or indirectly) to Nazism. In philosophical terms, Darwinism was a necessary, but not a sufficient, cause for Nazi ideology. But however logical or illogical the connections are between Darwinism and Nazism, historically the connections are there and they cannot be wished away.

Second, I need to stress that I am restricting my discussion about the impacts of Darwinism to its influence on ethical and social thought, especially on ideas about what is currently called biomedical ethics. I will focus primarily on Darwinian influence on eugenics, euthanasia, racial theory, and militarism in Germany. While these were fundamental features of Nazi ideology, I do not think it serves any purpose to label those holding these views as proto-Nazi, as though Nazism inevitably flowed from their views. Many of the figures I will discuss were liberals of some sort; some were socialists; some were pacifists; some were even Jews. Nevertheless, when we turn the spotlight away from political ideology and focus on ethics, the value of human life, and racial ideology, we often find that Darwinists who were poles apart politically had more in common than we may have suspected at first glance.

Another reason that the Darwinian devaluing of human life should not be treated as proto-Nazi is because similar ideas circulated in the United States, Britain, and other democratic countries. Ian Dowbiggin and Nick Kemp in their fine studies of the history of the euthanasia movement in the United States and Britain, respectively, both emphasize the crucial role played by Darwinism in initiating and ideologically underpinning the euthanasia movement. Dowbiggin states, "The most pivotal turning point in the early history of the euthanasia movement was the coming of Darwinism to America."[26] Kemp strongly supports this point, claiming, "While we should be wary of depicting Darwin as the man responsible for ushering in a secular age we should be similarly cautious of underestimating the importance of evolutionary thought in relation to the questioning of the sanctity of human life."[27]

Many studies on the eugenics movement in the United States, Europe, and elsewhere likewise show the importance of Darwinism in mediating a shift toward eugenics and other related ideas, including biological determinism, inegalitarianism, scientific racism, and the devaluing of human life.[28] The ideas expressed by Madison Grant, president of the New York Zoological Society, in *The Passing of the Great Race* (1916), sound ominously close to Nazi ways of thinking (and Hitler owned the German translation of Grant's book). Therein Grant wrote, "Mistaken regard for what are believed to be divine laws and a sentimental belief in the sanctity of human life tend to prevent both the elimination of defective infants and the sterilization of such adults as are themselves of no value to the community. The laws of nature require the obliteration of the unfit, and human life is valuable only when it is of use to the community or race."[29] Stefan Kühl has even explicitly shown the many links between the American eugenics movement and the Nazi eugenics program.[30] Thus, the Darwinian devaluing of human life was not just a German phenomenon, and it led to many human tragedies outside of Germany, such as the compulsory sterilization campaigns in the United States and Scandinavia. Nowhere did it reach the catastrophic level of Germany, however, since only in Germany did a dictator rule with the power to pursue his radical agenda under the cover of war.

Evolutionary theory in general and Darwinism in particular had a tremendous impact on German thought. Darwin wrote to Wilhelm Preyer in 1868, "The support which I receive from Germany is my chief ground for hoping that our views will ultimately prevail."[31] Already in the 1860s and 1870s, many young German biologists began promoting Darwinism, while some prominent biologists and other scholars, such as the famous theologian David Friedrich Strauss and the neo-Kantian philosopher

Friedrich Albert Lange, began appealing to Darwinism to support their political and social theories.[32] By the 1890s so many biologists and social theorists had tried to apply the Darwinian struggle for existence to human society that Ludwig Woltmann, who at that stage in his life criticized such attempts, began referring to them collectively as social Darwinists.[33]

Not only can the influence of Darwinism be gauged by the outpouring of books and articles in late nineteenth-century Germany, Austria, and Switzerland (henceforth when I refer collectively to the German-speaking lands I will use the term Germany as shorthand) discussing the social and ethical applications of Darwinism, but we also find it frequently in autobiographical testimony. Richard Goldschmidt (1878–1958), one of the leading geneticists of the twentieth century, captures some of the pathos of his encounter with Darwinian literature in his youth. At age 16, he explained, he read Ernst Haeckel's *Natural History of Creation*

> with burning eyes and soul. It seemed that all problems of heaven and earth were solved simply and convincingly; there was an answer to every question which troubled the young mind. Evolution was the key to everything and could replace all the beliefs and creeds which one was discarding. There were no creation, no God, no heaven and hell, only evolution and the wonderful law of recapitulation which demonstrated the fact of evolution to the most stubborn believer in creation. I was so fascinated and shaken up that I had to communicate to others my new knowledge, and this was done in the schoolyard, on school picnics, and among friends. I remember vividly a scene during a school picnic when I stood surrounded by a group of schoolboys to whom I expounded the gospel of Darwinism as Haeckel saw it.[34]

Goldschmidt claims that his experience of embracing this Darwinian worldview (*à la* Haeckel) was typical for educated young people of his day, and abundant testimony from his contemporaries confirms this. In 1921 the physiologist Max Verworn stated, "One can state without exaggeration that no scientist has exercised a greater influence on the development of our contemporary worldview than Haeckel."[35]

Ernst Haeckel, the most famous German Darwinist of the late nineteenth and early twentieth centuries, enthusiastically adopted Darwin's theory of natural selection and applied the struggle for existence to humans in many of his writings.[36] He believed the most important aspect of Darwinism was the animal ancestry of humans, which would "bring forth a complete revolution in the entire world view of humanity." The theory of human evolution would "necessarily penetrate deeper than every other advance of the human mind" and would help integrate all branches of knowledge.[37] Haeckel congratulated Darwin on his seventieth birthday for having "shown man his true place in nature and thereby overthrowing the anthropocentric fable," that is, the idea that humans are the center of the cosmos and history.[38] In his writings, he often criticized the "anthropocentric fable" as a religious idea no longer tenable in the light of Darwinian science.[39]

The physician Ludwig Büchner, a famous scientific materialist and one of the most important popularizers of Darwinian theory in late nineteenth-century Germany, agreed. He wrote to Hermann Schaffhausen, the anthropologist who had discovered the Neanderthal fossils:

> As the new conception of nature [i.e., Darwinism] gradually prevails, so with it is produced, as I believe, one of the greatest transformations and one of the greatest advances, which human knowledge has ever undergone . . . At the same time a clarity and simplicity never before suspected will enter our entire philosophy.[40]

Haeckel, Büchner, and many young men and women influenced by them saw Darwinism as more than merely a biological theory. For them it was a central ingredient of a new worldview that was locked in combat with traditional Christian religion and indeed any dualistic religion or philosophy. Alfred Grotjahn, professor of social hygiene at the University of Berlin and leading figure in the eugenics movement, fondly recalled the time in his youth when he read Büchner's book, *Force and Matter*, which stripped him of all traces of religious faith. Büchner's Darwinian materialism influenced not only him, but many of his generation, according to Grotjahn (who was born in 1869): "Like hundreds of thousands of other young people it swept my brain clear of metaphysical conceptions at an age decisive in the development of my world view and freed me up to receive positivist views and this-worldly ethical values."[41] Many other German scholars and intellectuals have similarly testified that in their youth their encounters with popular Darwinist writings—especially those by Haeckel and Büchner—were decisive in the formation of their worldviews.[42]

While Büchner defended a materialist worldview, where mind is merely a function of matter, Haeckel called his philosophy monism, viewing matter and mind as inextricably united everywhere. For Haeckel even inanimate matter and single-celled organisms possess psychical characteristics. Sometimes Haeckel called his view pantheism, which he considered a synonym for monism. He also admitted, however, that pantheism is the same as atheism.[43] In any case, whether materialist, monist, or positivist, the Darwinian worldview advanced by these and most other leading Darwinists—and certainly by most Darwinian social thinkers—was fully naturalistic, that is, it explained all phenomena, including religion, ethics, and human behavior, as products of natural causation subject to scientific laws.

Whatever Haeckel, Büchner, Carneri, and other leading Darwinists might have disagreed on, they agreed that natural processes could account for all aspects of human society and behavior, including ethics. They denied any possibility of divine intervention, heaped scorn on mind–body dualism, and rejected free will in favor of complete determinism. For them every feature of the cosmos—including the human mind, society, and morality—could be explained by natural cause and effect. Everything was thus subject to the ineluctable laws of nature. As a corollary to this, science became the arbiter of all truth. Not even ethics or morality could escape the judgments and pronouncements of science.

Almost all the thinkers I will discuss in this study embraced this naturalistic Darwinian worldview. There were, of course, many of their contemporaries who accepted the validity of evolution for biological processes, but denied it any influence on theology, ethics, or social thought. German theologians in the late nineteenth century, partly because of the onslaught of antireligious Darwinists, but even more in response to the rise of biblical criticism (which they generally embraced), drew a strict line of separation between God and nature. The latter was science's province, while the former belonged to theology. This made their theology impervious to scientific assault (but also less relevant to the real world).[44] Most German philosophers and many sociologists reacted to the encroachments of science on their domains in a similar manner, adopting the philosopher Wilhelm Dilthey's distinction between the natural sciences and the human sciences. Dilthey and his followers vehemently denied that the scientific method could be applied to the social sciences, agreeing with G. E. Moore that any attempt to derive morality from nature is to commit the "naturalistic fallacy."[45] However, despite those who resisted the naturalistic Darwinian worldview, many others embraced it zealously, especially scientists and physicians. Indeed, among scientists and physicians the voices of support for a naturalistic Darwinian worldview were far stronger than the voices of opposition. Most of the opposition came from the humanities. It is the naturalistic Darwinists, then, those who tried to apply biological evolution to ethics, rather than those denying its applicability, on whom we focus in this work.

The question then naturally arises: Are the intellectual developments I portray in this work the product of something specifically Darwinian or do they flow from a more general naturalistic (materialistic or monistic) worldview? This is a difficult question, for the historical connections between Darwinism and naturalism are incredibly complicated. Some historical evidence suggests that Darwinism may not have been all that influential in promoting philosophical naturalism; these include the following points: (1) naturalism was on the rise before Darwin published his theory (in Germany the three most prominent scientific materialists—Ludwig Büchner, Karl Vogt, and Jakob Moleschott—published their most famous works in the 1850s); (2) the related issue that many people embraced Darwinism after embracing naturalism; and (3) many people embraced Darwinism without embracing naturalism (e.g., neo-Kantians, Christian theologians). However, while these points demonstrate that there was not a necessary relationship between Darwinism and materialism or monism (one could and did exist in the absence of the other), nevertheless strong connections developed historically between Darwinism and naturalism in late nineteenth-century Germany that require some explanation. Factors linking the two include: (1) most materialists and monists adopted Darwinism with alacrity and argued that it supported their materialistic or monistic metaphysics; (2) many people claimed that Darwinism was the key factor converting them to materialism or monism; and (3) leading Darwinian biologists and ethical philosophers overtly argued that Darwinism implied psychological determinism and thus a materialistic view of the mind.

The historical picture is further muddied by the fact that some leading voices in spreading Darwinism made the controversial claim that they were promoting not just a biological theory, but a Darwinian worldview as well. Even though they may have embraced their metaphysical views before adopting Darwinism, it was convenient to claim scientific sanction for their metaphysics. So how can we make sense of this confusing and often contradictory landscape of Darwinism and metaphysics? I suggest that the reception of Darwinism in nineteenth-century thought was both influenced by and an influence on the spread of a naturalistic paradigm. To consider this counterfactually, I would assert that without Darwinism, materialism would still have increased during the late nineteenth century, but it would have been far less persuasive and thus would have gained fewer adherents than it actually did.[46]

Despite the sometimes tight relationship between Darwinism and naturalism, it is still relevant to return to the question: Which more directly influenced ethical and moral ideas—Darwinism or naturalism? Some of the ideas about morality I discuss are specifically Darwinian, drawing on elements of biological theory as justification. Others seem to have no Darwinian content, but rely on more general naturalistic principles. Interestingly, however, many naturalistic Darwinists argued that Darwinism justified these principles as well, so the situation is not at all clear-cut.

Since Darwinian naturalism permeated the early eugenics movement, we will examine in this study a number of prominent eugenicists. Not only did many leading Darwinists embrace eugenics, but also most eugenicists—certainly all the early leaders—considered eugenics a straightforward application of Darwinian principles to ethics and society. Darwin's cousin, Francis Galton, the founder of modern eugenics, developed his ideas upon reading Darwin's *Origin of Species*, and German eugenics leaders likewise relied heavily on Darwinian principles.[47] Alfred Ploetz, who founded both the German Society for Race Hygiene (the first eugenics organization in the world) and also one of the first journals devoted to eugenics, was decisively influenced by Haeckel in his youth. He informed a friend in 1892 that his main ideas about eugenics were drawn from Darwinism, and he often praised Haeckel as a key influence on his worldview.[48] He also told Haeckel that his newly founded journal "will stand on the side of Darwinism," which was abundantly clear from the advertisement he sent to prospective subscribers, saturated as it was with Darwinian terminology.[49] Furthermore, one of Ploetz's coeditors was Ludwig Plate, a Darwinian zoologist who took over Haeckel's professorship at the University of Jena when Haeckel retired. It was no surprise that Ploetz recruited the two leading Darwinists in Germany—Haeckel and August Weismann—to become honorary members of the Society for Race Hygiene when he founded it in 1905.

The Krupp Prize Competition, announced in 1900 and completed in 1903, illustrates clearly the close connection between Darwinism and eugenics. Friedrich Krupp, a prominent industrialist and avid amateur naturalist, anonymously funded a lucrative prize competition for the best book-length answer to the question, "What do we learn from the principles of biological evolution in regard to domestic political developments and legislation of states?"[50] Haeckel helped sponsor this prize competition and his protégé Heinrich Ernst Ziegler, a zoologist working with him at the University of Jena, was one of the judges. Winning the first prize of 10,000 marks—a handsome sum in those days—was the physician Wilhelm Schallmayer for his book, *Heredity and Selection* (1903), which expanded on his earlier eugenics pamphlet, *The Threatening Physical Degeneration of Civilized Peoples* (1891). Schallmayer's eugenics relied heavily on Darwinian theory, which he called the greatest discovery of the nineteenth century.[51] In a letter to another leading eugenicist he confessed that eugenics was indissolubly bound together with Darwinian theory.[52]

Eugenicists were not only avid Darwinists, but many were intensely interested in ethical issues. Indeed eugenics was a movement trying to develop a "scientific ethics" ostensibly based on Darwinian theory. Schallmayer wrote in the introduction of his prize-winning book,

"This view [Darwinism] had an especially powerful influence on ethics. It not only produced new views about the origin and evolution of ethical commands and thus new foundations for them, but it also led to the call for a partial alteration of presently valid ethical views."[53] Most leading eugenicists agreed heartily, as we shall see.

In what ways, then, did Darwinism impact ethical thought? First, Darwinism made philosophical materialism and positivism more respectable by providing a non-theistic explanation for the origin of ethics. Before Darwin some ethical theories denied (or simply ignored) the divine origin of ethics (e.g., Bentham's utilitarian views), but none could explain why humans have an innate moral sense or conscience, nor why humans act altruistically. Many simply assumed the existence of morality, but could not account for its origin. Kant, for example, presupposed the existence of morality and from this deduced the existence of God, immortality, and free will, but Darwinism undermined Kant's argument. Second, Darwinism contributed to the rise of ethical relativism by denying the timeless and transcendent character of ethics. Most Darwinists explained ethics as a product of nature, that, like all other natural phenomena, was constantly evolving. It was not carved on stone tablets, but written in the ever-changing sands of time. Third, Darwinism gave impetus to the view that the human moral sense is a biological instinct—or at least based on one—rather than a spiritual endowment (the traditional Christian position) or a purely rational function (Kant's view). Fourth, natural selection and the struggle for existence among humans influenced people's views on ethics. In his study on social Darwinism Hannsjoachim Koch states, "The concept of natural selection had an even greater effect on Darwin's contemporaries than the idea of evolution; it [natural selection] . . . called into question the validity of the hitherto existing ethical ideals in all areas of life, whether social, economic, or political."[54] Finally, Darwinism altered conceptions of human nature and the value of human life, which had far-reaching ethical (and political and social) implications.

Let us explore this last point a bit further. What was it about Darwinian theory that produced a change in thinking about the value of human life? First, Darwinism implied that humans arose from animals, and many interpreted this to mean that humans did not have the special position accorded them in Judeo-Christian thought. Instead of being made in the image of God and falling from a pristine state of perfection, humans ascended from some kind of simian. In explaining the evolution of human mental and moral traits from animals, Darwin and most Darwinists denied the existence of an immaterial and immortal soul, a central tenet of the Judeo-Christian worldview that undergirded the sanctity of human life.[55] Second,

Darwinism emphasized variation within species, which implied biological inequality. Applying this to humans, many biologists, anthropologists, and social thinkers used Darwinism to justify social and racial inequality. Third, natural selection and the struggle for existence in Darwin's theory—based on Malthus's population principle—also implied that death without reproductive success is the norm in the organic world, and that the death of multitudes of "less fit" organisms is beneficial and fosters progress. Death had previously been viewed by most Europeans as an evil to overcome, not a beneficial force. But Darwin perceived some good in this evil. In concluding *The Origin of Species* he wrote, "Thus, from the war of nature, from famine and death, the most exalted object which we are capable of conceiving, namely, the production of the higher animals, directly follows."[56] Darwin's theory was thus not just about biological change; it was a matter of life and death.

Darwin's theory thus raised many fundamental issues that intersected with traditional religious doctrines, including the foundation for ethics, the formulating of moral codes, and the meaning of life and death. Many recent studies on the reception of Darwinism by religious leaders have stressed accommodation, since many Christian theologians and pastors, even those of a conservative theological bent, were willing to embrace some form of evolutionary theory.[57] However, my study helps illuminate why the debate over Darwinism was so acrimonious at times. Also, it reminds us that regardless of how accommodating leading religious figures may have been to evolutionary theory, many leading Darwinists were not so accommodating to religion. Many not only used Darwinism to attack the traditional Christian understanding of miracles and the supernatural, but they also undermined many deeply cherished Christian values that had become entrenched in European culture. For many educated Germans in the late nineteenth and early twentieth centuries, as Detlev Peukert observes, science replaced religion "as the source of meaning-creating mythology."[58]

In chapters 1–3 we will examine the ways that Darwinists tried to explain and formulate ethics and morality in general. Then we will turn to the more specific moral question of the value of human life, examining how Darwinists devalued human life, especially the disabled and non-European races. The final chapter will show how these ideas contributed to the development of Hitler's ideology.

1. Laying New Foundations for Ethics

1. The Origin of Ethics and the Rise of Moral Relativism ✍

Darwin neatly summed up his view of ethics and morality in his *Autobiography*, stating that one who does not believe in God or an afterlife—as he did not—"can have for his rule of life, as far as I can see, only to follow those impulses and instincts which are the strongest or which seem to him the best one."[1] This was a radical departure from traditional ways of grounding morality, for Christianity relied on divine revelation, Kant and many Enlightenment thinkers grounded ethics in human rationality, and even those British moral philosophers basing ethics on moral feeling usually considered it an immutable part of human nature, whatever its origin. The philosopher David Hull in *The Metaphysics of Evolution* underscored the revolutionary nature of Darwin's theory for ethics, stating, "Because so many moral, ethical, and political theories depend on some notion or other of human nature, Darwin's theory brought into question all these theories."[2] Even before Darwin wrote his autobiography, many opponents feared the ethical consequences of Darwinism, and Darwin's comment about following our instincts surely did not soothe those worries.

However, while on the one hand radically departing from traditional views about the origin and justification of ethics, Darwin did try to allay the fears of his contemporaries. After making the above statement in his *Autobiography* about following our instincts, he hastened to add that in humans the social instincts or ethical feelings are stronger than selfish or hedonistic impulses. In *The Descent of Man* Darwin had argued that the human moral sense had arisen through the combined activity of social instincts and rationality. Darwin further explained that human social instincts and group selection had led quite naturally—with no need for supernatural intervention—to the Golden Rule: Do unto others as you would have them do unto you.[3] Thus, he assured his contemporaries, his theory was nothing to fear, for it confirmed one of the central tenets of Judeo-Christian morality.

Not all of Darwin's supporters, however, agreed with him about the moral implications of his theory. Darwinists debated among themselves whether Darwinism overthrew, modified, or confirmed traditional moral views. Though some stridently and contemptuously rejected all traditional ethics and morality as subjective and unscientific, most took a mediating approach, affirming the necessity of altruistic impulses and brotherly love, while rejecting many specific tenets of Judeo-Christian morality. These latter Darwinists hoped to salvage the kernel of religious ethics, while dispensing with those aspects they deemed superfluous religious trappings no longer necessary in a more enlightened age of science. More importantly, however, since they regarded ethics and morality as products of evolution, they considered all morality relative to the evolutionary stage of development and also relative to its ability to preserve the species. Thus, they denied any fixed moral code.

For Darwin explaining the origin of morality was no trivial side issue, but a key question he had to confront if his theory of human evolution were to be plausible. Most people throughout history, after all, viewed morality as a uniquely human phenomenon that elevated humanity far above the rest of the organic realm. First, Darwin needed to demonstrate that morality was not uniquely human. Second, he needed to explain the process or mechanism that produced it. Long before publishing his theory, Darwin wrestled with the origin of human ethical impulses. Darwinism "was always intended to explain human society," explain Adrian Desmond and James Moore, since Darwin integrated ideas about economics and social thought into his theory from the start.[4] His 1838 "M" notebook was filled with ruminations on human evolution and included many notations on morality.[5]

In *The Descent of Man* Darwin tried to demonstrate that all human traits—including moral behavior—are different in degree, but not in kind, from other organisms. He pointed out that other animals live in societies and cooperate, and the social instinct producing this cooperative behavior is heritable. In humans the social instincts have developed further than in most other species, and, harnessed together with expanded human cognitive abilities, produced what we call morality. The mechanism for producing the increase in social instincts was, according to Darwin, natural selection through the struggle for existence. Those groups with more cooperative and self-sacrificing individuals would out-compete (either directly through warfare or indirectly by increasing their population) those groups with more selfish individuals.[6]

By basing morality on biological instincts, Darwin's evolutionary explanation for ethics provided a rational, scientific account for the development of nonrational human impulses. On the one hand, his scientific theory embodied Enlightenment rationalism. Darwin presented his theory as the product of true Baconian empiricism, and his theory helped facilitate a shift in biology toward a positivist paradigm.[7] However, ironically Darwinism also helped undermine rationalism, especially in the field of ethics. The growing prestige of science in general and Darwinian theory in particular in the late nineteenth century helped foster biological determinism. Human mental and moral traits—and thus human behavior—were thus grounded in biological traits or instincts rather than in human rationality. Darwinists never denied human rationality, of course, but as they began explaining human behavior in Darwinian terms, they emphasized the importance of instincts for human behavior. If a Darwinian account of human behavior was possible, then humans must share many similarities with animals, implying that human reason was not the primary motivation behind human behavior.[8]

Darwin was by no means the first to portray morality as innate and nonrational. In fact, he received considerable impetus from previous British ethical philosophers. In eighteenth-century Britain, philosophers proposed various theories grounding morality in human sentiments. Shaftesbury was a pivotal figure in developing this moral sense philosophy, and he decisively contributed to the eclipse of reason in moral theory. Many prominent figures in this development, such as Francis Hutcheson and Joseph Butler, believed these moral sentiments were implanted in the human breast by God, though others, such as David Hume, could embrace moral sense philosophy without reference to any divine origin, simply grounding it on human nature. Since most eighteenth-century philosophers considered human nature static, most believed that moral sentiments were universal and immutable. Darwin was well acquainted with British moral philosophy.[9] In 1838, he read *Dissertation on the Progress of Ethical Philosophy* by James Mackintosh, his uncle by marriage. Mackintosh argued that morality is both innate, since it relies on moral motives, and also rational.[10] Darwin agreed in some ways with Mackintosh's approach, but decisively rejected his static view of human nature and morality.

Darwinian theory spawned numerous attempts to construct new systems of ethics on an evolutionary basis. Even before the advent of Darwinian theory, Herbert Spencer had already been trying to construct a "scientific morality," but in his 1851 formulation of ethics, he still retained God as the source of the human moral sense. After Darwin published his theory, Spencer dispensed with God and developed a fully naturalistic account of ethics. Besides Spencer, the agnostic intellectual Leslie Stephen was the best-known proponent of evolutionary ethics in late nineteenth-century Britain, but many lesser lights in Britain, the United States, and elsewhere tried to formulate ethical theories and morality on the basis of biological evolution. Of course, these forays into evolutionary ethics also spawned many critiques, including a prominent one by the Darwinian biologist Thomas H. Huxley.[11]

Darwinian explanations for the origins of morality found fertile soil in Germany, but German Darwinists did not always agree among themselves about the implications of Darwinism for the moral status quo. Some considered it revolutionary, completely overthrowing traditional ethical systems, especially Christian ethics. Others, like Darwin himself, tried to stress its harmony with existing morality. Most, however, took a more ambivalent approach, differentiating between some aspects of traditional morality that Darwinism supports and other aspects that Darwinism undermines.[12]

Haeckel's forays into ethical thought exemplify this mediating approach. Haeckel, like Darwin, grew up in a Christian moral climate where the Golden Rule was taken for granted. Even before embracing Darwinism, he wrote to his fiancé that the measuring rod for his life was the maxim, "Do unto others, as you would have them do unto you."[13] In 1877, six years after Darwin published his views on human morality in *Descent*, Haeckel articulated a view of morality identical with Darwin's. He claimed that every human has an internal drive to love others, to forego egoism in favor of the collective good. Like Darwin, he called this cooperative urge "social instincts." The task of evolutionary theory, according to Haeckel, was "not to find new [moral] principles, but rather to lead the ancient command of duty back to its natural-scientific basis." Science would provide an even firmer foundation for morality than religion had.[14]

So far, this sounds very compatible with Christian morality, and indeed Haeckel would continue to stress the Golden Rule as a fundamental component of his monistic ethics. However, when Haeckel began writing more extensively on ethics, beginning in 1892 in *Monism as Connecting Science and Religion*, his position toward Christian morality became more ambiguous. On the one hand, he praised Christian morality as the highest form of morality yet propagated (though he tempered this remark by claiming that the Golden Rule preexisted Christianity and also existed in Buddhism). He further claimed that his monistic ethics "does not contradict the good and truly valuable parts of Christian ethics." This statement implies—and his other works confirm—that he was willing to dispense with other parts of Christian ethics that he did not consider "good and truly valuable."[15]

Haeckel did not overtly criticize Christian ethics in *Monism as Connecting Science and Religion*, but he was not so restrained in his best-selling *Riddle of the Universe* (1899), which contains his most extended discussion of ethics. While agreeing that the Christian conception of love and compassion are basically valid, he criticized Christianity for committing "a great mistake, in that it one-sidedly exalts *altruism* to a command, while rejecting *egoism*. Our *monistic ethics* grants both the same *value* and finds the perfect virtue in the proper balance between love of neighbor and love of self."[16] Haeckel thus rebuked Christian ethics for ignoring the human instinct for self-preservation, which was just as important in evolution as the social instinct. Haeckel believed that an organism had to balance egoism and altruism in order to thrive biologically, and so Christianity erred by stressing the latter at the expense of the former.

Other aspects of Haeckel's Darwinism had important implications for ethics. First, he believed that Darwinism undermined free will in favor of strict determinism. It did so by providing a naturalistic explanation for the origin of human psychology and behavior, proving conclusively that free will is illusory. "The theory of evolution," Haeckel stated, "finally makes clear, that the '*eternal, iron laws of nature*' of the inorganic world are also valid in the organic and moral world."[17] Another point Haeckel mentioned, but did not stress very often, was that since morality developed entirely by evolutionary processes, it could change historically.[18] Thus, morality was neither eternal nor immutable, but constantly in flux.

Haeckel was no ethical philosopher, and he admitted, both in public and in private, that his treatment of ethics was the weakest part of his monistic philosophy.[19] Nonetheless, Haeckel thought Darwinism had five implications for ethics, and these points recur repeatedly in the writings of other naturalistic Darwinists when they applied evolution to ethics: (1) Darwinism undermines mind–body dualism and renders superfluous the idea of a human soul distinct from the physical body. (2) Darwinism implies determinism, since it explains human psychology entirely in terms of the laws of nature. (3) Darwinism implies moral relativism, since morality changes over time and a variety of moral standards exist even within the human species. (4) Human behavior and thus moral character are, at least in part, hereditary. (5) Natural selection (in particular, group selection) is the driving force producing altruism and morality. Not all Darwinian-inspired ethical theorists agreed with Haeckel on all five points, of course, but these were very influential ideas nonetheless. In fact, Haeckel's books published in the early twentieth century—especially *The Riddle of the Universe*, but also *The Wonders of Life* (1904) and *Eternity* (1917)—were probably the most popular nonfiction books in Wilhelmine Germany, and his ideas about ethics and morality featured prominently in them.

The first serious thinker in Germany to systematically relate Darwinism to ethics was Bartholomäus von Carneri, who was not an academic philosopher, but an aristocrat and a liberal politician in the Austrian parliament. He wrote extensively about the connections between Darwinism and ethics after jettisoning his erstwhile Hegelian views for, first a pantheistic, and later a positivistic monistic worldview. Though a private scholar, he received an honorary doctorate in philosophy from the University of Vienna in 1901 on his eightieth birthday to honor his work on evolutionary ethics. Carneri published his first major work on ethics, *Morality and Darwinism*, in 1871, the same year that Darwin's *Descent of Man* appeared. In this and in many subsequent books and articles Carneri explored the implications of Darwinism for ethics and morality, though he avoided deriving moral standards directly from Darwinian theory. He exulted that Darwinism had finally made it possible to develop a coherent scientific worldview, and only on the basis of this worldview could any valid ethical philosophy be built. Thus Darwinism was a central component of Carneri's worldview, not just one scientific theory among others. In order to develop his moral philosophy Carneri diligently studied Darwin and Haeckel, and he developed close relationships with several leading Darwinists, including Haeckel, whom he greatly admired.

Carneri thought that Darwinism made its most important contribution to ethical thought by demolishing the idea of free will. By undermining supernaturalism, dualism, and purpose in the cosmos, Darwinism places all human phenomena, including ethics, under the sway of the laws of nature. Carneri explained to Haeckel, "For me the value of Darwin is that the human no longer needs to have a supernatural soul, and that one no longer needs purpose to explain creation."[20] In Carneri's view, the Darwinian explanation for the evolution of humans implies that the causal law imperiously rules over all human affairs. "The human being," he stated, "is as subject to the universal law of causation, both mentally and physically, as the most unimportant cell, the most insignificant atom."[21] Part of Carneri's project was to explain how ethical ideals could have validity in a deterministic world.[22]

However, Carneri relativized morality, clearly rejecting the notion of inherent human rights and natural law morality, ideas that most European liberals in the nineteenth century cherished. In 1871 he stated, "An *ethic* consistent with Darwin's theory knows no natural or innate rights, and can therefore only speak of *acquired* rights, even in relation to tribes of people."[23] Human rights are thus historically malleable, not fixed and eternal. Specific moral codes vary from one society to another, and in a Darwinian world without purpose there is no standard by which to judge the validity of one form of morality over another. In his first book Carneri defended the idea—perhaps a remnant of his earlier Hegelianism—that there exists an "absolute good" that is universal, but he defined it in such a way that it did not constitute anything even close to an absolute moral code. For him, the "absolute good . . . appears as the identity or perfected harmony of the idea with its practical implementation without regard for the form under which it appears." Since "the form under which it appears" could vary according to time or place, the "absolute good" was not a fixed principle.[24]

When explaining how human ethical impulses arose, Carneri departed from Darwin's and Haeckel's model. He did not believe that humans had social instincts, nor natural biological predispositions to altruistic behavior. He considered the natural biological tendencies of humans basically selfish, with instincts tending toward self-preservation, not self-sacrifice. Instead of basing human morality on social instincts, he argued that ethics is based on the human drive for happiness (*Glückseligkeitstrieb*). This drive arose when humans were forced by their conditions to live in society. Thus, rather than being a biological trait, it is derived from human social life. As Carneri described it, the drive for happiness is by no means hedonistic, since true happiness does not consist in the pursuit of physical pleasures. Rather true happiness comes through the pursuit of higher goals, especially those involving the good of others. Carneri's reformulation (and rejection) of Kant's categorical imperative sounds similar to utilitarianism: "Always act such that the maxim of your desires can always serve simultaneously as the principle of the greatest happiness of the greatest number."[25] However, Carneri criticized utilitarianism for not taking into account moral feelings, which form the basis of human actions.[26]

If Carneri's description of the drive for happiness seemed fuzzy at times, even more puzzling is his discussion about its origin. Unlike Darwin's or Haeckel's account of the origin of morality, Carneri's theory completely ignored Darwinism as a mechanism. Though he embraced natural selection and the struggle for existence as key factors in biological evolution in his 1871 book, thereafter his enthusiasm for natural selection (but not for biological evolution) waned.[27] Rather than being the product of the struggle for existence, Carneri thought the drive for happiness is a way for humans to escape the struggle for existence, at least in its more repugnant forms. Carneri forthrightly refused to base his ethical theory on Darwinian theory, especially on the struggle for existence, though he insisted that any ethical theory must be consistent with Darwinism, by which he meant that it must be fully naturalistic and deterministic.[28] For Carneri Darwinism thus served more to destroy the status quo in ethical theory—especially Judeo-Christian and Kantian ethics—than to construct new ethical ideals.

Carneri's writings were influential, in part because his blend of high-minded ethical idealism and hard-nosed scientific determinism appealed to many of his contemporaries, especially scientists. Of the six major books he wrote on evolutionary ethics, by 1906 four of them had gone into second editions, and his most popular work, *Modern Man*, was in its seventh edition by 1902. Despite his influence, however, his specific theory of the drive for happiness never caught on. Nonetheless, many Darwinists thought he adequately demonstrated that Darwinism did not entail the abandonment of ethics and morality, which was important to Darwinists trying to assuage fears that Darwinism would lead to immorality or amorality.[29] He developed a close relationship with Haeckel and Friedrich Jodl, who expressed appreciation for Carneri's work, as did other Darwinian biologists, such as Oskar Schmidt and Wilhelm Preyer. Arnold Dodel, a Darwinian botanist at the University of Zurich, told Carneri that Carneri's first book, *Morality and Darwinism*, played a major role in the early 1870s in helping him cast off his religious worldview and formulate a new, scientific one.[30]

Next to Carneri, the economist Albert E. F. Schäffle was probably the most influential thinker to apply Darwinian thinking to ethical and social theory in the 1870s. His massive four-volume *Structure and Life of the Social Body* (1875–78) was one of the earliest systematic works in sociology in Germany. In this work and in numerous essays Schäffle explored the connections between Darwinian principles and social theory. Unlike Carneri, he insisted that the struggle for existence was an inescapable reality, even for humans. His analysis of society was saturated with biological analogies and metaphors, especially Darwinian ones. He stated, "Social evolution actually proceeds on the basis of unceasing variations, adaptations, and inheritance through the results of the struggle for existence."[31] However, though liberal in applying biological ideas to sociology, he did maintain that the human struggle for existence differed significantly from the struggle among animals. The human struggle for existence included violent episodes, including warfare, but it also included peaceful economic competition.

In fact, Schäffle did not construe the Darwinian struggle for existence as a brutal struggle among beasts or an amoral free-for-all, but on the contrary, he argued that morality was an essential element in the human struggle for existence. Like Darwin, he emphasized the collective struggle for existence among humans with tribe against tribe and nation against nation. Morality and law functioned to reduce conflict within societies, making those societies stronger than their neighboring societies and thus conferring a competitive advantage.[32] Thus the struggle for existence, far from contradicting morality, produced morality, according to Schäffle: "Law and morals necessarily arise in and through the selective struggle for existence, since they themselves are essential components of the power of collective self-preservation."[33] Evolutionary processes, then, tend to generate moral progress.[34]

Schäffle's explanation for the origins and status of morality was similar to Haeckel's in many respects. He agreed with Haeckel that morality is constantly in flux, undermining any system of fixed moral laws.[35] He also insisted that morality should not favor altruism so much that it excludes egoism. "Moral and legal self-preservation," he argued, "is, just like common sense, an absolute demand of a true ethic."[36] Self-preservation should by no means overrule morality, but rather everyone should maintain a balance of egoism and altruism.

Schäffle's ideas about evolutionary ethics exerted significant influence on other scholars, but many writers on evolutionary ethics reached a broader public. Max Nordau, a physician and leading Zionist, wrote several popular works discussing the relationship between evolution and morality. In one of his popular works, *The Conventional Lies of Our Civilization*, he urged moral reform by pointing out alleged inconsistencies between conventional morality and a moral code based on Darwinism. He clearly articulated his presuppositions:

> We believe that the evolution of the human species as well as all other species is perhaps only possible—and in any case is furthered—through natural selection, and that the struggle for existence shapes in its widest sense all of human history as well as the existence of the most obscure individual; and [the struggle for existence] is the basis for all phenomena of politics as well as social life. That is our worldview. From this flows all our principles of life and our conceptions of law and morality.[37]

This book, originally published in 1883, sold over 50,000 copies by 1903 and was translated into numerous foreign languages. Later Nordau published an entire book devoted to explaining the relationship between *Morals and the Evolution of Man* (1916).

However, despite his insistence on the ubiquity of the struggle for existence, Nordau departed from Darwin's explanation for the origin of morality, since he did not believe that it arose through struggle among humans. He believed, rather, that morality arose out of the need to live in society, thus staking out a position similar to Carneri's. Nordau, however, tried to explain scientifically why humans became social. He explained that it was the human struggle against the environment, not a struggle among humans, which gave rise to society and thus morality. He claimed that before the ice ages, humans led solitary lives, but scarcity and stress brought on by the ice ages forced humans to associate. Social life required morality, so morality became essential for human survival. Nordau wrote, "Morality must be regarded as a support and a weapon in the struggle for existence in so far as, given present climatic conditions on earth and the civilization arising therefrom, man can only exist in societies, and society cannot exist without Morality."[38]

Nordau was ambivalent about the standard Darwinian explanation of moral impulses as instinctual. Instead, he emphasized that morality was first and foremost an exercise of human intelligence. Reason dictated that humans needed to cooperate in order to survive, so morality arose as a rational response to environmental pressures. Morality originally functioned to restrain or limit egoistic instincts. However, Nordau believed that as evolution proceeded, the moral sense became permanently fixed in people's biological nature. Thus all people, except for some aberrations, possess a keen sense of sympathy that inhibits their selfish instincts.[39]

Though he thought morality had an instinctual basis rooted in sympathy for others, Nordau did not think that moral codes were permanent. He specifically rejected all supernatural ethics and Kantian ethics, since they generally imply fixed moral standards. Instead, he thought evolution implied moral change as well as biological change. "Good and bad," he stated,

> derive not only their existence but their measure and their significance from the views of the community. They are therefore not absolute but variable; they are not an immutable standard amid the ever-changing conditions of humanity, a rule by which the value of the actions and aims of mortals are indisputably determined, but are subject to the laws of evolution in society and therefore in a constant state of flux. At different times and in different places they present the most varied aspects. What is virtue here and now may have been vice formerly and at another spot, and *vice versa*.[40]

Moral relativism was thus an important component of Nordau's evolutionary ethics.

The two most influential academic philosophers in nineteenth-century Germany wrestling with the implications of Darwinism for ethical thought—Georg von Gizycki and Friedrich Jodl—followed the same general approach as Carneri, even though they differed with him on many details. Like Carneri, they used Darwinism to undermine previous systems of moral thought, but resisted applying Darwinism to the content of morality. Unlike Carneri, both Gizycki and Jodl embraced a form of utilitarian ethics, the main principles of which Bentham propounded before

Darwin was even born. Thus, their ethics had little or no evolutionary content.

In 1876, at the beginning of his career, Gizycki published *Philosophical Consequences of the Lamarckian–Darwinian Theory of Evolution*, and nine years later, he followed up with an article in a highbrow popular journal on "Darwinism and Ethics." In these works he argued that evolutionary theory has significant implications for ethical thought, even though one cannot derive morality from evolutionary biology. He stated, "The importance of the theory of evolution for morality lies less in deriving specific, special doctrines as its consequences, and much more in the more general *displacement of everything contrary or opposed to nature in the treatment of ethical questions* required by it."[41] For Gizycki this meant rescuing ethics and morality from its connections with religion, creating a this-worldly moral philosophy to replace the prevalent otherworldly conception. Gizycki further argued, as Carneri also had, that scientific theories of evolution—and here he included the Kant–Laplace nebular hypothesis, Charles Lyell's uniformitarian geology, and biological evolution—decisively proved that all phenomena are determined by natural causation in accord with scientific laws. These scientific theories "prove an unrelenting, all-ruling, immutable lawfulness, a firmly integrated order and unified continuity of all occurrences."[42] Darwinism had specifically shown that humans are subject to the same laws and do not stand separate from nature.

Even though Gizycki criticized Darwin for overemphasizing natural selection and competition in nature, he still thought natural selection had some validity. In fact, he invoked group selection to explain the persistence of moral traits in any given society. Selective pressures require a society to remain morally upright, for if it abandons morality, it would be destroyed by a competing society that practices greater selflessness. He thought that moral virtues tend to preserve society and thus life, since "Whatever has no life-preserving power is no virtue." Earlier in the same essay he asserted that Darwinism teaches us that we should "pay respect to the positive conceptions of morality of the most successful nations in the 'struggle for existence.'" This almost sounds as though the preservation of life in the struggle for existence is the ultimate arbiter of morality, but Gizycki had already explicitly denied this, since it smacked of teleology, which had no place in his naturalistic Darwinian paradigm.[43]

So, what is the proper standard by which to judge moral standards? Gizycki opted for a utilitarian ethics, in which the greatest happiness of the greatest number is the highest ethical principle. By no means did he derive this from Darwinism, but as we have seen, he believed that the Darwinian process would reinforce utilitarian ethics by selecting for those societies most in harmony with true ethical principles. Darwinism cannot provide any criteria for morality, though, since the mere existence of a moral trait or norm is no proof that it is adaptive. Adaptation is never perfect, and conditions are ever-changing, especially in highly complex situations, such as human society. Darwinism thus implies that morality is ever-changing, a point important for Gizycki, since as a socialist sympathizer he advocated far-reaching social reforms.[44]

Though Gizycki portrayed Darwinism as a fundamental influence on the formation of his philosophy, in his most important and influential works on ethical theory—including those written during the nine-year period between his two writings on Darwinism and ethics—he almost completely ignored evolutionary theory.[45] Darwinism may have facilitated his embrace of a secular ethics, specifically utilitarianism, but evolution did not really add anything fundamental to his utilitarianism. In a book aimed at a popular audience, *Moral Philosophy* (2nd ed., 1895), Gizycki only mentioned Darwinism in one passage, where he explicitly denied that morality can be derived from nature or natural laws, including Darwinism. However, he maintained in the same passage that immorality will never flourish, because Darwinian group selection will weed out groups that exploit the weak.[46] Gizycki's influence in Germany was profound, not only through his writings and lectures on ethical philosophy at the University of Berlin, but also through his position as cofounder of the German Society for Ethical Culture and editor of its weekly journal.

Jodl, professor of ethical philosophy at the University of Vienna (before 1896 at the German University of Prague) and editor of the *International Journal of Ethics*, emphasized even more than Gizycki the implications of Darwinism for moral relativism. His belief that morality had evolved led him to reject the notion that moral laws are fixed and immutable. After comparing the old idea of the fixity of species to the Christian and Enlightenment conception of the fixity of moral laws, he stated,

> Morality, too, is a product of evolution, and is in a state of continual transformation. . . . But all evolution—so teaches biology—is adaptation of the organic individual to the changeable conditions of its environment. The sum of the ethical principles or ideals, which at any time are current in any nation, presents *nothing else*, therefore, *than* the conception of all that is reciprocally required in a practical direction of its members, for the advantage and profit of the community and the individual persons in it.[47]

If morality is, as Jodl here alleges, *nothing else than* an adaptation to a changing environment, then morality has no fixed reference point. Moral laws or principles that are adaptive for one place and time are maladaptive in a different situation. Thus, moral principles are not fixed or objective, but constantly in evolutionary flux.[48]

Not only were moral principles constantly changing, but moral beliefs in any given society "are always a step behind the times," claimed Jodl, since it takes time to adjust them to changing conditions.[49] This means that the status quo cannot provide guidance about the validity or viability of any particular moral principles. Jodl thus provided Darwinian legitimation for attempts at moral and social reform. However, in an 1893 essay he specifically rejected the idea that humans have any inherent rights, so the social reform he advocated did not aim at bringing society into congruence with any universal principles.[50]

Jodl's ethical philosophy comprised part of a positivist worldview heavily influenced by Darwinian thought. He not only studied Darwin and Haeckel intensively and forged close relationships with Haeckel and Carneri, but he also wrote major works on the Enlightenment skeptic David Hume and the materialist Ludwig Feuerbach, both of whom he admired. He told his friend, Wilhelm Bolin, that he hoped his *International Journal of Ethics* would "become a stronghold in the struggle against the theological spirit, which strides through the land murdering the soul."[51] Not surprisingly, Jodl sympathized with the renegade Left-Hegelian theologian, David Friedrich Strauss, in his turn toward Darwinian materialism. Jodl informed his friend Carl von Amira that he wanted to present "a new world view for the people: to do well, what [D. F.] Strauss did in a mediocre way. That is the highest goal of my ambition, of my work."[52]

Strauss, in his sensational book, *The Old Faith and the New* (1872), had tried to replace Christianity with a naturalistic scientific worldview containing large doses of Darwinian science. His earlier book, *The Life of Jesus* (1835) had sent shock waves through the Christian world by portraying the Gospels as mythological rather than historical. He completely abandoned Christianity in his 1872 work, which sold so rapidly that ten editions appeared within seven years. Just one year after Darwin published *Descent* Strauss was already using Darwinism to defend psychological determinism. The human soul, he claimed, is nothing more than the physical brain, and ethical relations are merely useful adaptations in the struggle for existence. Strauss argued that even the Ten Commandments lose their sanctity, once one recognizes that they are merely tools useful to humans in the course of evolutionary competition, rather than divinely ordained commands.[53] *The Old Faith and the New* was immensely popular and won the praise of Haeckel, despite receiving widespread criticism within academic circles.[54]

Jodl was also sympathetic with the Darwinian ethnologist Friedrich Hellwald, whose *History of Culture* (1875) explained human history within a Darwinian framework. Hellwald's book was so well received that he expanded it into a two-volume work, which went into its fourth edition in 1896, by which time other contributors updated and expanded on Hellwald's work, but usually in the same spirit as the original author. Hellwald claimed that all of human culture, including morality and religion, function as weaponry in the universal human struggle for existence. Denying any objective morality, Hellwald believed in only one source of right: power and the success it brings in the struggle for existence. While Darwin had scoffed at a newspaper article accusing him of advocating the principle that might makes right, Hellwald espoused exactly such a position, stating, "The right of the stronger is a natural law."[55] He further stated,

> In nature only One Right rules, which is no right, the right of the stronger, or violence. But violence is also in fact the highest source of right, in that without it no legislation is thinkable. I will in the course of my portrayal easily prove that even in human history the right of the stronger has fundamentally retained its validity at all times.[56]

Jodl objected to Hellwald's coarse depiction of the human struggle for existence, but he agreed with Hellwald that Darwinism undermined the notion of inherent human rights or objective moral standards.[57]

Hellwald's views do, indeed, seem crass, but he was not the only Darwinian social thinker to dismiss human rights with contempt and exult in the right of the stronger. Under the influence of Haeckel and Carneri, Alexander Tille tried to develop a consistent evolutionary ethics in the 1890s. Though he was a professor of German language and literature at the University of Glasgow until 1900, he considered his work on evolutionary ethics his real calling. In 1896 he told Haeckel that he hoped to change fields to philosophy, so he could devote his full attention to evolutionary ethics.[58] Lacking financial resources, he had to abandon this idea, and after leaving his teaching post in Glasgow, he returned to Germany and served as a business representative for industrialists.

Tille agreed with Carneri that in the light of Darwinism, the idea of innate human rights was no longer tenable. He stated, "Darwinism knows no inborn human rights, but only earned ones, and the modern view built on Darwinism knows no other earned rights, than those earned through one's own labor."[59] Tille believed that many modern ideals, such as freedom, equality, and peace, were inconsistent with evolutionary theory, since they were based on the idea of innate human rights. These ideals would have to give way before the right of the stronger in the struggle for existence, for "Against the right of the stronger, every historical right is completely invalid."[60] Tille not only agreed with Carneri and other leading Darwinian ethical thinkers that Darwinism implies some kind of moral relativism, but he also agreed with them about the strong link between Darwinism and determinism. In 1895 he asserted, "The denial of free will, on which the mythological ethics believed, the denial of the responsibility of the criminal, on which cruel punishments like mutilation are founded, have been the direct consequences of the penetration of the theory of evolution into the fundamental problems of ethics."[61]

Though Schallmayer as a socialist sympathizer was in some respects more humane in his views than Tille, he agreed with Tille about the implications of Darwinism for human rights. In his Krupp Prize–winning book on the implications of Darwinism for legislation, he insisted that ethics, morality, and law, as well as all other human cultural achievements were merely weapons in the inescapable human struggle for existence. Thus they have no universal or eternal validity. He wrote, "But the right of the stronger, that asserts itself in the victory of the better adapted forms over the less perfect, reigns not only in nature, but also in human social history."[62]

Another influential Darwinist stressing moral relativism as a consequence of Darwinian theory was August Forel, the world-famous Swiss psychiatrist who devoted much of his life to moral and social reform. In an autobiographical essay he confessed that his whole worldview had been dramatically transformed by reading Darwin's *Origin of Species* in 1865, because he recognized that Darwinism undermined mind–body dualism and human free will.[63] Evolutionary theory played a central role in his book, *The Sexual Question* (1905), wherein he asserted, "Morality is therefore relative, and our reasoning ability never allows us to recognize something as absolute good or absolute bad."[64] Nonetheless, Forel, as many Darwinists in the late nineteenth and early twentieth centuries, did not think morality was simply individually or even socially constructed. Like Darwin, he believed that morality was based on hereditary biological instincts. Before becoming a psychiatrist, Forel had become an expert on ants, observing their instinctual social behavior. Based on his belief in human evolution, he assumed that human behavior was similar to that of ants—determined by instincts, above all by social instincts.[65]

In *Hygiene of Nerves and Mind* Forel argued that feelings of sympathy were the basis for ethics and morality. These feelings "are innate or instinctive in human beings. Any one who does not possess them is a monster, a moral idiot, a born criminal." Forel, like Darwin and Haeckel before him, saw human moral character as primarily instinctual and hereditary. He stressed the supremacy of biological factors over environmental factors in determining human behavior. In a 1910 speech, he claimed that environment or education only affects the superficial aspects of human character, while biology determines the foundational aspects. He stated, "The inherited, inborn ethical feelings cannot in itself be inculcated; it is a drive, an instinct, that at most can be strengthened through practice and weakened through neglect."[66]

The idea that human moral behavior is determined largely by biological instincts inherited from one's forebears became widespread in scientific and medical circles by 1900, partly through the growing influence of Darwinism and partly through the failures of psychiatry to treat many mental illnesses. Darwin and Haeckel had both expressed the view that character traits, such as diligence, thrift, honesty, and sobriety, just like intelligence, were primarily biological and thus hereditary. Darwin's cousin, Francis Galton, and most eugenicists, such as Forel, adopted this view of the heritability of moral traits.

So did Büchner, who devoted an entire book to the subject of *The Power of Heredity and Its Influence on the Moral and Mental Progress of Humanity* (1882). In this work Büchner claimed that the discovery of the power of heredity was one of the greatest discoveries of the nineteenth century. Though he credited Darwinism with laying the groundwork for this breakthrough, he relied heavily on the French psychiatrist Théodule Ribot's book, *Heredity* (1875) to back up his claims. The main thrust of Büchner's book was to demonstrate that not only physical traits, but also mental and moral traits are hereditary: "Habits, inclinations, drives, aptitudes, talents, instincts, and aesthetic sense are likewise passed on through heredity, just as feelings and passions, temperaments and character, intellect and moral sense."[67] Not only are tendencies for virtues and altruism inherited, but also tendencies for vice or crime. Environment plays a distinctly subsidiary role in shaping human behavior, according to Büchner. Nonetheless, he still considered education important, since he did not think the specific content of morality was hereditary.[68]

Büchner's stress on the "power of heredity" is especially intriguing, because Büchner—like Haeckel and Forel—remained committed to the inheritance of acquired characteristics (Lamarckism), while simultaneously upholding Darwinian selection theory.[69] Because so many scholars have focussed attention on the intellectual struggle in the late nineteenth and early twentieth centuries between Weismann and Lamarckians over the inheritance of acquired characteristics (Weismann denied it), they often draw a strict dichotomy between proponents of "hard heredity" (Weismann and his followers) and "soft heredity" (Lamarckians). While this dichotomy can sometimes be helpful in sorting out the debates over the process of evolution, it often leaves the mistaken impression that those accepting Lamarckism believed that heredity was extremely plastic. But Büchner, Haeckel, and Forel do not fit very easily into this dichotomy, for while rejecting Weismann's theories in favor of Lamarckism, their position on the "power of heredity" seems closer to "hard heredity."

In fact, many nineteenth-century Darwinists (including Darwin himself) embraced a position on heredity that lies somewhere between the polar opposites of "hard" or "soft." They could do this because of their stress on gradualism. Forel provides an excellent example of this, for his stress on the power of heredity played a pivotal role in the introduction of eugenics into Germany. However, he remained convinced that acquired characteristics can be inherited, as he confessed to Richard Semon, a former student of Haeckel's who developed a Lamarckian evolutionary theory in the first decade of the twentieth century. He explained to Semon, however, that any modifications that occur through the inheritance of acquired characteristics must proceed very slowly over many generations, since species are relatively constant.[70] Forel's position, then, seems somewhere between "hard" and "soft" heredity. For our purposes it is important to understand that most evolutionists—whatever their understanding of the causes of evolution—believed that mental and moral traits were heritable, and many stressed the primacy of biological inheritance over environmental factors.

Stress on heredity of mental and moral traits became especially pronounced in the field of psychiatry, and not only through Forel's influence. This was a significant shift from the early nineteenth century, when the budding field of psychiatry, imbued with liberal notions of progress and human malleability, hoped for therapeutic breakthroughs to rid society of mental illness and the resulting errant behavior. One important theorist wedding Darwinian theory to psychiatry was the Italian psychiatrist Cesare Lombroso, whose influence extended throughout Europe. Heavily influenced by Darwin and Haeckel, Lombroso developed his theory of the "born criminal" in the 1870s. Lombroso believed that certain biological types are predisposed (or perhaps even constrained) by their heredity to commit crimes. Hoping to identify a link between physical and mental or moral characteristics, he studied the physical traits of known criminals, especially their cranial and facial characteristics. He theorized that the "born criminals" in modern Europe were atavistic, that is, throwbacks to people at an earlier stage of evolution, who were more savage and less morally refined. Darwin had upheld a similar view in *The Descent of Man*, stating, "With mankind, some of the worst dispositions, which occasionally without any assignable cause make their appearance in families, may perhaps be reversions to a savage state, from which we are not removed by very many generations."[71] Lombroso's theory spawned a new field, criminal anthropology, and though many psychiatrists and criminologists rejected his theory of atavism, many embraced the notion that criminality is partly or even primarily a hereditary phenomenon.[72]

The psychiatrist Hans Kurella, editor of the important psychiatric journal, *Centralblatt für Nervenheilkunde und Psychiatrie*, was one of the most influential disciples of Lombroso in Germany, writing prolifically on criminal anthropology in the 1890s and thereafter. In his biography of Lombroso and elsewhere he stressed the Darwinian influence on Lombroso's thought.[73] When describing Lombroso's theory linking criminals to "primitive" peoples, Kurella explained the link between Lombroso's theory and evolutionary ethics:

> The psychic analogies rest essentially on the evolutionary theory of morality, especially Darwin's and Spencer's hypothesis, that the presently reigning views of morality and law depend on the inheritance of feelings, which are acquired through adaptation to the requirements of the rising culture in very long periods of time. Without full understanding of evolutionary ethics one can scarcely appreciate this side of Lombroso's hypothesis; in this sense the criminal would reproduce a stage of evolution, on which many primitive peoples still stand today.[74]

Kurella not only embraced Lombroso's view that "born criminals" were reverting to earlier stages of human evolution, but he stressed the "primatelike" traits of criminals (see figure 1.1).[75] For Kurella the task of criminal anthropology was to transform moral and legal theory by bringing it under the banner of science, especially Darwinian science.[76]

Another German psychiatrist sympathetic to Lombroso's ideas was Robert Sommer, a professor of psychiatry at the University of Giessen who organized the Seventh Congress of Criminal Anthropology in Cologne in 1911. Sommer agreed with Lombroso's stress on the hereditary nature of criminal behavior, but did not see the "criminal type" as atavistic. Though fully supporting the idea of the "born criminal," he warned against possible dangers that might arise if the theory was misapplied to politics. His foreboding seems especially prescient in light of subsequent Nazi practices. In responding to those who thought the state would be threatened by the doctrine of the "born criminal," Sommer cautioned against the opposite

Figure 1.1   Illustration in psychiatrist Hans Kurella's book, *Die Grenzen der Zurechnungsfähigkeit und die Kriminal-Anthropologie* (1903), showing the allegedly ape-like features of an Italian criminal

danger, which he considered an even greater threat. It is becoming ever clearer, he asserted,

> that with uncritical application this theory runs the danger of becoming a dangerous means of a *police state*. . . . The theory of the born criminal in the hand of dogmatic proponents of state order can become a fearful weapon against the *personal freedom of the individual*. Not in the direction of *psychiatrization*, but in that of the *coercive state* with *detention ad libitum* lies the true danger of this scientifically unavoidable theory with its possible improper application. The Committee for Public Safety of the French Revolution with unlimited power over elements dangerous to the present state is the form of state order, for which the theory of the born criminal is most suitable.[77]

While Sommer considered detention of some "born criminals" a necessary way to protect society, he wanted to limit its extent as much as possible.

Lombroso's ideas remained controversial in Germany (and elsewhere), but the idea of the "born criminal," or at least that heredity plays an important role in moral and immoral behavior steadily gained ground in late nineteenth and early twentieth-century German psychiatry. One of the most prominent psychiatrists in Germany, Emil Kraepelin, professor at the University of Munich (and earlier at Heidelberg), acknowledged in his memoirs the influence of Lombroso on his early intellectual development.[78] In his influential textbook on psychiatry he promoted the idea that "moral insanity" is a hereditary condition, in which a person has no moral feelings to offset their selfishness. Neither education nor training can remedy such a person's condition, since it is biologically determined.[79] In the 1904 edition of Kraepelin's text, he altered the heading "Moral Insanity" to "The Born Criminal," which underscores even more the hereditary character of immoral and criminal behavior.[80] The psychiatrist Otto Binswanger agreed with Kraepelin that "moral insanity" is a hereditary condition beyond the reach of psychiatric treatment. In an 1888 article (wherein he makes clear the importance of Darwinism for his views), he called these "morally insane" people "atrophied, defective people born through hereditary degeneration." They almost always end up criminals, he claimed.[81]

In 1916 Eugen Bleuler, Forel's successor as director of the Burghölzli Psychiatric Clinic in Zurich, summed up the consensus of the psychiatric community in his *Textbook of Psychiatry*, which was widely used in universities to train the next generation of psychiatrists. When discussing the origins of mental illnesses, he began with what he considered the most important cause—heredity. He explained that mental problems run in families, and thus one's biological constitution is the most important factor causing mental illness. One does not inherit specific mental illnesses, but rather only the tendency or propensity toward that mental illness, so not everyone with mental weaknesses actually display pathological symptoms. Some of Bleuler's categories of mental illnesses reflected the prevailing view among German psychiatrists at that time that morality is a biological trait. He listed spendthrifts, vagrants, and "moral idiots" as examples of those suffering from mental illnesses. He agreed with Lombroso that "moral idiots" or those having "moral insanity" are predisposed to criminality, and they are also usually antisocial and work-shy. While Bleuer and most German psychiatrists in the early twentieth century did not believe that specific moral commands are innate, they did believe that the aptitude to behave morally was an innate, biological trait. Those lacking a moral faculty are aberrant and a danger to society.[82]

Psychiatrists' ideas about the heritability of moral traits found fertile soil in the budding eugenics movement in the 1890s and early 1900s. Many psychiatrists, including Forel, Kurella, Sommer, and Bleuler, became prominent advocates of eugenics. Schallmayer, Ploetz, and most other eugenicists used the specter of the "born criminal" or "moral defective" to motivate their contemporaries to adopt eugenics policies. Another leading eugenicist, Felix von Luschan, professor of anthropology at the University of Berlin, believed that eugenics was the key to abolishing crime. In his view, "the Criminal should only be considered as a pure lunatic, who is not responsible for his ill doings. . . . Crime in the great majority of cases is a hereditary disease generally caused by drunkenness of the parents or ancestors." Since crime is a hereditary disease, the way to solve the problem is by "*the complete and permanent isolation of the criminal.*"[83]

The idea that morality and immorality are primarily hereditary biological traits had tremendous implications for social policy, including education, justice, and penal reform, marriage policy, and the control of reproduction. Besides issues of marriage and sexual reform, to which we will return in later chapters, one of the crucial questions emerging from the rise of biological determinism concerned personal responsibility. How can society hold individuals responsible for their behavior if that behavior—or at least predispositions toward that behavior—is programmed into their biological constitution? By rejecting human free will, many psychiatrists argued that responsibility as it had been traditionally conceived was misguided. However, they generally argued that society had a right to protect itself against morally "aberrant" individuals, so they were not necessarily advocating reduced penalties for criminals. However, they thought penalties should be determined on the basis of the threat of the individual to society, not based on the particular crime. The medicalization of criminal justice could sometimes result in longer sentencing, but for "medical" reasons, not retributive justice.[84]

Whether or not Darwinism actually implies materialism or determinism is a philosophical question beyond the scope of this work, but I have clearly demonstrated that historically many people thought it did. Evaluating Darwinism's contribution to the rise of moral relativism is even trickier. Certainly many Darwinists proclaimed the death knell for Christian, Kantian, or any other fixed system of ethics, and they contended that moral relativism was a logical consequence of a Darwinian view of morality. They completely rejected the natural law tradition of morality that had been so influential in the Enlightenment. Thus, it seems safe to say that Darwinism did play a role in disseminating ideas about moral relativism.

However, Darwinism was only one factor among others in the trend to historicize ethics and undermine natural law morality in nineteenth-century Germany. There were other forms of historicism (both before and after Darwin) prominent in Germany (and elsewhere in Europe), which contributed to the sense of ethical crisis in the late nineteenth and early twentieth centuries. Historicism—the idea that everything is in flux and phenomena can only be understood as part of a historical process—was a feature common to most major systems of thought in nineteenth-century Europe, including Hegelianism and Marxism.[85] The historian James Kloppenberg has shown that the intellectual move toward greater uncertainty of knowledge and historicizing ethics involved those rejecting the application of science to ethics. The philosopher Wilhelm Dilthey, with his strict separation between the natural sciences (including Darwinism, of course) and the human sciences (including ethics), rejected any kind of evolutionary ethics, as did most neo-Kantian philosophers.[86] Hegel's, Marx's, or Dilthey's moral relativism owed nothing to Darwinism, but rather flowed from their historicist worldviews. Darwinism was only one form of historicism among others, but it fostered moral relativism by providing scientific sanction for it. For some audiences this was more important than all the philosophizing Hegel or Dilthey could do.

2. Evolutionary Progress as the Highest Good ⌒

In 1904 one of the leading German Darwinian biologists, Arnold Dodel, proclaimed, "The new world view actually rests on the theory of evolution. On it we have to construct a new ethics . . . All values will be revalued."[1] Though proponents of evolutionary ethics did not always agree on the specifics of the new morality, they all agreed that present moral norms had to be examined anew in light of evolutionary theory. Some traditional moral norms might be valid, but others must be revised or completely overthrown. Their moral relativism implied that some moral values might have been valid in the past, but may no longer apply under modern conditions.

Thus evolutionary ethics faced a daunting question: What would replace the old values? While Carneri, Jodl, and Gizycki did not think that evolution provided much guidance on this matter, many proponents of evolutionary ethics disagreed. One common attitude of those writing about evolutionary ethics in the 1890s and early 1900s was poignantly expressed on a postcard sent by Willibald Hentschel, a former student of Haeckel's, to Christian von Ehrenfels, a philosopher and eugenics enthusiast: "That which preserves health is moral. Everything that makes one sick or ugly is sin."[2] Indeed, though many Darwinists insisted that morality was not fixed, but historically changing, and though many emphasized the relativism of morality, one factor still remained constant: the evolutionary process itself. Thus many writers on evolutionary ethics exalted evolutionary progress—and everything that contributes to it—to the status of the highest moral good. Health and sickness became criteria for making moral judgments, since they influence evolutionary progress. This is obviously a tremendous shift from Christian or Kantian ethics, where health and sickness may be influenced by moral choices, but by no means do they provide moral guidance.

Proponents of evolutionary ethics and evolutionary progress could proceed in two main directions. The first was that favored by most Darwinian social thinkers before the 1890s, who considered natural selection the most sure path toward evolutionary progress. Since untrammeled competition in the natural world had already produced such remarkable progress, people must only make sure they do nothing to hinder it. If a particular society has already adopted "unnatural" institutions that prevent natural selection from operating, it should quickly get rid of them. This position is usually known as social Darwinism. Many social Darwinists thought competition occurred simultaneously on two levels, between individuals and between social groups, though some social Darwinists placed greater stress on one or the other form.[3]

The second way to promote evolutionary progress was to engage in artificial selection. Rather than get rid of "unnatural" institutions and return to "barbaric" forms of competition, a society should retain its higher culture, while offsetting the disadvantages of reduced competition. As rational creatures, humans should make reproductive choices to further evolutionary progress. This second method of promoting evolutionary progress is called eugenics. Social Darwinism and eugenics are not mutually exclusive positions, for natural and artificial selection can both operate simultaneously. Indeed, many eugenics proponents also advocated measures for increasing competition among humans.

Most social Darwinist thinkers broached the topic of ethics and morality, and certainly, their views had significant moral implications, but many did not treat the subject systematically. One of the more prominent social Darwinists who did was Gustav Ratzenhofer, a retired Austrian military officer who wrote extensively on the social implications of Darwinism. Like other social Darwinists, Ratzenhofer believed that humans were locked in an ineluctable struggle for existence, but he thought that the human struggle was mostly between societies, not individuals.[4] His book, *Positivist Ethics* (1901), tried to formulate an ethics on the basis of a monistic worldview. Darwinism was a key component of Ratzenhofer's monism, and his book was loaded with biological terminology and examples. Though he did not believe in timeless or universal moral precepts, he argued that good and evil are not completely relative. He asserted that actions can be judged by an absolute standard, which is their impact on the evolution of humans. That which promotes evolutionary progress is good, while that which hinders it is evil. He stated, "The ethical nature of humans is directed at nothing other than at the flourishing of the species, and this rests on the mutual dependence of all humans."[5] By stressing human interdependence and the interests of the species, Ratzenhofer avoided an ethics of individual egoism. Just like Darwin, he believed that morality arose as individuals replaced egoism with concern for others in society, which would help their society emerge victorious in the struggle for existence with other societies.

Another social Darwinist thinker, Tille, agreed with Ratzenhofer that evolutionary progress was the highest good, but his social Darwinism was more individualistic than Ratzenhofer's. His evolutionary ethics was a synthesis of Darwinism and Nietzscheanism. While Tille applauded Carneri's efforts at applying Darwinism to ethics, he criticized Carneri for not going far enough. Carneri failed, he thought, to appreciate the central role of competition and selection, and "the increase of the abilities of the species is not yet the reigning ideal" in his ethics.[6] Like Nietzsche, Tille forthrightly declared battle against Christian and humanitarian ethics, which he considered contrary to an ethic based on Darwinian evolution. The moral ideal of evolutionary ethics, according to Tille, must be "the elevation and more excellent formation of the human race." Because of this, morality must judge actions by their effects on the physical and mental prowess of coming generations. If an act contributes to biological decline, it is immoral, even if it fulfills the Christian command of love and compassion.[7] Tille suggested replacing the Christian command to honor one's parents with a principle derived from evolutionary ethics: " 'Honor your child, that it may become fit and accomplish its work in life,' teaches ethics on a scientific foundation."[8] Tille embraced both social Darwinist competition and eugenics as twin means to the goal of evolutionary progress.

Tille also interpreted Nietzsche as a proponent of evolutionary ethics, even though he admitted that Nietzsche was not always conscious of the evolutionary underpinnings for his ideas. Tille claimed that Nietzsche did not reject all morality, but only slave morality, such as Christianity or humanitarianism. But Nietzsche promoted a morality that would produce the Superman, which Tille interpreted as a higher stage of human evolution. According to Tille, "A physiologically higher condition of future humanity is his [Nietzsche's] final moral goal, his moral ideal, and thereby he draws the final ethical consequence from Darwinism, thereby he expands the theoretical world view of the theory of evolution to the ethical world."[9] Tille exulted in Nietzsche's rejection of compassion and sympathy, for they will not lead to evolutionary progress. Rather, Tille asserted,

> Even the most careful selection of the best can accomplish nothing, if it is not linked with a merciless elimination of the worst people. . . . And the proclamation of social elimination must therefore be one of the supreme features of every ethics, which elevates as its ideal the goal that the theory of evolution has demonstrated. . . . Out of love for coming generations . . . Zarathustra preaches: Do not spare your neighbor! For the person of today is something that must be overcome. But if it must be overcome, then the worst people, the low ones, and the superfluous ones must be sacrificed. . . . Therefore this means becoming hard against those who are below average, and in them to overcome one's own sympathy.[10]

Tille, like Nietzsche, thus reverses much of Christian morality by spurning moral responsibility toward the weak and sick in favor of the strong and healthy. Tille not only rejected love and compassion as moral precepts, but in many cases he considered them immoral.

Many philosophers today object to Tille's biological interpretation of Nietzsche, pointing out that Nietzsche consistently posed as an anti-Darwinist and rejected positivism and scientific materialism.[11] His penchant for freedom set him in opposition to the rising tide of biological determinism in the late nineteenth century, and Nietzsche's vision of the Superman was not of a biologically superior specimen who would out-reproduce his fellow men, but rather of a creative genius who would rise to dominate others, but not necessarily leave behind more plentiful offspring. Nietzsche had contempt for the "herd," while admitting that mediocrity had an advantage in the struggle to survive.[12]

Nonetheless, many historians studying eugenics have noted the influence of Nietzsche on the budding eugenics movement.[13] Why did eugenicists, most of whom based their entire worldview on Darwinism, look favorably on such an anti-Darwinist as Nietzsche? Part of the answer, of course, is that Nietzsche's writings are often ambiguous and offered something for everyone—at least everyone who, like him, rejected Christianity and its morality. Many feminists were willing to overlook his misogyny, anarchists and some socialists conveniently forgot about his contempt for the masses, and nationalists bent his philosophy for their own purposes. Eugenicists were not the only ones to selectively read Nietzsche.[14]

However, completely dismissing the connections between Nietzsche and evolutionary ethics or Nietzsche and eugenics as a facile misinterpretation of a complex thinker does not adequately grapple with Nietzsche's intense engagement with biology, especially with evolutionary ethics. Rüdiger Safranski, Gregory Moore, and Jean Gayon have all recently argued that the biological content of Nietzsche's philosophy must be taken seriously, a position Steven Aschheim has long upheld.[15] Indeed, Nietzsche formulated much of his moral philosophy in direct response to evolutionary ethics. Already in 1873 Nietzsche confronted evolutionary ethics in his *Untimely Meditations* on Strauss's *Old Faith and the New*. Nietzsche strongly criticized Strauss's moral philosophy, not because Strauss tried to derive morality from Darwinism, but precisely because he did *not* draw the correct ethical implications from his Darwinian worldview. Strauss, Nietzsche groaned, had merely justified presently accepted moral standards, while a "true and seriously consistent Darwinian ethic" would "derive moral precepts for life from the *bellum omnium contra omnes* [the war of all against all] and the prerogatives of the stronger." Nietzsche further showed affinity with Darwinism by noting that Darwinism implied inequality, since the strong suppress the weak and bring about their demise.[16]

Nietzsche's keen interest in ethical philosophy was stimulated further in the mid-1870s by his close friend, Paul Rée, who lived with Nietzsche for a time while composing his book, *The Origin of Moral Feelings* (1877). Upon publishing this work, Rée wrote to Nietzsche, "To the father of this work, with the deepest thanks, from its mother."[17] Nietzsche politely distanced himself from Rée's remark, and in the 1880s Nietzsche would drift further apart from his friend's attempt to formulate a positivistic moral philosophy. However, Nietzsche's work *Human All-Too-Human*, was composed in the late 1870s under the heavy influence of Rée, and many of Nietzsche's friends criticized Nietzsche's turn toward positivism in that work.[18]

The fundamental idea animating Rée's work, *The Origin of Moral Feelings*, is that Darwin's and Lamarck's theories of biological evolution prove that "moral phenomena—just as well as physical [phenomena]—can be reduced to natural causes," and thus undermine any transcendent origin for morality. Rée argues that human behavior is completely determined, because all moral qualities of people are biologically inherent. Humans are essentially egoistic, since they behave in such a way as to maximize their survival. Even an individual's "unegoistic" behavior is simply an adaptation helping him or her survive in the struggle for existence. Thus, altruism can also be reduced to egoism. Rée therefore denied that anyone could be properly praised or blamed for moral or immoral behavior, since such behavior is programmed into their biological nature.[19] Of course, Nietzsche did not agree with all of Rée's points, and much of his later moral philosophy was a refutation of some of Rée's ideas. However, he did agree with Rée's dismissal of transcendent morality and also embraced determinism.

As Nietzsche developed his moral philosophy in the 1880s, he read extensively the literature on evolutionary biology and especially on evolutionary ethics. He was heavily influenced by non-Darwinian theories of evolution, especially by reading Karl Nägeli's *Mechanical–Physiological Theory of Evolution* (1884), Wilhelm Roux's *Struggle of Parts in the Organism* (1881), and William Rolph's *Biological Problems: Toward the Development of a Rational Ethic* (1882). Nietzsche heavily annotated his copy of Rolph's book, which shaped his critique of Darwinism. Rolph—and likewise Nietzsche—clearly accepted biological evolution, but he thought that Darwin's theory of natural selection was misguided. Humans do not struggle for existence in a situation of scarcity, as Darwin's theory postulated, but rather humans struggle for supremacy and mastery in conditions of superfluity and abundance. Nietzsche heavily marked a passage from Rolph's book explaining how his view of struggle differed from the Darwinian struggle for existence:

> Furthermore, the life-struggle is then no defensive struggle, but rather a war of aggression . . . But growth and reproduction and perfection are the consequences of that successful war of aggression . . . While the Darwinists hold that no struggle for existence takes place where the survival of the creature is not threatened, I believe the life-struggle to be ubiquitous: it is first and foremost precisely such a life-struggle, a struggle for the increase of life, but not a struggle for life![20]

Nietzsche certainly reveled in struggle, and he incorporated Rolph's idea into his philosophy by positing the human will to power as the driving force in life.[21] In some instances he referred to the will to power as an instinct, thus seeming to give it biological content.

Nietzsche intently studied much of the important literature on evolutionary ethics in the 1880s, including Herbert Spencer (whom he opposed) and Francis Galton, as well as some more obscure works, such as Georg Heinrich Schneider's *The Animal Will* (1880) and Erdmann Gottreich Christaller's *The Aristocracy of the Mind as the Solution of the Social Problem: Handbook of Natural and Rational Selection of Humanity* (1885).[22] Even though Nietzsche did not agree with many of the ideas of these writers on evolutionary ethics, he certainly did incorporate into his writing some of their key themes, especially ideas about degeneration and breeding. Further, many passages in Nietzsche exalt health and physical vitality above the traditional Christian virtues of renunciation and self-denial. Nietzsche also explicitly promoted many eugenics ideas, including health certificates before marriage. In many passages in *The Will to Power* Nietzsche manifests his eugenics ideas. In a representative passage he stated,

> There are cases in which a child would be a crime: in the case of chronic invalids and neurasthenics of the third degree. . . . Society, as the great trustee of life, is responsible to life itself for every miscarried life—it also has to pay for such lives: consequently it ought to prevent them. In numerous cases, society ought to prevent procreation: to this end, it may hold in readiness, without regard to descent, rank, or spirit, the most rigorous means of constraint, deprivation of freedom, in certain circumstances castration.—
>
> The Biblical prohibition "thou shalt not kill!" is a piece of naiveté compared with the seriousness of the prohibition of life to decadents: "thou shalt not procreate!"—Life itself recognizes no solidarity, no "equal rights," between the healthy and the degenerate parts of an organism [this refers to Roux's evolutionary theory]: one must excise the latter—or the

whole will perish.—Sympathy for the decadents, equal rights for the ill-constituted—that would be the profoundest immorality, that would be antinature itself as morality![23]

As the quote demonstrates, Nietzsche was more radical than most of his contemporaries in drawing moral conclusions from his views on evolution and eugenics. He clearly rejected the Christian conception of the sanctity of human life. According to Safranski, "this naturalization of the mind and the consequent relativization of the special status of man, which was in effect a disparagement of man, is only one of the two major aspects of the effects of Darwinism [on Nietzsche]."[24]

Indeed, Nietzsche's pronouncements on human breeding often carried murderous overtones, though, as Safranski has argued, Nietzsche usually shrouded his views in ambiguity to dodge responsibility and ward off criticism. Zarathustra, for example, encourages people to die "at the right time," when one is in victory and when one wills to die, rather than hanging on to life while one declines. Nietzsche not only seems to favor suicide in this passage, but also expresses more perverse hopes, stating, "Far too many keep living and hang much too long on their branches. May a storm come, which shakes all this rotten and worm-eaten fruit from the tree."[25] In a section of *The Gay Science* entitled, "Holy Cruelty," a "saint" gives advice to a father to kill his disabled child, asking, "Isn't it more cruel to allow it to live?"[26] In *Twilight of the Idols* Nietzsche included a section entitled "Morality for Physicians," in which he called sick people parasites having no right to life. He enjoined physicians to cultivate a "new responsibility" by fostering "the *ascending* life," while demanding the "most ruthless suppression and pushing aside of the *degenerating* life."[27] Finally, in *Ecce Homo* Nietzsche discussed the future if his ideas triumph:

> If we cast a look a century ahead and assume that my assassination of two thousand years of opposition to nature and of dishonoring humans succeeds. That new party of life, which takes in hand the greatest of all tasks—the higher breeding of humanity, including the unsparing destruction of all degenerates and parasites—will again make possible that *superfluity of life* on earth, from which also the Dionysian condition must again arise.[28]

It may be true that Nietzsche did not explicitly advocate killing the disabled in these passages, but he certainly gave more than enough hints to make his views clear. Many leading eugenicists, in any case, embraced Tille's interpretation of Nietzsche and thought he supported their cause.

Eugenicists generally believed that ethics and morality needed to be rewritten in light of evolutionary theory. Evolutionary ethics undergirded—sometimes overtly, sometimes implicitly—the whole enterprise of eugenics as it expanded rapidly in the early twentieth century. Eugenics was, after all, the attempt to find practical measures to improve human heredity. Its adherents often claimed scientific status for the enterprise, but because of their stress on psychological determinism, most of the early leaders also claimed that all the human sciences were subject to the natural sciences. Just like their mentor Haeckel, they tried to bring ethics and morality under the purview of science.

Few were more insistent on this point than Schallmayer. Beginning his university studies in philosophy, he became disillusioned because of the many different answers proposed by philosophers, so he abandoned philosophy for natural science as a more reliable path to true knowledge.[29] In 1904–05 he wrote two articles—one for Ploetz's eugenics journal and the other for an academic philosophy journal—arguing that the human sciences needed to embrace the same methodology as the natural sciences.[30] Schallmayer insisted that ethics and morality were no exceptions, but could and should also be brought under the sway of natural science. He asserted that if our views on moral righteousness are not in harmony with natural laws, then we must amend our views, "for we are subject to natural laws, and they do not ask about our crazy ideas."[31] Ethics was so integral a component of Schallmayer's eugenics program that he often referred to eugenics as "generative ethics."[32] In his most important book, *Heredity and Selection*, which went through four editions by 1920, he opened and closed the book with remarks about the necessity of altering present ethics. Darwinism would form the basis for this new ethics, since, according to Schallmayer, "Theoretically the theory of evolution leads undeniably to the demand for the continued development of ethics in the sense of evolutionary ethics."[33]

Schallmayer believed the function of ethics was to help social organisms triumph in the struggle for existence between societies. Thus, the measuring rod for morality was the survival and reproduction of the greatest number. He criticized utilitarian ethics, since happiness and pleasure may or may not correspond with survival and reproduction. He pointed out that hedonism sometimes leads to biological and social decline, as it did for the Greeks and Romans.[34] In an article on the relationship between ethics and eugenics, he summed up his criteria for evaluating moral precepts:

> The value of existing or intended rules of social life and especially of sexual life may therefore be measured first and foremost according to their contribution to the success of the human evolutionary process, secondly according to their usefulness for society, and thirdly according to their adaptation to the individual desire for happiness and avoidance of pain.[35]

Not only was the evolutionary process the standard by which to judge moral norms, but natural selection would ensure that the highest form of evolutionary ethics would ultimately prevail. The first society to adopt evolutionary ethics would have an advantage in the struggle for existence and would likely prevail over those retaining more traditional morality.[36]

Schallmayer attacked Christian morality as an impediment to evolutionary progress. He asserted that "the views of Christianity, insofar as they are at all influential, do not have the tendency to improve selection, either consciously or unconsciously, but rather—naturally unconsciously—has the opposite tendency." He deemed some of the ideals of Christian morality, such as humility and despising earthly goods, harmful in the struggle for existence and thus irreconcilable with true morality.[37] How, then, did Christian morality become so prevalent, if it conferred such disadvantages to its bearers? Schallmayer suggested that some aspects of religious ethics can be advantageous in the struggle for existence. However, those aspects of Christianity most disadvantageous, such as loving one's enemy, survived only because they remained a dead letter, never being put into practice.[38]

Ploetz was just as aware as Schallmayer that his eugenics program entailed significant changes in morality. In the foreword to the first issue of his journal he stated that "the modern natural scientific-biological view offers even moral philosophy new points of departure."[39] In the first sentence of the first article in his journal he set forth the principle that would guide his ethical philosophy: "Everyone who would like to participate with the work on the development and realization of human ideals, and who struggles to find a clear guiding principle for this work, will again and again encounter the elementary fact, that all spiritual values of the beautiful, true, and good are firmly tied to the living body."[40] Ploetz explained that monism and psychological determinism were the philosophical foundation for his entire program. Thus, it is not surprising that he—like so many other eugenicists—thought morality must ultimately be linked to physical life.

Ploetz's formulation of the highest moral principle was quite similar to Schallmayer's and Tille's. In his 1895 book—the only full-length book Ploetz wrote—he stated, "The first measuring rod of all human activity is the preservation of the healthy, strong, flourishing life." This is not an individualistic enterprise; for Ploetz believed that race hygiene—his term for eugenics—should promote the greatest welfare for the greatest number of people.[41] Ploetz was not thereby supporting utilitarianism, for—like Schallmayer—he saw welfare not in terms of happiness or pleasure, but as survival and reproduction. Ploetz's stress on promoting the welfare of others may seem odd in light of his frequent criticisms of the contraselective effects of humanitarianism. However, Ploetz saw his eugenics program as a way to harmonize the seeming contradiction between humanitarianism and Darwinism. As a young man he was committed to socialist ideals, and the subtitle of his book was *An Explanation of Race Hygiene and Its Relationship to the Humane Ideals, Especially Socialism.* Rather than dispensing with humanitarianism or socialism, Ploetz considered eugenics the only way to preserve humanitarian ideals without producing biological degeneration. Eugenics could foster evolutionary progress through rational artificial selection, thus obviating the need for the cruel, laissez-faire struggle for existence that brings progress to the rest of nature.[42]

Like Haeckel, however, Ploetz thought Christianity went too far with its ethic of love and compassion. We need to recognize, he declared, "that it does not suffice to turn the left cheek, when the right cheek is struck, but *that in the moral view of the future besides love for one's neighbor the preservation and increase of fitness, manliness, and beauty of people must also be fully guaranteed.*"[43] Christian love, therefore, needs to be balanced with self-preservation and a concern for biological improvement.

Ploetz's brother-in-law and coeditor, the sociologist Anastasius Nordenholz, largely shared Ploetz's views on the goals of morality. According to Nordenholz, the highest moral principle is: "Everything that promotes the increased reproduction of the more fit racial elements, even if [it is] at the expense of the unfit."[44] He was fully aware of the harsh implications of his moral views, claiming,

> Solidarity has its extent and its limits in the need of a progressing society to get rid of its inferior elements or those no longer sufficiently able to adapt. Otherwise, the progress of society would soon become lame through such ever-accumulating ballast; indeed society itself would be strangled by it. Solidarity, altruism, love for one's neighbor, and humaneness are all only social [i.e., moral], to the extent that the progress of society is fostered through them.[45]

These views obviously flew in the face of traditional Christian or humanitarian ethics, but Nordenholz countered their objections by asserting that in the long run his own morality was more humane. He compared the harshness of his views with a painful operation, which inflicts pain only to bring greater good in the end.[46]

Judging from the outpouring of articles and books on the subject, interest in the ethical implications of Darwinism and eugenics exploded in the first decade of the twentieth century. The vast majority of those promoting evolutionary ethics and eugenics were scientists or physicians, including biologists, anthropologists, psychiatrists, and medical professors.

Most German philosophers and many sociologists, on the other hand, opposed the application of science to ethics. The philosopher Wilhelm Dilthey wielded considerable influence in German academe in the late nineteenth and early twentieth centuries. He had argued that the natural and human sciences were incommensurable, since they necessarily employed different methodologies.

The only prominent German philosopher to promote an ethic based on Darwinism was Christian von Ehrenfels, who accepted the chair in ethical philosophy at the German University of Prague in 1896. Ehrenfels is better known today for his early work on Gestalt theory, but he devoted much of his career to promoting eugenics, which he considered the logical consequence of evolutionary ethics. His first article on evolutionary ethics appeared in 1892 in a popular journal. In it he confessed that despite the many similarities between his ethic and Nietzsche's, he had not derived his ideas from Nietzsche, but rather from Darwinian science. He overtly rejected Christian and utilitarian ethics, preferring instead a morality that relies on healthy instincts to further human evolution. This evolutionary ethic would strive to increase life and growth, beauty and strength, in order to produce a more noble kind of humanity in future generations.[47]

Nine years later Ehrenfels privately published a book spelling out the implications of his evolutionary ethics, which he distributed to friends and colleagues, apparently wanting feedback before openly divulging all his views. In this book, Ehrenfels drew a sharp distinction between an organism's fitness and its value. Fitness can only be determined by the survival of an organism, while Ehrenfels defined the value of an organism as its ability to perform mental functions. Thus, humans are the highest organisms. Later in the same book, however, Ehrenfels asserted that greater value is usually linked to greater physical strength and health, but it is unclear why this would be so, unless he thought that mental and physical vigor normally coincided. In any case, Ehrenfels believed that morality should sanction whatever promotes health, especially mental vigor, but also physical prowess.[48]

Because more valuable organisms were not always more fit, according to Ehrenfels, biological decline or degeneration was possible. He feared that in human society this process of degeneration was already underway, since various social institutions provided "less valuable" people with reproductive advantages over the "more valuable" ones. Ehrenfels constantly distinguished between "inferior" and "superior" people in his writings, and his primary goal was to suggest ways to favor the "superior" over the "inferior" in order to improve the human species. While he suggested several kinds of reforms, including abolishing the inheritance of capital, the centerpiece of his proposal was an alteration of sexual morality to favor polygamy for the "most valuable" males (see chapter 7).[49]

In 1903, Ehrenfels began publishing his views in earnest, and for the next 15 years he wrote tirelessly about Darwinian ethics and eugenics. In an article on "Evolutionary Morality," Ehrenfels favored a replacement of Christian and humanitarian ethics with an ethic squarely based on Darwinism. According to Ehrenfels, "Evolutionary theory arose, and a generation after its founding the freest spirits, the most far-sighted pioneers, proclaim the ideal of the regeneration of our blood as the highest ethical goal, as the most pressing demand of evolution."[50] Over the next decade, he published many articles promoting biological regeneration through moral reforms. Like most other advocates of evolutionary ethics, Ehrenfels argued that morality is not a fixed system, since what is biologically beneficial under some circumstances may not be so in other situations. Like Jodl, he suggested that morality is often a step behind present conditions, necessitating a constant reform of morality. His task as a moral philosopher was to try to find moral precepts that would match current conditions, thus fostering evolutionary progress.[51]

Under the influence of Darwinian biology and eugenics, the idea that health should become one of the highest moral principles became prominent in the medical profession in the early twentieth century. Felix von Luschan, professor of physical anthropology at the University of Berlin and an early leader of the eugenics movement in Berlin, stated in a 1909 lecture to the Society of German Scientists and Physicians, that "every means is good, if it raises the fruitfulness of the fit and limits that of the unfit." He hoped that anthropology would contribute to this enterprise and thus fulfill a practical function, in addition to its theoretical purposes.[52] Indeed, Luschan wanted not only anthropology, but the entire university to contribute to the cause of promoting eugenics. He stated in a lecture that the "Duty of Health" must come to dominate the university, then the school system, and finally, the whole nation.[53]

A pathology professor at the University of Bonn, Hugo Ribbert, not only thought that promoting physical health was a moral imperative, but he also linked health and morality even more tightly by claiming that those who are healthy are necessarily moral. On the face of it, this seems a rather odd position to take, but Ribbert stated it quite boldly:

> The man who is thoroughly healthy in every respect simply cannot act badly or wickedly; his actions are necessarily good, necessarily, that is to say, properly adapted to the evolution of the human race, in harmony with the cosmos. . . . The healthy human being knows nothing of evil because he is

capable of no other actions than those that are good, that is to say, in harmony with the aims of human evolution.[54]

Ribbert could take this position, because he—like Darwin and most eugenicists—believed that morality rested on social instincts that would prevail in any healthy person. He optimistically asserted that altruism is the natural state of a healthy individual. But since one's moral character determines one's health, and health determines morality, Ribbert's position seems tautological. Any individual with immoral character is automatically defined as unhealthy. Nonetheless, Ribbert insisted that if humans could conquer disease, immorality would vanish.[55]

These ideas spread rapidly, and not just among scientists and physicians. Theodor Fritsch, a highly influential anti-Semitic publicist, developed similar ideas about Darwinian ethics and eugenics. In one article he confronted head-on the question of the origins of morality, concluding that "Morality and ethics arise from the law of preservation of the species, of the race. Whatever ensures the future of the species, whatever is suited to raise the species to an ever higher level of physical and mental perfection, that is moral."[56] For Fritsch this was not just a minor side issue, but rather a central doctrine of his anti-Semitic worldview. Therefore, he continually stressed the priority of eugenics and health consciousness for maintaining the vitality of the German people. He stated, "The preservation of the health of our race (*Geschlecht*) is one of our highest commands," and he did not recoil from the harsh implications of this view, for he continued:

> We do not approve of any *false humanity*. Whoever seeks to preserve the degenerate and depraved limits space for the healthy and strong, suppresses the life of the whole community, multiplies the sorrows and burdens of existence, and helps rob happiness and sunshine from life. Where human power cannot triumph over sorrow, there we honor death as a friend and redeemer.[57]

The idea that Darwinism exposed the futility of Christian and humanitarian compassion for the downtrodden, and that characterized concern for the weak as "false humanity," became ever more widespread as the eugenics movement (and Nietzschean philosophy) became ever more popular in early twentieth-century Germany.

The constant stress on physical health in Darwinian and eugenics circles contributed to various movements promoting healthy lifestyles, which often took on the character of a moral crusade. Fritsch, for example, wanted to abolish the use of tobacco, coffee, and tea in his utopian eugenics community, and many other eugenics enthusiasts abstained from these and other unhealthy substances. The importance of Darwinism and eugenics in influencing the budding health reform movement is debatable, but it is clear that they played a significant role. Even advertisements reflected this, suggesting that these ideas must have been widespread. In a 1908 issue of a journal promoting sex reform and eugenics, a company promoted its malt coffee-substitute as a way to win the Darwinian struggle for existence: "One can only move forward in the struggle for existence, if one possesses a healthy body and healthy nerves. Therefore beware of everything that can destroy your most precious possession in life, your health."[58]

Eugenics also played a key role in the temperance movement. Gustav von Bunge, a physiologist at the University of Basel, became convinced that alcohol was unhealthy, not only to the individual drinker, but also to his or her offspring. He became the leading figure in the early temperance movement in Germany, recommending complete abstinence from alcohol to avoid hereditary degeneration. Many biologists in the early twentieth century agreed, since scientific evidence suggested that alcohol damaged gametes, and this allegedly would harm the hereditary health of the following generations. Bunge's views on temperance became rather influential in the eugenics movement, which he fully supported, accepting honorary membership in the Society for Race Hygiene when it was founded in 1905.[59]

Forel, who deserves to be called the grandfather of the German eugenics movement, was also a tireless supporter of temperance. As a psychiatrist, he saw the deleterious effect of alcohol on many of his patients, and he was concerned that alcohol might cause hereditary diseases in drinkers' children. In 1888, after a two-year experiment in abstinence, he decided to permanently abstain from alcohol.[60] He became president of the Swiss section of the Knights Templar, a temperance society, and led the Swiss branch to break away from the international parent organization, when it opposed his stance on religious neutrality.[61] Nonetheless, he was a tireless organizer, founding new temperance lodges, not only in Switzerland, but also in France and North Africa.[62] Through his direct influence other leading figures in the eugenics movement, including Ploetz, vowed to abstain from alcohol. Ploetz was so serious about proving the debilitating hereditary effects of alcohol that he daily fed several thousand rabbits alcohol to observe the effects. After spending a huge sum on this endeavor (1.5 million marks, if his son's memory is correct) and finding no discernible signs of degeneration, he abandoned the experiment.[63]

The moral crusade for health reform—including temperance, but also abstinence from other harmful substances—seems to flow logically from the exaltation of hereditary health, physical strength, and mental prowess to new moral imperatives in the Darwinian worldview. However, though the various health reform movements were controversial, most were not a serious threat to the moral order, and some were welcomed enthusiastically by those upholding traditional morality. More serious threats to the moral status quo would emerge as Darwinian and eugenics ideas were applied to other areas of morality, especially sexual morality and medical ethics. After sketching out the attempts to organize evolutionary ethics in the early twentieth century in chapter 3 (in part to show the influence of these ideas), we will discuss these more controversial topics at length in subsequent chapters.

3. Organizing Evolutionary Ethics ⤳

By the 1890s and early 1900s Germany had reached a state of ethical and moral crisis. Many factors contributed to the increasing sense of malaise and disorientation in the realm of morality. German (and European) intellectual life had become increasingly secularized during the nineteenth century, a process that Darwinism furthered. Though the anti-clerical philosophers of the eighteenth-century Enlightenment had generally retained the fundamental tenets of Judeo-Christian ethics, many intellectuals of the nineteenth century no longer found traditional morality satisfying. Nietzsche's rejection of Christian morality resonated with many young thinkers around the turn of the twentieth century. The rise of historicism in the nineteenth century brought many leading intellectuals in Europe to abandon moral certainty.[1] However, though many intellectuals pressed for significant reforms in morality, few wanted to abandon morality altogether. Rather they sought a secular replacement for Judeo-Christian ethics. Darwinism would play a prominent role in this search for a secular ethics and morality, especially among the scientific and medical elite.

Urbanization was another factor contributing to the sense of dislocation and disorientation, causing many new urbanites to jettison traditional religion and ethics. The famous French sociologist Emile Durkheim coined the term anomie to describe this rising sense of unease among urban dwellers. The German sociologist Ferdinand Tönnies analyzed the same phenomenon, noting that traditional social ties in communities (*Gemeinschaft*) were being displaced by more impersonal urban society (*Gesellschaft*). The growth of crime and prostitution in urban areas aroused concerns about the widespread breakdown of morality.

The German Society for Ethical Culture, founded in October 1892, was the first attempt to organize secular ethics in Germany. Impetus for establishing the Society came from Felix Adler, an American professor who had already organized a similar society in the United States. In 1892 he visited

Berlin and encouraged his colleagues in Germany to create an organization dedicated to promoting the divorce of ethics from religion. The time was propitious for such an endeavor, since Prussia passed a new school law in 1892, which tried to enforce greater religious orthodoxy, thereby arousing the ire of secular intellectuals. The new Society was able to capitalize on this dissatisfaction, quickly recruiting 2000 members and establishing local societies in many German cities in its first few months of existence.

The Ethical Culture Society was not overtly Darwinian, so it could not properly be considered an organization promoting evolutionary ethics. Indeed it refused to endorse any particular worldview. Obviously, those upholding more traditional religious views were staunch and vocal opponents, since they were unwilling to separate ethics from religion. Nonetheless, among secularists the Society tried to be as inclusive as possible, refusing to take sides in debates over metaphysics. It was precisely this agnosticism on metaphysical questions that alienated those who believed that ethics and morality could not be understood apart from a more comprehensive worldview. Haeckel, for example, attended the founding meeting of the Society in Berlin, but he refused to join, since he insisted that ethical discussions had to be based on a proper worldview (his monistic philosophy).[2]

Though snubbed by Haeckel, other leading Darwinian-inspired philosophers and social reformers were more sanguine about the possibilities of the Ethical Culture Society. Even though Darwinism was not a central point of discussion of the Society, many (but not all) of the intellectuals and social reformers involved in the Society embraced Darwinism and believed that it had implications for ethics and morality. One of the most important figures in the founding of the Society was Gizycki, a moral philosopher influenced by evolutionary theory (see chapter 1). Gizycki hoped the Society would supplant religion and become the "church of the future."[3] He also became editor of its weekly journal, *Ethische Kultur*.

Besides Gizycki, Wilhelm Foerster, an astronomer at the University of Berlin, played a leading role in establishing and guiding the Ethical Culture Society. He presided at its founding meeting, and in his closing remarks to that gathering, he revealed how evolutionary principles affected his ethical thought. He stated

> In the deepest sense of the biological theory of evolution, we are of the conviction that the human soul is the ultimate aim of this evolution on the earth, insofar as the refinement of its senses and the ennobling of its thinking will produce the feeling and the knowledge of a more comprehensive fellowship of happiness and unhappiness, of pleasure and pain, and the unifying consciousness of humanity.[4]

Foerster, just like Gizycki, thus considered human morality the product of the evolutionary process. The feeling of human solidarity, which he believed must be the basis for all morality, arose as a product of biological evolution. However, it is unclear how much Foerster relied on biological evolution in actually deriving any of his ethical principles.

Foerster's eldest son, Friedrich Wilhelm, whose 1893 dissertation explored Kantian ethics, also became a leader in the early years of the Ethical Culture Society, coediting the Society's journal for several years in the 1890s. In his memoirs he explained the powerful influence Darwinism exerted on him and other young people in the 1880s and 1890s. He was raised in a thoroughly secular family, whose home was "a place of worship for the natural–scientific worldview." Darwin's *Origin of Species*, which he read at age eighteen, had a tremendous influence on him, convincing him that humans were "an animal among other animals." Thereafter he imbibed many works advocating a "scientific worldview," including Haeckel's *Natural History of Creation* and Strauss's *Old Faith and the New*, and he even studied some biology under Weismann at the University of Freiburg. Foerster ultimately abandoned the search for a secular ethics and quit the Society for Ethical Culture after converting to Catholicism in 1899. His growing doubts about the ability of Darwinism to explain the origins of the human conscience was one factor that drove him away from a naturalistic worldview.[5]

The first secretary of the Ethical Culture Society, Rudolph Penzig, who frequently lectured at meetings of the Society throughout Germany and later became editor of its journal, discussed the connections between Darwinism and ethics at greater length than the Foersters. First of all, he argued that biological evolution undermines any religious foundations for morality. Rather morals are autonomous, deriving their sanction from people, whether individually or socially. Thus, evolution underpinned Penzig's call for the secularization of ethics, as it did for many members of the Ethical Culture Society. But evolution served yet another role in the formulation of ethics, according to Penzig. Morality produces human solidarity, which benefits the species, allowing it to survive. Therefore, "ethics derives its law from biology, the *universal* science of life: preservation of the species through preservation, adaptation, and replication of the individual." Despite this reference to the individual, Penzig believed that individuals could only flourish biologically when the community takes precedence over the individual. He stated, "One may certainly without exaggeration characterize as the most essential task of our century [the task] to place the relationship of the individual to the species on a new foundation." He believed that Christianity had unduly exalted the value of the individual, while Darwinism showed the importance of the national and racial community.[6]

Jodl, the leading ethical philosopher in Germany under the influence of Darwinism, eagerly joined the Society for Ethical Culture, hoping that it would provide a platform for his moral philosophy (see chapter 1). Even before the founding of the Society, he was a friend of Gizycki's and had contact with the Ethical Culture Societies in the United States and Britain. As Gizycki and Foerster hammered out the agenda for the Society, Jodl persuaded them to alter the original document, since it seemed to welcome with open arms those with various religious persuasions. Jodl wanted to keep religious partisans away, so that the organization could be resolutely secular in its approach.[7] Jodl joined the leadership of the budding organization in 1893, and the following year co-founded the Austrian Ethical Society.[8] Jodl later withdrew from the German Ethical Culture Society, because he opposed the growing influence of socialist ideas in its ranks, especially through the influence of Gizycki and his wife, Lily (more famous under her later name, Lily Braun). After Gizycki's death in 1895, his widow withdrew from leadership in the Society to devote her full energies to promoting Social Democracy, but Jodl refused to rejoin the Ethical Culture Society, though he continued to be sympathetic with its goals.[9]

Another important figure in the early days of the Ethical Culture Society, Moritz von Egidy, was a charismatic and zealous moral reformer who gained quite a following in Berlin in the 1890s. Egidy, a retired cavalry officer, promoted sweeping social reforms by appealing to religious sentiment. However, though he advocated a "United Christendom" and used some Christian terminology, his worldview was far removed from orthodox Christianity. Rather his religious position seems much closer to Haeckel's monism or pantheism. He explained,

> In the place of the contemporary conception of God, a personal God, God as a Spirit, God as a Being, Triune, or even a unitary God, the conception of a "holy law of evolution" will emerge. . . . The thought of pure materialism cannot satisfy; we need something that will meet our desire for imagination and that does not contradict serious and honest thought. We have this in the conception of a "holy law of evolution," a concept, which we piously call "Providence."[10]

Egidy believed that only by replacing a personal God with the "holy law of evolution" could religion avoid contradicting science, and he, like so many of his contemporaries, thought that opposing anything bearing the name of science was tantamount to committing intellectual suicide. Egidy's influence among German social reformers was substantial, but ultimately fleeting, for his circle dispersed upon his death in 1898.

Many other leading figures promoting a Darwinian worldview participated in the Ethical Culture Society. Carneri, the leading writer on evolutionary ethics, was a founding member of the German Society and also
participated in the Viennese branch, but because of age and infirmity,
he was never particularly active in the Society. He did, however, suggest
that the Society concentrate on moral instruction in the schools, and
he advised the Viennese branch to follow his friend Jodl's path (rather than
following the Berlin Society's socialist inclinations).[11]

By the first decade of the twentieth century, the Ethical Culture Society
was in decline, and it had obviously failed to create the "church of the
future," as Gizycki had hoped. Many other attempts to establish and institutionalize nonreligious ethics ensued in the first decade of the twentieth
century, many of them more overtly Darwinian than the Society for Ethical
Culture. Many of these organizations had overlapping goals and overlapping memberships, thus competing with each other for the time and
financial resources of their members.

One of the more ambitious attempts to organize evolutionary ethics was
spearheaded and financed by Albert Samson, a Berlin banker who retired
to Brussels in the early 1890s after acquiring a fortune. Intensely interested
in science, he studied medicine at the University of Berlin, though he did
not pursue a career as a physician. In the course of studying natural science,
he became enthusiastic about Darwinism, agreeing largely with Haeckel's
monistic philosophy. He wanted to use his wealth to promote scholarly
research—especially in the natural sciences—related to evolutionary ethics.
In 1899 he approached Haeckel with an offer to fund an institution for the
purpose of investigating evolutionary ethics. Haeckel was excited about the
prospect and began discussing with Samson the founding of an Academy
of Physiological Morality, which would investigate the biological bases of
ethics and morality.[12] While Haeckel hoped to establish the academy at his
home institution, the University of Jena, Samson proposed that it be situ-
ated in Brussels, since he wanted it to be international in scope.

In March 1900 Haeckel sent his assistant, the zoologist Heinrich Ernst
Ziegler, to Brussels to negotiate with Samson about the proposed institute.
Because of Samson's misgivings, they ultimately agreed that he would provide 5,000 marks to found a more modest international organization of
scholars committed to investigating evolutionary ethics.[13] Haeckel quickly
recruited three prominent scientists to help lead the fledgling organization:
Wilhelm Waldeyer, an anatomist and physical anthropologist at the University
of Berlin (who knew Samson personally, since Samson had attended his
classes); Paul Flechsig, a neurophysiologist at the University of Leipzig, and
Hermann Munk, a neuroanatomist at the University of Berlin. In 1901
Haeckel sent letters inviting prominent scholars interested in the ethical

and social implications of Darwinism to join the Ethophysis Society. According to its statutes, the society's aim was "the formation of a natural-scientific worldview and of an ethics founded on it," especially by investigating the anatomy and physiology of the brain, physiological psychology, and the ethical implications of other scientific findings.[14] The Ethophysis Society apparently experienced a stillbirth, for it disappeared without a trace.

Thereafter Samson sought other avenues for funding research in evolutionary ethics. Sometime before 1904 he set up a research foundation with the Bavarian Academy of Science, agreeing to endow it with 500,000 marks upon his death. The purpose of this foundation was: "Scientific research and explanation of the morality of individuals and of social morality in light of the results of scientific and historical research, especially of empirical and experimental psychology; further, discovery of the implications of the results of this research for the life of individuals and society."[15] Apparently he believed in spreading his money around, for by 1904 he was negotiating with Waldeyer to set up a similar foundation with the Prussian Academy of Science for the investigation of "the natural, biological foundations of morality."[16] When Samson died in 1908 the Bavarian Academy received their 500,000 marks, while the Prussian Academy received almost a million marks.[17] Both Waldeyer and Munk, leaders of the defunct Ethophysis Society, were named to the Samson Foundation board. Despite Samson's massive funding, his hopes remained unfulfilled, for most of the money from the Prussian foundation funded projects only distantly related to ethics. The most important project supported from the Prussian Samson Foundation was an observation station for apes on Tenerife. Most of the scientific work at Tenerife—such as Wolfgang Koehler's famous work on Gestalt psychology—focussed on cognition rather than ethics.[18]

It is not clear what Samson hoped to accomplish through his funding. He obviously was devoted to the idea that science could not only provide insight into the origins and nature of morality, but could also give practical direction in formulating morality. We do not know much about his own stance on ethics, but we do know that he considered the Bible the highest expression of moral teaching and highly esteemed Jesus' teachings on morality. Waldeyer replied to Samson that the Bible does contain a kernel of moral truth, but he asserted that "we cannot set up moral laws valid for all times," since morality evolves, just as everything else in the cosmos. Waldeyer also warned that one must not take Jesus' teaching literally, and he specifically criticized Jesus for not emphasizing "healthy egoism" enough in his moral doctrine.[19] Though Waldeyer did not overtly mention Darwinism as the justification for his ethical views in this correspondence, the context of the discussion was evolutionary ethics, and Waldeyer's position certainly resonated with Haeckel's ethical views.

Despite massive funding, Samson's projects largely failed in propagating evolutionary ethics. However, other efforts early in the twentieth century were more successful. First of all, simultaneous with his negotiations with Samson, Ziegler was helping organize the Krupp Prize Competition under the auspices of Haeckel. Since the prize question focussed on political applications of Darwinism, participants could hardly avoid dealing with the ethical implications of Darwinism. In his expanded explanation of the prize question, Ziegler forthrightly included an ethical dimension to the problem. He advised all participants to consider two main points in their submitted work: heredity and adaptation. Here Ziegler was emphasizing Haeckel's view that heredity and adaptation were the two key elements of evolution—the latter producing biological changes and the former stabilizing species. In discussing heredity Ziegler included the inheritance of human character traits, including "egoistic instincts, family instincts, social instincts, etc." This implies that some kind of moral characteristics are inherited biologically. But these instincts are not immutable, for Ziegler noted that adaptation brings about gradual changes of laws and morals.[20]

In light of Ziegler's guidelines it should come as no surprise that Schallmayer and other contestants wrestled with the ethical implications of Darwinism in their works. Schallmayer considered his eugenics proposals as a straightforward ethical application of Darwinism to society. Not all prize-winning authors agreed with Schallmayer that ethics could be based on science. Arthur Ruppin and Albert Hesse, two of the three second-place winners, both maintained that Darwinism could not provide ethical guidance. The difference between Schallmayer's views and those of Ruppin and Hesse may reflect their educational background, as Schallmayer was a physician and Ruppin and Hesse studied social science, especially economics. However, even Ruppin and Hesse argued that moral traits were hereditary, and both endorsed eugenics. Hesse even argued that if a society's morality conflicted with the laws of evolution by hindering natural selection, such a society would fall prey to societies whose morality did not contravene the laws of nature. Moral codes of societies could be naturally selected, just as biological traits of individuals.[21] By publishing ten of the best entries to the Krupp Prize Competition, Ziegler powerfully advanced the cause of social Darwinism and eugenics.

The rise of social Darwinism and eugenics in early twentieth-century Germany helped spawn several organizations seeking to apply Darwinism to ethics and society. First and foremost was the Monist League, founded in 1906 at the behest of Haeckel. Haeckel had been interested for some time in giving organizational form to his monistic philosophy. The primary purpose of the Monist League was to replace religious and dualistic world-views—the targets here were primarily Christianity and Kantian philosophy—with a monistic worldview. Naturalistic ethics based on evolution featured prominently in this monistic worldview. One of the theses Haeckel formu-lated to serve as the basis for the Monist League called for a monistic ethics based on evolutionary theory, in which altruism would be balanced with egoism.[22] A leading article in the first issue of the Monist League's journal also made clear that one of the most important reasons for the League's existence was to promote secular moral instruction for youth.[23]

The ethical implications of monistic philosophy and particularly of evo-lutionary theory became one of the favorite topics of the Monist League in the first decade of its existence, judging from the outpouring of journal articles and speeches on this subject. One of the more important popular-izers of evolutionary ethics among the Monists was Johannes Unold, a Munich physician who served as vice-president of the League and for sev-eral months in 1910–11 as president. He wrote many books and articles promoting a monistic ethics based on science, especially Darwinian theory. Unold sharply criticized any ethics based on human happiness, such as util-itarianism, since science demonstrates that the "*first law and the most uni-versal purpose* of the entire organic world is the *preservation of the species* through *preservation, adaptation and reproduction of the individual.*" Ethics, according to Unold, must reflect this scientific insight by placing the preservation of the species above happiness or any other value: "To the question: 'What *should* I do? How do we order our life?' the scientific ethics answers first of all: '*Do everything that contributes to the preservation of your-self, your people, and your species!*'"[24] Because the evolution of higher life forms depends on the "suppression and destruction of the lower" forms, Unold criticized Christian compassion.[25] Unold's ideas, derived in part from Ratzenhofer's, reinforced the growing social Darwinist discourse of the early twentieth century and propagated evolutionary ethics within the Monist League.

Though Jodl's utilitarian ethical philosophy was quite different from Unold's, after recovering from his disappointment in the Society for Ethical Culture, Jodl hoped he could infuse the Monist League with his ethical and social concerns. In 1911 Jodl spoke to the First International Monist Congress in Hamburg on "Monism and the Cultural Problems of Today." Jodl noted that "modern monism is first of all a child of natural science, of evolutionary theory," and it seeks to apply evolutionary theory to all fields. Because of this, many wrongly think monism abandons ethics and morality in favor of the struggle for existence and survival of the fittest. Jodl argued that this was a misunderstanding. Rather, biological evolution has produced ever higher forms of morality, since humans as social animals have banded together and produced human culture. Human morals, however, are not fixed, according to Jodl, but continue to evolve, adapting to ever-changing conditions.[26]

Jodl's speech struck a responsive chord with his audience, which included representatives of the Ethical Culture Society and other social reform organizations. The ethical concerns he aroused dominated the next two annual meetings of the Monist League, which from that time on became more active in pressing for social reform.[27] After hearing Jodl's speech, the president of the Monist League, Wilhelm Ostwald, asked Jodl if he would author a textbook on moral instruction from a monistic standpoint that could be used in schools. Jodl refused, but Ostwald's request shows how important the issue of morality was for the Monist League.[28] Another Monist encouraged by Jodl's speech (which he read later) was Ziegler, who thought the Monist League had been negligent by not promoting a positive ethical program. He applauded Jodl for calling upon fellow Monists to begin formulating a positive approach to moral and social reform.[29] Jodl died less than three years after giving that speech and did not participate much in the Monist League after his 1911 speech, but his influence was lasting. Ostwald later claimed that Jodl's speech provided a new direction for the Monist League and set the tone for its subsequent activities.[30] Of the organizations pressing for the secularization of ethics in Wilhelmine Germany, the Monist League was probably the most influential, numbering 6,000 members in 1914.[31]

One prominent member of the Monist League, the world famous psychiatrist August Forel, was not completely satisfied with the League's approach to ethics. Forel was an especially ardent crusader for moral reform, who retired in 1898 at age 50 from his career in psychiatry to campaign for moral and social reforms. In 1908, two years after refusing Haeckel's offer to become president of the Monist League, Forel founded the International Order for Ethics and Culture. Though he claimed it would complement rather than compete with the Monist League, the two organizations overlapped considerably in function and membership. Forel believed that the Monist League and other freethinker organizations would ultimately fail to win many segments of the populace to their worldview, because they appealed almost exclusively to reason and did not captivate human feelings. He wanted to build a community of ethical monists by patterning some aspects of his Order on the Christian churches, whose rituals appealed to people's emotions. For instance, the local groups within the Order were supposed to hold weekly meetings for the purpose of moral edification and exhortation and also to conduct weddings and funerals for members. Forel's hope that he could build a "well-organized army with strict discipline" to supplant the Christian churches was obviously not fulfilled, since his organization remained even smaller and less influential than the Monist League.[32]

Another influential figure in the drive for a new, secularized ethics, Helene Stöcker, helped found the League for the Protection of Mothers in 1905 and edited its official journal. She also had connections with the Monist League, lecturing at their annual meetings and writing articles for their journal. In her persistent call for a "new ethics," Stöcker appealed to Nietzsche's overthrow of conventional morality, while promoting a Darwinian-inspired eugenics program. Her synthesis of Nietzsche and Darwin was not as idiosyncratic as it may seem at first, for other of her contemporaries—including her erstwhile lover, Tille, also interpreted Nietzsche as a proto-eugenicist, as we have seen (see chapter 2). As many of Stöcker's writings make clear, the League's practical activity on behalf of single mothers was part of its wider concerns with reforming ethics and morality. When Werner Sombart suggested in a 1908 meeting of the League's leadership that they restrict their program to practical activity and forget about theoretical concerns related to reforming ethics, Stöcker and other leading members of the League replied that their practical activity could not be separated from its theoretical underpinnings. Thus the League for the Protection of Mothers played an important role in the movement to secularize ethics in Wilhelmine Germany.[33]

Many other secular organizations formed around 1900 to promote various health or social reforms. Some aimed at overcoming the perceived biological decline in German society. Others focussed their efforts on counteracting moral decay. With the rise of biological determinism in the late nineteenth century, many believed that biological and moral decline were integrally related, so both problems had to be tackled simultaneously. Thus many organizations, such as Ploetz's Society for Race Hygiene, promoted biological renewal as a means to moral rejuvenation. Ploetz's organization was extremely influential among physicians, but it never appealed to the masses, since he favored an elitist approach in organizing the eugenics movement. In addition to Ploetz's organization, many other organizations promoting social reform had at least some inclination toward eugenics, and for some eugenics was the guiding principle.[34] The antialcohol movement in Germany, for example, was driven by eugenics concerns, since many physiologists feared that alcohol caused biological degeneration in one's gametes, thus resulting in various kinds of hereditary illness, especially mental illnesses. Many—perhaps most—German psychiatrists also considered alcohol a key cause of mental illness, not only for the user, but for the offspring as well (see chapter 2).[35]

A quite different kind of organization aimed at moral renewal and improving the health and vitality of the German people was the German Renewal Community, founded by Theodor Fritsch, a prominent anti-Semitic publicist. Fritsch wanted to promote moral regeneration through establishing garden communities, that is, utopian settlements in the countryside that would practice eugenics and health reform. In 1908–09 he actually bought a landed estate and tried to launch his experiment in communal living.[36] Fritsch's whole worldview, and particularly his stress on health and moral reform, centered on his understanding of Darwinism and its implication for ethics. This reliance on evolutionary ethics and eugenics is reflected in the "Fundamental Principles of the [German] Renewal Community":

> *The preservation of the health of our generation* belongs to our highest commands. . . . We do not approve of *false humaneness*. Whoever aims at preserving the degenerate and depraved, limits the space for the healthy and strong, suppresses the life of the whole community, multiplies the sorrow and burden of existence, and helps rob happiness and sunshine from life.[37]

With his rejection of humanitarian ethics in favor of an ethics based on health and biological vitality, Fritsch's views reflected the thought of many other thinkers applying Darwinism to ethics. Nonetheless, his organization did not achieve a widespread following.

Though many of these organizations did not survive more than a couple of decades, the emergence of so many organizations devoted to evolutionary ethics and eugenics around 1900 shows the popularity of Darwinian-inspired social and ethical thought at that time. The membership of these organizations included many leading professors, physicians, and writers, who zealously spread their views throughout Germany. However, the demise of these organizations came about in part because they could not agree on what the ethical implications of Darwinism were. Evolutionary ethics was not a coherent philosophy, but rather, attempts to formulate ethics on the basis of evolutionary theory produced a cacophony of voices promoting contradictory visions of moral or social reform. Each, however, tried to give his or her own particular agenda scientific imprimatur by claiming harmony with the laws of evolution. In addition to disagreements on ethical issues, they also could not agree on the organizational form best suited to win the German people to their cause. The multiplicity of organizations reduced their effectiveness, since they competed for membership and duplicated efforts.

Despite these institutional weaknesses, the Monist League and the League for the Protection of Mothers both persisted until the Nazi seizure of power. However, the Monist League's influence declined significantly after World War I, largely because of Ostwald's resignation from the presidency in 1915 and Haeckel's death in 1919. Despite their support for eugenics, both these organizations were hostile to many other aspects of Nazism, such as its suppression of individual freedom. They both tilted decidedly toward the Left politically, supporting pacifism and other policies diametrically opposed to Nazism. The Monist League's publications overtly criticized the Nazis before their seizure of power, and in 1932 it joined with other freethinking organizations in signing a statement opposing the rise of Nazism.[38] Because of their staunch opposition to Nazis, neither organization was able to survive the Nazi "coordination" (a euphemism for Nazification) of institutions. The Society for Race Hygiene, however, not only survived, but thrived under Nazi rule. Ploetz had opposed Nazism before Hitler came to power, but he was won over to the Nazi regime by their eugenics program.[39]

The Nazi suppression of the Monist League was not a function of a fundamental change in the Monist League's orientation during the Weimar period, as Gasman has argued, but rather reflected significant differences between Haeckel and Hitler. Haeckel and the Monist League promoted many social reforms that were anathema to Hitler, such as homosexual rights, feminism, and pacifism.[40] Gasman's Haeckel-to-Hitler thesis ultimately failed, in part because he ignored the many areas of sharp disagreement between Haeckel and Hitler. However, while acknowledging the many differences, we should not ignore the many features of Monist ideology that featured prominently in Hitler's worldview, such as eugenics, euthanasia, and social Darwinist racism. Haeckel and the Monist League were very prominent in promoting these ideas, which we shall now explore in greater depth in the succeeding chapters. Despite the many disagreements among Darwinian thinkers about ethics, social reform, and politics, Darwinism clearly did make a difference in how many people thought about the value of human life.

2. Devaluing Human Life ✍

4. The Value of Life and the Value of Death ✍

Darwinism was a matter of life and death. No one understood this better than Darwin did. Immediately after explaining that each organism "has to struggle for life, and to suffer great destruction," he closed his chapter on "The Struggle for Existence" on a more comforting note: "When we reflect on this struggle, we may console ourselves with the full belief, that the war of nature is not incessant, that no fear is felt, that death is generally prompt, and that the vigorous, the healthy, and the happy survive and multiply." This put a rather positive spin on the struggle for existence, the "law, leading to the advancement of all organic beings, namely, multiply, vary, let the strongest live and the weakest die."[1] Even while overtly denying any purpose or goal for evolution, Darwin could not resist the mid-Victorian cult of progress, as these passages illustrate with their vision of increasing health, strength, and even happiness.

One of the alluring features of Darwinism, it seems to me, was that it offered a secular answer to the problem of evil and death. Indeed, it was more than an answer—it gave Darwinists hope and inspiration that suffering and death would ultimately spawn progress. Darwin clearly viewed death and destruction as an engine of evolutionary progress, as we see in the penultimate sentence of *The Origin of Species*: "Thus, from the war of nature, from famine and death, the most exalted object which we are capable of conceiving, namely, the production of the higher animals, directly follows."[2] Darwin's jubilation at the power of natural selection to wrest victory from the jaws of death is reminiscent of the biblical promise, "Death is swallowed up in victory."[3] In one respect, then, Darwin's theory of natural selection was a secular answer to Judeo-Christian theodicy (the justification of a benevolent God in a world of evil), since it provided an explanation for the existence of evil and promised that evil would ultimately fulfill a good purpose.

In a speech honoring Darwin's hundredth birthday in 1909 Max von Gruber, a famous professor of hygiene at the University of Munich, expressed exactly this point. He opened his speech by countering the common misconception that nature is peaceful, harmonious, and idyllic. Rather it is "filled with pitiless, gruesome struggle, with torment and death . . . Not only do animals murder animals, but plants murder plants." Darwin, Gruber exulted, had discovered a rationale behind all this seemingly meaningless misery:

> The never-ceasing struggle is, according to him [Darwin], not useless. It constantly clears away the malformed, the weak, and the inferior among the generations and thus secures the future for the fit. Thus only through the inexorable extermination of the negative variants does it provide living space for the strong and its strong offspring, and it keeps the species healthy, strong, and able to live.[4]

Suffering and death, then, were not gratuitous, but fulfilled a higher purpose—the preservation and advancement of all living beings. Even though Gruber thought human reason and pity could and should mollify the struggle among humans, Darwinism helped him find purpose and meaning in the mass destruction of other organisms.

Perhaps the promise of evolutionary progress gave comfort in the face of death and destruction, but, on the other hand, Darwinism increased the sting of death, at least in a quantitative sense. Darwin formulated his theory of natural selection after reading Thomas Robert Malthus's famous *Essay on the Principle of Population*. Malthus based his population principle on the biological observation that most organisms produce far more progeny than can possibly survive. He argued that, like other organisms, the human population tends to increase faster than the food supply, unless checked by other restraints (disease, war, etc.). Because of this imbalance between reproduction rates and food supply, Malthus believed that the vast majority of people must die without reproducing. Death—indeed mass death—was thus central to the Malthusian vision that Darwin appropriated and then propagated. Adrian Desmond in his biography of T. H. Huxley is not exaggerating when he claims that according to Darwin's theory, "only from death on a genocidal scale could the few progress."[5]

To be sure, the struggle for existence among organisms is more often peaceful competition than bloody combat, but Darwin recognized that killing—even within species—is also a normal part of the struggle:

> It may be difficult, but we ought to admire the savage instinctive hatred of the queen-bee, which urges her instantly to destroy the young queens her

daughters as soon as born, or to perish herself in the combat; for undoubtedly this is for the good of the community; and maternal love and maternal
hatred, though the latter fortunately is most rare, is all the same to the
inexorable principle of natural selection.[6]

Darwin may have admired the queen bee's instinctive hatred and infanticide (if I may indulge in anthropomorphizing here), but he certainly did
not see it as a model for human conduct, since he thought human instincts
tended more toward love and altruism. But what if humans had aggressive
instincts that were more powerful than the altruistic ones? Darwin consistently denied this possibility, but not all Darwinists would follow suit.

The Darwinian idea of death as a natural engine of evolutionary
progress represented a radical shift from the Christian conception of death
as an unnatural, evil foe to be conquered. This shift would bring in its train
a whole complex of ideas that would alter ways of thinking about killing
and the "right to life." Before Darwinism burst onto the scene in the mid-
nineteenth century, the idea of the sanctity of human life was dominant in
European thought and law (though, as with all ethical principles, not
always followed in practice). Judeo-Christian ethics proscribed the killing
of innocent human life, and the Christian churches explicitly forbade murder, infanticide, abortion, and even suicide.[7] The sanctity of human life
became enshrined in classical liberal human rights ideology as "the right to
life," which according to John Locke and the United States Declaration of
Independence, was one of the supreme rights of every individual. Until the
second half of the nineteenth century, and to a large extent even in the
twentieth century, almost all Christian churches and most anticlerical
European liberals upheld the idea of the sanctity of human life, even
though they may not have used that terminology. This was reflected in
European legal codes, which strictly forbade assisted suicide, infanticide,
and abortion. According to the historian Udo Benzenhöfer, no one advocated assisted suicide in medieval and modern Europe before the second
half of the nineteenth century (the only possible exception is Thomas
More, but Benzenhöfer argues that More's treatment of euthanasia was
parody, not advocacy).[8]

Only in the late nineteenth and especially the early twentieth century did
significant debate erupt over issues relating to the sanctity of human life,
especially infanticide, euthanasia, abortion, and suicide. It was no mere
coincidence that these contentious issues emerged at the same time that
Darwinism was gaining in influence. Darwinism played an important role in
this debate, for it altered many people's conceptions of the importance and
value of human life, as well as the significance of death. Some historians

writing about euthanasia and eugenics in Germany have commented on this phenomenon, but few have analyzed it in any depth.[9]

What aspects of Darwinism brought about this transformation in thinking about the value of human life? First, Darwinism altered some people's conceptions of the human position in the cosmos and in the organic world. T. H. Huxley had dubbed this the question of "Man's Place in Nature," and many German Darwinists, including Ernst Haeckel, considered this one of the most important aspects of Darwinism.[10] The traditional Christian view of the value of human life was one idea Haeckel wanted to revise in the light of evolution. In his 1904 book, *The Wonders of Life*, he remarked that "the value of our human life appears to us today, on the firm foundation of evolutionary theory, in an entirely different light, than it did fifty years ago."[11] How did Haeckel think it had changed? Stated succinctly, Haeckel did not think that human life was particularly valuable, nor did he think that all people had the same value. This point comes through in many of his writings, and he expressed it quite clearly already in 1864 to his devout Christian father:

> I share essentially your view of life, dear father, only I value human life and humans themselves much less than you. . . . The individual with his personal existence appears to me only a temporary member in this large chain, as a rapidly vanishing vapor . . . Personal individual existence appears to me so horribly miserable, petty, and worthless, that I see it as intended for nothing but for destruction.[12]

Haeckel and many other German Darwinists fought incessantly against all dualistic views of humans, which endued human life with much greater value than animals. For Haeckel and most German Darwinists, humans were not much different from animals, and they often criticized Christians and other dualists for insisting on significant qualitative distinctions between humans and animals.[13]

In rejecting mind–body dualism Haeckel explicitly denied the existence of an immaterial human soul. Haeckel contended that all the activities traditionally ascribed to the human soul were nothing more than material processes originating in the central nervous system.[14] He even admitted that his psychology was materialistic, since he believed that psychology could ultimately be reduced to physiology.[15] Despite his slippery use of religious terminology, Haeckel was clearly a reductionist who denied free will and insisted on mechanistic explanations for everything, including the human soul. Though Darwin (at least in his published works) was never as explicit as Haeckel in denying mind–body dualism, Darwin did nonetheless embrace reductionism by providing natural explanations for all human characteristics, including those traditionally considered unique aspects of the human soul or spirit, such as rationality, emotions, conscience, morality, and even religion.

Forel's psychiatry—and his eugenics—was based squarely on his rejection of body–soul dualism, which, he claimed, had been undermined by Darwinian theory. Darwinism was, according to him, the greatest achievement of the nineteenth century, since it "gave birth to the germ of the discovery of the identity of the human soul with the brain, and therewith dealt the deathstroke to the dualism of body and soul." According to his own account, when he read Darwin's *Origin of Species*, "it was as though scales fell from my eyes" and it converted him to the belief that human psychology can be reduced to brain physiology.[16] At his habilitation defense (a second dissertation qualifying one to teach in German universities) Forel defended the Darwinian thesis: "All characteristics of the human soul originate from those of higher animals."[17] Forel explained to Haeckel that in his view monism is the "scientific proof of the essential identity of the psychological activities of humans and animals and their neurophysiological side."[18] By undercutting the Judeo-Christian and Kantian claim that humans had unique moral status based on an immaterial soul, Haeckel, Forel, and other Darwinists helped undermine the idea that human life is intrinsically sacred and inviolable.

Most leading German psychiatrists adopted views of the human mind similar to Haeckel's and Forel's. They rejected body–soul dualism and embraced a deterministic view of the human psyche. However, it is not always clear what role Darwinism played in the formation of these views. Most German psychiatrists did not leave us a detailed autobiography of their intellectual development. One who did, Kraepelin, revealed in his memoirs that his interest in biological evolution developed already in his childhood. Unfortunately, though, he never explained how or if it influenced the development of his views on psychiatry. What we do know, however, is that Kraepelin embraced a worldview including Darwinian evolution and a deterministic view of mind. He criticized the traditional notion of body–soul dualism as an impediment to the scientific investigation of the mind.[19] Many other prominent German psychiatrists upheld similar views, which denied special status to the human mind or soul.

Whether or not Darwinism entails the rejection of body–soul dualism, as Haeckel and Forel insisted, and whether or not the denial of such dualism entails a devaluation of human life I will leave to the philosophers.[20] What is interesting historically is that Haeckel was by no means alone in his sentiments about evolution devaluing human life. In 1880, the zoologist

Robby Kossmann (who later became a professor of medicine) explained the implications of Darwinism for the significance of human life to a popular audience in his article, "The Significance of the Life of an Individual in the Darwinian World View." Like Haeckel, Kossmann argued that Darwinism should revolutionize one's entire worldview. Evolution had huge implications for the significance of human life, because it "tore down the boundaries between the animal and human world."[21] The Darwinian worldview, according to Kossmann, subordinated the individual to the community, since all individuals necessarily perish—indeed myriads die before reproducing—but the species continues. This means that the value of an individual's life can only be measured by its contribution to the welfare of the community. Kossmann pursued this logic relentlessly, explaining,

> We see that the Darwinian world view must look upon the present sentimental conception of the value of the life of the human individual as an overestimate completely hindering the progress of humanity. The human state also, like every animal community of individuals, must reach an even higher state of perfection, if the possibility exists in it, *through the destruction of the less well-endowed individual*, for the more excellently endowed to win space for the expansion of its progeny. . . . The state only has an interest in preserving the more excellent life *at the expense of the less excellent*.[22]

Kossmann was thus declaring war on the traditional idea of the sanctity of human life, since he thought only some human lives were worth protecting. Later Kossmann wrote an entire book applying the same logic to state politics. In short, he thought politics should be subordinate to Darwinian principles, including eugenics considerations.[23]

Even though Dodel was far more humane in his ethical views than Haeckel or Kossmann, he, too, believed that Darwinism stripped humanity of the special status that religion had accorded it. Dodel, like Darwin and most early Darwinists, recognized that in order to persuade his contemporaries that humans had evolved from animals, he would have to reduce the distance between the two. Humans (especially "primitive" people) had to become more animal-like, and animals more human-like. After examining the similarities of humans and animals in anatomy, embryology, and other fields, Dodel posed the question, "Is the human something special?" The answer, "founded on the scientific results of the last couple of decades," he assured us, was "decisively: No!"[24]

Many Darwinists agreed with Haeckel and Kossmann that humans could be reduced to animals, and quite a few reduced animals to their physical and chemical components. This kind of Darwinian reductionism was strongest among scientists and physicians, to be sure, but it had severe consequences for the value of human life when applied to human affairs. Eugenicists, for example, often compared the selective breeding of animals, which they saw as rational and scientific, with human reproduction, which seemed irrational and arbitrary. The clear implication was that humans would be better off if they would treat each other the way they treat animals, at least in the area of reproduction. Sex was thus reduced to a mere biological function. The jurist and eugenics proponent Hans von Hentig, for example, stated, "The idea, though today it disgusts us, that one could breed humans, like we have bred other animals for the sake of certain useful characteristics, will become important, familiar, and fruitful." Humans are, after all, the most useful creatures around, continued Hentig, so why not act "scientifically" and breed them for desired characteristics. Hentig claimed his entire book, *Penal Law and Selection*, was an attempt to subsume humans under nature and scientific laws.[25]

Hentig's approach was rather common among eugenicists, for eugenics was all about breeding better humans. Otto Ammon, a freelance anthropologist and early eugenics proponent, compared humans to animals with even more ominous overtones. He explained that "in every herd there are badly developed individuals." After noting that animal breeders kill these individuals to keep their herd strong and healthy, he wrote in a passage dripping with irony:

> With *people* a planned selection of this kind is not possible. We practice humanity, in that we chase the unfortunate individual out into the wide world and, pursued from place to place, let them perish gradually, or put them in institutions where they cannot cause any immediate damage. The *prevention* of their reproduction is an important interest of society, which may be opposed neither by legislation nor administration nor through private charity.[26]

The irony is even more apparent in the original German, where the words "chase" and "pursued" were words used commonly for hunting game. In this passage and elsewhere in his writings, Ammon portrayed humanitarianism as misguided and even cruel, a position not at all uncommon among social Darwinists and eugenicists.

Not only did the general idea of biological evolution impact the way people thought about the value of human life, but Darwin's particular theory of evolution by natural selection—with the Malthusian population principle embedded in it—contributed to a devaluing of human life, too. Many German Darwinists, including Kossmann, argued that the mass destruction of organisms, including humans, showed that individual human lives were not really so important. In his 1878 Darwinian diatribe against socialist egalitarianism, Haeckel—basing his arguments forthrightly on the Malthusian element in Darwinian theory—argued that most humans necessarily perish in the struggle for existence. The more fit ones survive and reproduce, while the less fit die. Haeckel recognized that this vision of struggle might rub some people the wrong way, but he affirmed it nonetheless:

> The cruel and unsparing "struggle for existence," which rages—and naturally must rage—everywhere in the biosphere, this unceasing and inexorable *competition* of all living creatures, is an undeniable fact; only the chosen minority of the privileged fit ones is in the condition to survive successfully this competition, while the great majority of the competitors must necessarily perish miserably. One can deeply lament this tragic *fact*, but one can neither deny it nor alter it.[27]

Haeckel underscored his equanimity about the plight of unfit organisms, including the vast majority of humans, by ironically quoting the Bible. "Many are called," he quipped, "but few are chosen!" Haeckel's vision of evolutionary progress (just like Darwin's) required incredible sacrifice—including multitudes of human sacrifices—since the survival of the chosen few means the "destruction of the majority."[28]

The physiologist Wilhelm Preyer, a colleague of Haeckel at the University of Jena, argued forcefully for the application of the Darwinian struggle for existence to human society. The Malthusian element of Darwin's theory underlay his analysis of "Competition in Nature," an article published in a popular journal in 1879. Because of scarcity, "the human's greatest enemy is another human," and "one part of humanity was, is, and always will be poor and sick, another part rich and healthy." Most of this article, as well as an earlier one on "The Struggle for Existence," exuded optimism about the progress produced by competition. He admitted that competition was "life-destroying," but found comfort in the thought that it was also "life-bringing." Predictably, Preyer emphasized the beneficial aspects of competition much more than the death and destruction it wrought. Death, poverty, and misery were perhaps regrettable, but they had a purpose, for ultimately they produced progress.[29]

Büchner's writings also reflected this Darwinian view of death. He agreed with Haeckel that Darwinism had delivered the deathblow to the "anthropocentric fable," that is, the notion that humans are the centerpiece of the cosmos. Büchner contended that the vast expanses of time involved in evolution reduced the significance of the individual. "The individual is nothing in relation to the course [of time]," he wrote in 1882, "the species is everything; and history as well as nature mark every step forward, even the smallest, with innumerable piles of corpses."[30] In Büchner's vision of Darwinian evolution, then, multitudes die, and an individual's death only has significance inasmuch as it promotes progress for the species.

Like Haeckel, Preyer, and Büchner, the Darwinian ethnologist Hellwald applied the struggle for existence to humans in his influential book, *The History of Culture* (1875). His Darwinian view of history decisively shaped his view of human life and death. Hellwald saw the human struggle for existence as "the motive principle of evolution and perfection, in that the weak are worn down and must give place to the strong; so in world history the extermination of weaker nations by the stronger is a postulate of progress." Hellwald evinced little sympathy for the downtrodden losers of the Darwinian struggle, for death is a fact of nature. Progress will come as the victors in the human struggle "stride across the corpses of the vanquished; that is natural law."[31] Thus for Hellwald and many other Darwinists death was no longer an enemy, as Christianity portrayed it, but a force for progress. Carneri displayed this attitude when he called death "nothing but the inexhuaustible source of continuous rejuvenation."[32] Gizycki agreed, calling death "good," since it keeps the world young and vigorous.[33]

Not only did death foster progress, but, according to many Darwinists, the more death, the better! Some Darwinists only implied this, but others, like Haeckel, clearly explained the Darwinian logic behind it. Natural selection can only function if there are variations, and the more individuals that are produced, the more variations there are likely to be. Also, more individuals competing among themselves tend to heighten the selective pressure. Thus high reproduction rates should bring about more rapid evolutionary progress. But, the greater the population pressure, the more individuals will necessarily perish before reproducing. By this logic, death is beneficial, since more deaths mean more progress. This mentality led many Darwinists and eugenicists to promote population expansion. Just before World War I, as German population growth was decelerating (the population was still increasing, but not as rapidly), leading eugenicists, such as Gruber and Grotjahn, led a chorus of worried voices calling for measures to fight the declining reproduction rates.[34]

The idea expressed by Büchner and Kossmann, as well as by Darwin (though in his case only in relation to bees), that the individual is far less important than the species was a common theme in the writings of German Darwinists around the turn of the century. It resonated with the growing popularity of collectivism and the decline of liberal individualism. This was an important move in devaluing the life of individuals, for their life was now considered valuable only to the extent that it contributed to the well-being of the entire community, which might mean all of humanity or might mean a particular race, depending on the particular evolutionist applying the principle. Tille, in his zeal to synthesize Darwin and Nietzsche, stated the principle this way: "Humans belong to nature, just like plants and animals, and nature knows no pity. It brutally sacrifices the individual, in order to preserve the species."[35] Forel concurred, stating that "the interests of the whole [species] must be placed above the interests of the individual. . . . In many cases the life of a single human is more important than that of several others."[36] Unold also expressed this point repeatedly, for evolution demonstrates "the overriding importance of the lasting community (the species) against the highly transitory individual."[37] For these Darwinists individual life thus had no importance in and of itself. The individual's welfare was subservient to that of the species.

The famous Darwinian biologist August Weismann gave his own special spin to this idea, when he argued that death is an evolutionary adaptation benefiting the species. In an 1881 lecture and essay, "On the Duration of Life," he argued that in examining the duration of biological life, *only the interest of the species* comes into consideration, not *that of the individual*." An individual serves the species by reproducing, but after reproducing, the individual "ceases to have value for the species," so it dies. Death serves an important biological function, keeping a species vigorous, since it clears away injured individuals, "who are worthless for the species and even harmful."[38] A few years later he wrote "On Life and Death," an extensive essay expounding the same idea. Here he once again argued that death was beneficent for multicellular species, since it rids each species of injured individuals, who have become "worthless and even harmful."[39] While the physical body must die, the reproductive cells of a species continue, being passed along from one generation to the next, so reproduction is in a sense the only path to "eternal life." Weismann did not directly mention humans in these essays, but his arguments clearly presage the ideas of many eugenicists, for whom the individual's interests are subservient to those of the species. Weismann also lent support to the eugenics movement by joining Ploetz's Society for Race Hygiene as an honorary member when it was founded in 1905.

One leading eugenicist, Schallmayer, referred explicitly to Weismann's views on death and the relationship of the individual to the species to support his eugenics ideology. Schallmayer contended that evolution shows that individual interests are only significant inasmuch as they contribute to the welfare of the species: "This natural law, the complete subordination of the individual interests under those of the species, must also be valid for human evolution." Thus, the individual has no purpose in and of itself. Those having "no value" for the species perish. Schallmayer criticized European culture for laying too great a stress on the value of the individual, which sometimes damages the interests of the species. Schallmayer and many other eugenicists thus devalued the lives of people, making their value dependent on their contribution to evolutionary progress.[40]

In his book, *Moses or Darwin?*, Dodel expressed much the same thought as Weismann. After discussing the Malthusian population principle and the resultant struggle for existence, he stated, "Death is the end of the individual, but it is also the greatest benefactor for the whole. Without death [there is] no progress, and progress is life; so the death of the individual is the condition of life for the whole." He applied this principle to humans as well as other organisms. He further maintained that a proper understanding and relationship to nature—which he called "our mother"—would help people overcome their fear of death.[41] In an earlier book, Dodel had discussed the need for some animals—including "barbarian" people—to engage in violent competition for mates in order to reproduce. "So nature destroys," he remarked, "in order to reproduce."[42]

By the beginning of the twentieth century, these Darwinian ideas about the value of life and death found fertile soil among scientists, physicians, and some social theorists, taking root and springing up as the eugenics movement. As all leading eugenicists in the late nineteenth and early twentieth centuries confessed, the core idea of eugenics derived from Darwinism. The physician and eugenicist Eduard David, a Social Democratic member of the German parliament, explained succinctly the connections between Darwinism and eugenics in terms reminiscent of Weismann:

> A strong counterbalance to the degenerative effect of this imbalance and atrophy of a people's organic condition is the rapid death of the damaged individual, as well as any of its weak progeny. This process of natural selection is frustrated through institutions of social assistance, which aim at preserving the life of damaged organisms, allowing them to reproduce and also preserving the lives of their progeny with inferior health.[43]

David's fear that modern institutions, especially those motivated by compassion or humanitarianism, would produce biological degeneration was a commonplace lament among eugenicists.

Haeckel was one of the earliest German Darwinists to warn about the biologically deleterious effects of modern institutions. Since he viewed natural selection through the struggle for existence as a beneficent force in human society, he worried that helping the weak, sickly, and unintelligent might have ill effects, favoring them over the strong, healthy, and intelligent. Many Darwinists and eugenicists labeled any such tendencies contraselective, since they selected the "wrong" people. These Darwinists ignored (or some forthrightly denied) the fact, of course, that any value judgments about who "should" survive cannot be derived from empirical science. Strictly speaking, the word contraselective makes no sense in the light of Darwin's definition of fitness, since by definition those who survive are more fit. In any case, in 1870 Haeckel identified several causes of contraselection: modern medicine, clerical celibacy, and modern warfare. All three, according to Haeckel, were artificial institutions either disadvantaging those with "good" biological traits or aiding those with "bad" characteristics.[44] However, Haeckel was optimistic about the prospects for evolutionary progress and never lapsed into the gloom-and-doom of the *fin-de-siècle* prophets of biological degeneration. He believed that ultimately natural selection was a strong enough force to overcome these contraselective institutions.

Weismann shared Haeckel's general optimism that natural selection would counteract many of the ill effects of contraselective forces, but nonetheless he also contributed significantly to the rising tide of anxiety about biological degeneration.[45] He wrote an important essay in 1886, "On Regression in Nature," where he pointed out that evolution does not always bring progress, since many organisms lose functioning parts and thus regress, as they adapt to different environments. Weismann explained that when an organism no longer needs a particular organ to survive and reproduce, there is no selective pressure for the organism to retain that organ, so over many generations, it gradually disappears. For example, a blind cave fish species did not lose its sight, according to Weismann, from the direct influence of the environment or from disuse, but rather because its forebears didn't need eyesight to survive and reproduce. This allowed individual fish with poorer and poorer eyesight to reproduce, ultimately leading to loss of function.

In applying these biological insights to humans, Weismann claimed that uncivilized peoples have better senses of hearing, seeing, and smelling than do civilized peoples, who rely more on their mental acuity and technology. Further, reliance on technology can be biologically harmful. For example, Weismann argued that wearing glasses encourages nearsightedness and dentistry promotes the development of weak teeth, by allowing those with poor eyesight and weak teeth to reproduce better than they could if left to their own devices. Weismann concluded that "in many respects the physical condition of civilized people has been worsened through civilization and will likely be worsened even more."[46]

Psychiatrists working within the Darwinian paradigm contributed significantly to the growing concern about hereditary degeneration, as well. Even though the originator of the idea of psychiatric degeneration, the French psychiatrist Augustin Morel, published his theory two years before the advent of Darwinism, by the 1880s and 1890s most discussions of degeneration (both within and outside the psychiatric community) sought support in Darwinian principles.[47] Kurella, following the lead of Lombroso, forthrightly synthesized Morel's degeneration theory and Darwinism, criticizing Morel's model for its pre-Darwinian character.[48] In a 1908 article Kraepelin blamed biological and psychiatric degeneration on the deleterious effects of civilization, which by providing many conveniences and comforts had counteracted the beneficial effects of natural selection. He noted that social welfare measures were now keeping alive those weak and sick individuals who would have perished in previous ages.[49] By the 1890s, psychiatrists were frequently voicing their concerns about degeneration.

As psychiatrists increasingly expressed their fears about degeneration and the ill effects of modern civilization on mental health, German authorities were taking measures to reduce the threat to the safety and welfare of society. Between 1885 and 1900 the state of Prussia, for example, expanded the number of mental asylums from 71 to 105. Since the new asylums were generally larger than most of the earlier ones, and because they also expanded the size of some of the existing institutions, the population in mental asylums increased 429 percent in these 15 years, while the Prussian population increased 48 percent. These figures are not all that different from statistics from other German states and also other European nations in the same period. The phenomenal increase in institutionalized individuals was caused by the increased prestige of medicine, including psychiatry, as well as trends toward greater bureaucratization and increasing social control. It did not demonstrate an increase in mental illness, but this was not apparent to many at the time, especially since many psychiatrists were warning about degeneration. It is even possible that the fear of degeneration helped motivate the increase of asylum inmates.[50]

Psychiatry played a fundamental role in spreading fear of biological degeneration to a wider audience, especially to those already embracing Darwinism.[51] The two leading figures in the German eugenics movement— Ploetz and Schallmayer—were both physicians decisively influenced by their contact with psychiatry (we have already discussed the influence of Darwinism on them in the Introduction). In 1885, at the end of a year's internship in a mental asylum in Munich, Schallmayer wrote a dissertation on "The Rejection of Food and Other Disorders regarding Food Intake by the Insane." The following year he wrote the first draft of his book, *On the Threatening Physical Degeneration of Civilized Humanity* (published in 1891), which was the first book entirely on eugenics published in Germany. Underlying Schallmayer's analysis of degeneration was his view that some social institutions further "the improving selection of the struggle for existence," while others hinder it. Under the influence of Haeckel, he identified modern medicine and the military as two sources of contraselection, but unlike Haeckel, he also criticized the capitalist economic system as a degenerative factor.[52]

Like Schallmayer, Ploetz developed his views on degeneration under the influence of psychiatry. While studying medicine in Zurich in the late 1880s, Ploetz interacted with Forel, who won him over to the cause of eugenics. Like many other psychiatrists in the late nineteenth century, Forel stressed the hereditary character of mental illness. He believed that as civilization was advancing, the beneficial effects of natural selection diminished, causing mental illness to increase.[53] Forel not only preached eugenics as the solution to biological degeneration, he also took Ploetz and his friends on tours of the Burghölzli Psychiatric Clinic, so they could witness the specter of mental illness firsthand. Forel also exerted considerable influence on other early eugenicists, such as the psychiatrist Ernst Rüdin and the physician Agnes Bluhm, both of whom were classmates of Ploetz in Zurich, as well as Ploetz's friend, Gerhart Hauptmann, whose first successful play, *Before Dawn* (1889), centered on biological degeneration and eugenics (many believe the protagonist—a crusader for eugenics—was modeled on Ploetz).[54]

As we have seen, Darwinism contributed to new ways of thinking about life and death in the late nineteenth and early twentieth centuries that often led the most avid Darwinists in Germany to devalue human life. This is not to say that everyone who embraced Darwinism denied the value of human life. Ideas about the sanctity of human life, ascendant for centuries in European thought, could not be swept away that easily. One leading popularizer of Darwinism, Wilhelm Bölsche, even protested against the devaluing of human life that he saw in the writings of some of his fellow Darwinists.[55] Thus, some contemporaries recognized the trend in Darwinian circles to devalue human life, even if they opposed it. However, among leading Darwinists who saw Darwinism as the centerpiece of a new scientific worldview, Bölsche's views on the value of human life did not predominate.

More Darwinists, in fact, took the opposing view, though few were as extreme as the racial theorist and Nietzsche enthusiast Heinrich Driesmans, who exulted in Darwinism as a mephistophelean liberation from stultifying nineteenth-century humanitarianism in his book, *Demon Selection: From Theoretical to Practical Darwinism* (1907). Driesmans called Darwinian selection a "scientific demon," since it functions "to eliminate gradually and to exterminate those who become weak." According to him, Darwinism "brought us knowledge, that if not all, at least much of the human misery that we tried to help, was *declining life*, determined by nature to be eliminated, in order to make room for the healthier, and that one does a service neither to the latter nor to the former if one prolongs its sickliness."[56] The lesson Driesmans drew from Darwinism was that the healthy should eliminate the unhealthy. How to eliminate the "unfit" was a key problem confronting eugenicists, and we will return to it later, but before tackling that issue, they needed to figure out who the "unfit" are. In chapters 5 and 6 we will engage this latter issue.

5. The Specter of Inferiority: Devaluing the Disabled and "Unproductive" ⌒

When the French social Darwinist Georges Vacher de Lapouge wrote the introduction to the 1897 French edition of Haeckel's *Monism as Connecting Religion and Science*, he rang the death knell for the three main ideals of the French Revolution: liberty, equality, and fraternity. To replace these allegedly outmoded and unscientific liberal dogmas, he argued that the Darwinian revolution had introduced a new, improved triad: determinism, inequality, and selection.[1] Lapouge agreed with Clemence Royer, who wrote in her 1862 preface to her French translation of Darwin's *Origin of Species*, "What is the result of this exclusive and unintelligent protection accorded to the weak, the infirm, the incurable, the wicked, to all those who are ill-favored by nature? It is that the ills which have afflicted them tend to be perpetuated and multiplied indefinitely; that evil is increased instead of diminishing, and tends to grow at the expense of good."[2] Lapouge's views found more resonance in Germany than in his native France, probably due in part to the stress on biological inequality in the writings of other German social Darwinists and eugenicists, who prepared the soil in Germany for the spread of Lapouge's ideas.

In the late nineteenth and early twentieth centuries many German scientists and other scholars recognized that biological inequality was a fundamental component of Darwinian theory. Evolution could not occur, after all, without significant variations. Some individuals were more fit than others and thus survived and reproduced, while others were less fit and perished without reproducing. This emphasis on biological inequality stimulated many scientists and physicians to categorize people as "inferior" or "superior," "more valuable" or "less valuable." Many eugenicists groped for a way to scientifically categorize people, some by measuring heads and other body parts. However, the categories of inferior and superior were more often than not highly subjective and somewhat nebulous, though they generally focussed on intelligence and health.

Since biological inequality was a key presupposition of Darwinian theory, stressing natural inequality served an important rhetorical purpose. In order to convince contemporaries that organisms—including humans—evolved, Darwinists in the late nineteenth century needed to demonstrate that organisms varied considerably. To make human evolution seem plausible, they claimed that humans were really not as far removed from the closest related species as most people thought. Further, they needed to show that significant variations exist within the human species. Darwinism, after all, implied that populations gradually diverge in order to form subspecies and later, distinct species. For these reasons, Darwinism provided powerful ammunition for those arguing for human inequality, and many Darwinists readily used it.

Haeckel regularly marshaled Darwinian arguments in support of inegalitarianism. In *The Natural History of Creation* (1868) he explained that

> *between the most highly developed animal soul and the least developed human soul there exists only a small quantitative, but no qualitative difference,* and that this difference is much less, than the difference between the lowest and the highest human souls, or as the difference between the highest and lowest animal souls.[3]

It may be hard for us today to imagine that a serious scientist could actually believe that the differences within the human species are greater than the differences between humans and other animals, but this was indeed Haeckel's position, which he reiterated in many publications.

Not only did Darwinian inegalitarianism serve an important function in convincing contemporaries of the validity of Darwinism, but it was also a handy weapon in the intellectual struggle against socialism in the late nineteenth century, and quite a few Darwinists entered the fray against socialism. To be sure, most socialists embraced biological evolution with alacrity, since it comported with their materialist worldview and aided in their campaign against religion. However, many socialists did not embrace Darwin's theory of natural selection, and most denied that Darwinian theory could be applied to society. The famous pathologist and politician Rudolf Virchow, one of Haeckel's professors, but later an opponent of Darwinism, stirred up debate over the relationship between Darwinism and socialism. In an 1877 speech to the Society of German Scientists and Physicians he warned that trying to introduce Darwinism into the public schools, especially in the dogmatic fashion of Haeckel and his followers, might produce a backlash against scientists in general. In the course of his speech, he insinuated that Darwinism leads to socialism, hoping to arouse fears in his audience about the dangers of teaching Darwinism to the public.[4]

Haeckel, who had previously lectured on Darwinism to the same congress, but had not stayed for Virchow's speech, responded with indignation. Provoked by Virchow, he published a contemptuous rebuttal, complete with a tirade against socialism. Rather than benefiting from Darwinism, Haeckel argued, Darwinism actually refuted socialism. In his opinion, Darwinism was the best antidote against socialism, since it proved the necessity of inequality and competition. This applied to human society, he explained, just as it did to nature. Thus, if Darwinism supported any political system, it must be in some sense aristocratic. By no means was Haeckel supporting the landed aristocracy, for which he had nothing but contempt, but—in keeping with his more liberal political views—he favored an aristocracy of talent or a meritocracy.[5]

Attempts to refute socialism on the basis of Darwinian theory were by no means uncommon in late nineteenth-century Germany. Usually the point of dispute was the socialist doctrine of human equality. Like Haeckel, the Darwinian biologist Oscar Schmidt was outraged that Virchow should connect Darwinism and socialism, so at the 1878 Conference of the Society of German Scientists and Physicians he delivered a rebuttal. In the midst of his attempt to refute socialism on Darwinian grounds, he asserted, "The principle of evolution is certainly the abolition of the principle of equality."[6] Another prominent biologist, Wilhelm Preyer, who lectured and wrote on the implications of Darwinism for economic competition, agreed with Haeckel and Schmidt that Darwinism demonstrated "that the inequality of humans is a natural necessity."[7] Likewise Hellwald spurned the egalitarian idea that "wants to treat everything that lives under the name 'human' on earth in the same way."[8] Heinrich Ernst Ziegler, a Darwinian biologist who wrote extensively on the social applications of Darwinian theory in the 1890s and early 1900s, wrote an entire book devoted to refuting socialism on the basis of Darwinism. In this book and in other writings he constantly stressed human inequality as a logical consequence of Darwinian theory.[9]

Not all Darwinists, however, agreed with the antisocialist stance of Haeckel. Some even argued that socialism was more consistent with Darwinism than was capitalism. Büchner, for example, considered himself a Leftist politically and wrote to Haeckel that he enjoyed reading his book rebuking Virchow, except for the part about socialism.[10] However, even though he continually pressed for greater socioeconomic equality, Büchner agreed with Haeckel that Darwinism implied biological inequality. In fact, one of the things Büchner found so objectionable about the capitalist system was that it gave advantages based not on biological inequalities, but on economic considerations. As he explained to Haeckel, capitalism bequeaths property to those who are not always biologically superior. Thus it skews the struggle for existence in favor of those with property, regardless of their biological makeup.[11] "The root of evil," Büchner stated, "lies in the incredibly great inequality of the weapons or means (external as well as internal) with which each individual is required to fight the inescapable struggle for existence." In order to level the playing field and make the struggle for existence more fair, Büchner favored the abolition of inheritance and greater social welfare measures. In Büchner's meritocracy the more intelligent and talented would reap greater economic rewards, but could not pass them on to their children.[12] Biological inegalitarianism was thus an integral aspect of Büchner's synthesis of Darwinism and socialism.

Another prominent Darwinian popularizer rejecting human equality was Nordau, who contemptuously dismissed the idea of human equality as a "fable" and a "delusion of ivory-tower scholars and dreamers," because it contradicted the most essential tenets of Darwinian science. In a book that sold fabulously he wrote that equality

> stands in contradiction to all the laws of life and evolution in the organic world. We, who stand on the ground of the scientific world view, recognize in the inequality of living things the impetus for all evolution and perfection. For what is the struggle for existence, this source of the beautiful variability and the many forms of nature, other than a constant confirmation of inequality? A better equipped organism makes its superiority felt by the other members of its species, diminishes their portion at the meal provided by nature, and stunts their possibility for the full development of their individuality, in order to win more space for its own [progeny]. . . . The least perfect individuals will be destroyed in the struggle for first place and will disappear. . . . Inequality is therefore a natural law . . .[13]

Nordau thus not only justified inequality as an inescapable part of nature bringing evolutionary progress, but he also recognized that the struggle for existence meant death for multitudes of less well-endowed individuals.

As Darwinism expanded its influence in the late nineteenth and early twentieth centuries, the biological inegalitarianism inherent in it won many adherents and became a cornerstone of the eugenics movement, underpinning their whole doctrine. Ploetz continually railed at Christianity, humanitarianism, and democracy for their egalitarian ideals.[14] When he founded the Society for Race Hygiene, he hoped to recruit members only from those who were biologically superior, ranking in the upper one-fourth of the population. This proved impractical, as he had no way of objectively measuring biological superiority, but members were required to pledge to undergo a medical exam before marrying to determine their fitness to reproduce.[15] Schallmayer dismissed egalitarianism with scorn. "Making the unequal equal," he wrote, "can only be an ideal of the weak."[16] Ehrenfels expressed a sentiment widely shared by his fellow eugenicists when he declared, "The liberal-humane fiction of the equality of all people was surely one of the most arbitrary, ignorant things that the human mind has ever devised."[17]

Darwinian inegalitarianism became so pervasive by the early twentieth century that it even infiltrated the ranks of socialists and other political radicals. In fact, the earliest leaders in the eugenics movement—Forel, Ploetz, Schallmayer, Ehrenfels, Grotjahn, Max von Gruber, and Woltmann, among others—tilted decidedly toward the Left politically. Some called themselves socialists, and several even joined socialist political parties, though most remained aloof from party politics. The Darwinian botanist and socialist Dodel upheld a view of inequality and the struggle for existence similar to Büchner's.[18] The socialist physician Alfred Blaschko wrote an article for a leading socialist journal opposing Ammon's attempt to use Darwinian arguments to refute socialism. In this article, however, he staked out a position that is perilously close to Haeckel's and Ammon's:

> It cannot be denied that the Darwinian theory is an eminently *aristocratic* theory, aristocratic on the one hand, because it proclaims the inequality of all who bear a human face, and on the other hand, because proceeding from this inequality, it preaches the right of the stronger, of the one better equipped for the struggle for existence.[19]

But, while admitting the inevitability of biological inequality, Blaschko did not see this as a reason to maintain social or economic inequality, as Haeckel, Ammon, and other social Darwinists did. Emil Reich, a professor at the University of Vienna, also stressed human biological inequality in another socialist periodical, stating, "From an evolutionary standpoint equality is not to be understood as a principle of the equal value of individuals, but only as the equal right of each individual to be allowed to develop his abilities."[20] Even Karl Kautsky, the leading Marxist theoretician after Engels' death in 1895, supported eugenics, though he believed it was only possible in a socialist economy.[21]

Nor was Lily Braun, a leading feminist and socialist, completely committed to equality. She appealed to nature, especially evolutionary theory, to vigorously oppose strict egalitarianism, but at the same time—like

Büchner—demanded greater socioeconomic equality, so that only biologi-cal differences would influence success in the struggle for existence. In her memoirs she faulted socialists for teaching the masses "equality of all in the sense of the same value and the same developmental ability." In nature, she continued, every blade of grass is different, and nature has given humans even greater variability. Thus, in order to act in harmony with nature and evolution, we need to respect natural differences instead of encouraging "every Joe Fool to call Goethe his brother."[22] Braun formulated her per-spective on biological superiority and inferiority under the influence of both Darwinism and Nietzschean individualism.

The penetration of Darwinian inegalitarianism into socialist ranks helps explain the increasing acceptance of eugenics in socialist circles in the early twentieth century. Even Karl Kautsky, the leading Marxist theorist in Germany, embraced eugenics and used the terminology of biological inequality in his writings, though he lay far greater stress on introducing a socialist economy to improve the human condition.[23] Grotjahn, professor of social hygiene at the University of Berlin and a leading eugenicist, joined the Social Democratic Party (SPD) as a student in the 1890s, and after dropping out of the party for a time in the first decade of the twentieth century, he later rejoined and served as a member of parliament for the SPD. Eduard David was another prominent figure in the SPD promoting eugenics.

Despite his acceptance of eugenics, Kautsky and most socialist eugeni-cists were often more humane than their antisocialist counterparts. One antisocialist eugenics proponent who spelled out the implications of Darwinian inequality with brutal frankness was Tille. In 1895 he wrote,

> From the doctrine that all men are children of God and equal before him, the ideal of humanitarianism and socialism has grown, that all humans have the same right to exist and the same value, and this ideal has greatly influ-enced behavior in the last two centuries. *This ideal is irreconcilable with the theory of evolution.* . . . It [evolution] recognizes only fit and unfit, healthy and sick, genius and atavist.[24]

In an earlier book he criticized Jesus and the "humanitarian fanatics" of the eighteenth century for their stress on human equality.[25] Tille, like many of his colleagues, rarely defined the vague categories of fit and unfit, but they were sure that some humans were less valuable than others based on their physical and mental traits. In the light of evolution, according to Tille, these inferior individuals do not even have the "right to exist."

By the early twentieth-century Darwinian inegalitarianism was becoming manifest through the increasing use of the German term "minderwertig" (properly translated as "inferior," but literally meaning "having less value") to describe certain categories of people. Aside from non-European races (whom we will cover in chapter 6), two overlapping categories of people were generally targeted as "inferior" or "unfit": the disabled (especially the mentally ill) and those who were economically unproductive. In a 1909 speech to the Society for German Scientists and Physicians the anthropologist and eugenicist Felix von Luschan answered the question, "Who is inferior?" with the following list: "The sick, the weak, the dumb, the stupid, the alcoholic, the bum, the criminal; all these are inferior compared with the healthy, the strong, the intelligent, the clever, the sober, the pure." For Luschan the most important practical task for anthropology was to figure out what to do about the "inferior" elements of society.[26]

The first German biologist to apply Darwinian inequality to the disabled was Karl Vogt, a political exile in Switzerland because of his participation in the Revolutions of 1848. Vogt, professor at the University of Geneva, was one of the earliest German biologists to embrace Darwinism. In his two-volume work, *Lectures on Man* (1863), which is considered such a classic work in anthropology that it was republished in 2003, he asserted that some mentally disabled people (he used the term "idiots") were closer to apes in their brain function and mental abilities than they were to the lowest normal humans. He claimed an "idiot" is biologically closer to an ape than to his or her own parents.[27] In 1867 Vogt argued that microcephalic persons were evidence for Darwinian evolution, since they are atavistic throwbacks to earlier phases of evolution. He saw them as a kind of contemporary missing link between apes and humans. He noted that their brains are about the same size as a spider monkey, and he further claimed that they generally had excellent dexterity in climbing! (Darwin, by the way, agreed with Vogt's views on this matter).[28] Though Vogt stopped short of advocating infanticide for the disabled, he did argue in *Lectures on Man* that morality is relative and supported his contention with a telling example: "If it is a capital offense in the civilized world to kill one's old lame father, there are Indian tribes in which this is considered an entirely praiseworthy deed of a son."[29] Vogt's relativizing of killing the weak and sick, together with his claim that some mentally ill people are closer to animals than humans would blossom and bear fruit later in the eugenics movement.

Vogt was not alone in characterizing certain atypical individuals as atavistic. In fact, this idea circulated widely in popular culture through the "freak shows" that were a prominent form of entertainment in Germany, and not just for lowbrow audiences. Some of the "freaks" on display in Germany had a rare genetic condition that causes excessive hair growth all over the body. Circus companies not only exploited their odd looks to draw audiences, but they also degraded these individuals by billing them as "missing links" in the evolutionary chain between apes and humans. Some physicians interpreted other abnormalities, such as protuberances at the base of the spine, as tails, thus displaying signs of earlier evolutionary stages.[30]

Around the turn of the twentieth century the eugenics movement focussed even more attention on the evolutionary significance of disabilities and other atypical human traits. One does not have to read very many writings by German eugenicists before it becomes apparent that they assigned differing values to different people. They constantly referred to the mentally and physically handicapped as inferior, lower, useless, burdens to society, and even worthless. Erwin Baur, a prominent Mendelian geneticist and supporter of eugenics, for example, at the end of his widely used text on genetics, drew distinctions between people who are inferior and superior.[31] The famous neuro-physiologist Kurt Goldstein wrote a book on eugenics, in which he argued that population increases are beneficial, but only if "every individual has value." This implies that some individuals have no value, a point Goldstein made explicit immediately thereafter, stating, "The increase of worthless individuals is indeed rather harmful."[32] Similar expressions about the inferiority and even worthlessness of some individuals—usually the disabled—abound in the writings of biologists, psychiatrists, and physicians around 1900.

Some eugenicists even claimed that individuals with physical or mental disabilities were not only worthless, but of negative value. Gizycki, for example, whose support for eugenics predated the emergence of the eugenics movement in Germany, exemplified this attitude already in 1883. He admonished his society to curb the conception of children who will be "worth less than nothing," because they cause more pain than happiness—a mortal sin against his utilitarian ethics.[33] Ironically, Gizycki was disabled and confined to a wheelchair, though he still had his full intellectual powers. Hugo Ribbert, professor of pathology at the University of Bonn, used similar language, when he wrote,

> The care for individuals who from birth onwards are useless mentally and physically, who for themselves and for their fellow-creatures are a burden merely, persons of negative value, is a function altogether useless to humanity, and indeed positively injurious.[34]

Ribbert's rhetoric seems rather shocking to us today, but it was rather commonplace among eugenicists in the early twentieth century and not just in Germany. Grotjahn also reflected this concern with biologically "inferior" elements, estimating that one-third of the entire population was physically or mentally "inferior" or "defective."[35]

German feminists, though striving for greater gender equality, were also not impervious to the Darwinian emphasis on inequality and inferiority. In fact, many leading feminists—Ruth Bré, Helene Stöcker, Henriette Fürth, Adele Schreiber, and Gertrud Bäumer, to name a few—were staunch supporters of eugenics. Like so many other eugenicists, they freely used the rhetoric of inferiority and disparaged the disabled in ways that would be embarrassing today.[36] Schreiber, for example, a leading figure in the League for the Protection of Mothers and later a Social Democratic member of the German parliament, argued that only through birth control "could higher breeding and selection, exclusion of worthless and bad elements, [and] the creation of a society free from the oppressive burden of the unfit of various kinds, be achieved."[37]

Bré, original founder of the League for the Protection of Mothers, was incensed when Stöcker and her followers toned down the eugenics goals of the League for the Protection of Mothers. In 1905, less than two months after adopting Bré's declaration as the basis for the new organization, Stöcker and the majority of members in the organization expressed disapproval of Bré's original document. Among other changes, they wanted to drop the word "healthy" and all references to eugenics from it, thereby implying that they would help all women, regardless of health. Bré felt betrayed and indignantly responded that in the founding declaration of the League, first published in Woltmann's journal (which promoted a racist form of eugenics), the eugenics goals were explicit. The founding declaration, according to Bré, "emphasized the breeding of the 'healthy,' rather than the coddling of the sick and ailing." She railed at her colleagues for departing from eugenics by helping the "inferior." Bré was so upset about the changes that she quickly withdrew and formed a short-lived rival organization.[38]

Whatever disagreements existed on other issues, Bré was probably overreacting on the eugenics issue. Stöcker and her supporters were not abandoning eugenics at all. They were simply concerned that restricting help to "healthy" mothers would lead to arbitrary decisions about who qualified for help. In May 1905 the League's advisory committee discussed the reasons for dropping eugenics rhetoric from the League's program. They unanimously agreed that despite altering the official program, in their practical activity—such as founding maternal homes for unwed mothers—the League should support only the "fit mothers and children, but those who suffer from an infectious or hereditary illness, especially syphilis or tuberculosis, will not be allowed to receive care." They further requested Ploetz, who was still a member of the League's advisory committee, to provide practical suggestions about how the League could implement eugenics.[39]

Even after Ploetz and a few other eugenicists resigned from the League, Stöcker, Schreiber, and other members continued to promote eugenics. Stöcker told the Monist League's international conference in 1913 that those embracing a scientific worldview could not evade the question of who should be born: "Because we want higher humans, we need eugenics and race hygiene."[40] In a 1909 speech to the General Assembly of the League for the Protection of Mothers she expressed the hope that Galton's vision of a religion of eugenics would be fulfilled in a future where everyone would be born healthy and happy. She underscored the need to educate people about "generative ethics" that would elevate humanity by eliminating the births of the disabled. Though the main point of her speech was to promote the legalization of abortion, eugenics played a key role in her justification for abortion, since "children from parents with infectious diseases, or children of the chronically ill, as well as children of those with heart or mental illnesses should not be permitted to be born."[41] In many publications Stöcker stressed the need to improve human heredity by hindering the reproduction of the disabled, "whose existence has no value and no enrichment for the whole [of society]."[42] Statements such as these by Stöcker—and similar ones by Bré and Schreiber—contributed to the devaluing of the disabled and stripped them, at least in part, of their human dignity.

Nowhere did the specter of inferiority—linked as it was with the notion of degeneration—loom larger than in psychiatry. Leading psychiatrists constantly described their patients as "inferior," "degenerate," or "defective." In 1888 Julius Koch had introduced the term "psychopathic inferiority" ("*psychopathische Minderwertigkeiten*") to describe mental problems that bordered on mental illness, but were not as severe.[43] This terminology became standard among psychiatrists, who began applying the term "inferior" not only to the condition, but also to the people afflicted with the problem. The influence of psychiatry rose steadily in late nineteenth and early twentieth-century Germany, as more mental institutions were erected and many more people were institutionalized. This contributed to public fears that mental illness was increasing, but it was probably more the result of greater scrutiny by government officials and physicians. In any case, this misperception helped psychiatrists gain even greater prestige and powers of social control.[44] They had an attentive audience for their warnings about degeneration and inferiority and their suggestions for remedying it.

Darwinian inegalitarianism was not the only factor contributing to the devaluing of the disabled. Another consideration was money. Many eugenicists lamented the economic burden placed on others by the mentally and physically handicapped. In 1911 a prominent science journal, *Umschau*, awarded a cash prize of 1,200 Marks—a considerable sum in those days— for the best essay on the topic, "What do the bad racial elements cost the state and society?" The instructions for the contest stressed the economic burden of the biologically "inferior," who "would have been better not to have been born." However, even though eugenicists regularly mentioned the economic burden of the "inferior" in their writings, no one had ever compiled statistics to measure it. Judges for the competition were the editor of *Umschau*, along with two eugenicists, Gruber and A. Gottstein.[45] Gruber, a professor of hygiene at the University of Munich who was world-renowned for his discoveries in serology, considered the disabled not only an "enormous burden," but a "constant danger for the healthy."[46] The prizewinning essay by Ludwig Jens, which appeared in a prominent journal on social hygiene in 1913, calculated the cost of institutionalizing "inferior" people in the city of Hamburg alone at 31,617,823 Marks annually.[47] The lesson was obvious: Something must be done to reduce this burdensome expenditure. An anatomy professor at the University of Vienna, Julius Tandler, drew exactly this conclusion. After citing Jens's study in a 1916 speech, he remarked, "As cruel as it may sound, it must be said, that the continuous ever-increasing *support of these negative variants is incorrect from the standpoint of human economy and eugenically false.*"[48]

Stimulated by the *Umschau* contest question, Ignaz Kaup, a professor of social hygiene who worked closely with Gruber at the University of Munich, wrote an article on "What Do the Inferior Elements Cost the State and Society?" Kaup investigated the cost of "inferior" youth, and concluded that society needed to segregate the "inferior," that is, the disabled, so they could not reproduce. He warned against false compassion for the "inferior," since "our healthy offspring have the right to be protected from decay through those who are genetically pestilent (*Keimschädlinge*), and every progressive nation has the duty to reduce the ballast of the costs of inferiority."[49] The term I have translated as "pestilent" (*Schädlinge*) in this quotation often means pests, parasites or vermin. Though sometimes it is also used to describe people, it usually refers to evil people who exert noxious influences on others. Using this term—as well as "ballast"—to describe the disabled shows the utter contempt toward the disabled that reigned in eugenics circles.

Kaup was not the only eugenicist to describe the disabled as "parasites." In an 1895 article promoting eugenics Kurella argued that not only physical, but also mental and moral traits are largely hereditary. Kurella's assertion that "psychopathic inferiority," as well as drunkenness, laziness, and criminality, are inherited traits was widely shared by many of his contemporaries, especially those in the eugenics movement. In describing these persons with "anti-social" traits, Kurella called them "parasites," as well as persons with "socially useless and harmful [biological] constitutions."[50] In his earlier book, *The Natural History of the Criminal* (1893), Kurella discussed at length those who shirked work, repeatedly labeling them "parasites."[51] The image of the physically and especially mentally disabled sucking the life juices out of their "host" society, harming and maybe even destroying their "host" in the process, was widespread, even among those who did not resort to calling them "parasites" overtly.

Heinz Potthoff, a member of the Monist League and a National Liberal member of the German parliament with expertise in socioeconomic affairs, provides a further example of how Darwinian inequality meshed with economic considerations to devalue the lives of the disabled. In response to Ploetz's speech at the first Congress of German Sociologists, Potthoff insisted that many in Germany were perishing, not because they were biologically inferior, but because of economic inequalities. However, he fully agreed with Ploetz that those who are biologically "unfit" should not receive state help, which he considered a luxury Germany could no longer afford. He specifically criticized social welfare provisions that supported "idiots and cripples." Rather it is more economically responsible, according to Potthoff, to spend those funds on the impoverished, but not on the biologically "unfit." The protocol to this meeting indicates that Potthoff's remarks calling for an end to financial support for the disabled were greeted with applause rather than indignation.[52]

Potthoff expanded on these ideas in an article, "Protection of the Weak?" in which he affirmed that "not the weak are the most valuable in the state, but the strong, those capable of life and performance; [and] a social policy that wants to protect the weak, must lead the state down the wrong track." He reiterated that all social welfare provisions should favor those who are economically disadvantaged, but not those who are biologically weak. Since so many are dying as a result of poverty, he stated,

> we do not have the right to withdraw from them, what we spend on those members who are incapable of life, useless, and unhappy. Unhappy! The heroes of humanity should not forget, that they would not only create infinitely more practical advantage, but also infinitely more happiness and welfare, if their humanitarianism would finally advance so far that they would give death to the dying and life to those capable of living![53]

Though Potthoff explicitly denied that anyone should actively intervene to kill the "weak," he recognized that withdrawing support from them was a death sentence. But, since these "weak" members of society were "useless and unhappy," he welcomed their demise. Potthoff clearly shared Darwin's vision of happiness increasing through the destruction of those "unfit" in the struggle for life. Potthoff's views seem rather harsh and extreme to us today, but they were not all that uncommon in the decades around 1900. Many eugenicists expressed similar ideas.

Economic considerations influenced discourse on the disabled in yet another related way. Many eugenicists began using economic productivity as a measuring rod to determine human value. Forel claimed that humans should be ranked in value according to their ability to be productive citizens. He defined a "fit" individual as one who contributes more to society than he or she receives.[54] "I maintain," stated Forel, "that two hereditarily fit children constitute a social *plus*, and two bad ones are a social *minus*. The former will by all means press ahead and then *produce* more than they consume, but the latter will do *vice-versa*."[55] This statement implies that we can mathematically calculate the value of individuals' lives by determining their economic contribution or liability to society. Schallmayer likewise thought that, for practical purposes, the "hereditary value" of individuals could best be determined by their economic productivity.[56]

The influential Viennese sociologist Rudolf Goldscheid, a leader in the Austrian branch of the Monist League, developed his entire sociology around the idea of a "human economy." Goldscheid was deeply concerned about ethics and the value of human life. "Perhaps never before in history," he stated, "has ethics been valued so little as in our days." By stressing the evolutionary and economic value of his fellow human beings, he hoped to elevate their value and status. Goldscheid was grieved and incensed at the capitalist oppression of the working classes, and since religious and moral teachings had proven impotent to curb it, he hoped that his scientific sociology could provide the rationale to end oppression. Once societies understood that humans have economic value, that they are capital not to be squandered, society would no longer tolerate capitalist oppression, Goldscheid thought, and socialism would triumph. "The evolutionary value of humans in their present form is thus *social property*," he explained. "Whoever wantonly damages it, *wrongfully destroys social capital*."[57]

Goldscheid insisted that his sociology was scientific, and evolution was an integral component of it. However Goldscheid's views on evolution differed from the strict Darwinism of many other eugenicists. While honoring Darwin for his scientific achievements, Goldscheid rejected the Malthusian element in Darwinian theory. Because of this, he insisted that human populations do not need to keep expanding, nor does natural selection need to be intensified. On the contrary, competition and selection should be reduced to increase the quality of humans. While many leading eugenicists insisted that increasing rates of reproduction would lead to greater biological improvement, Goldscheid argued the opposite. According to him, "with declining natality, just as with increasing usefulness, the biological and economic value of the individual human life is continually elevated."[58] Unlike most other eugenicists, he stressed the importance of the environment much more than heredity in shaping human character.

While Goldscheid opposed some key elements of mainstream eugenics, he shared many of their basic presuppositions and goals. He hoped that his sociology would help improve people's biological quality and thus produce evolutionary progress. These concerns led him to join the Society for Race Hygiene in 1908 or 1909.[59] Also the Vienna Sociological Society, founded and led by Goldscheid, had a Department of Social Biology and Eugenics. The famous Lamarckian biologist Paul von Kammerer and a professor of hygiene, Julius Tandler, headed up this department, which took a more Leftist orientation than the Society for Race Hygiene.[60] Goldscheid's views were surely more humane and compassionate than those of most eugenicists, but his stress on improving biological quality and the value of humans, and especially his focus on the economic value of humans, presupposes that some humans are more valuable than others. It also raises the troubling question: What about those humans who have no "economic value," such as the disabled? It seems odd that Goldscheid never broached this question, for he was closely connected to the eugenics movement, where this issue was raised repeatedly. Perhaps he could not bring himself to confront the issue, since when applied to the disabled, his rational conception of human economy seems in conflict with his humane disposition. As champion of the oppressed, Goldscheid hoped to elevate the value of human life—a noble goal—but it seems to me that by making the value of human life dependent on economic value, his presuppositions contributed instead to devaluing human life.

Many Darwinists promoted the idea that inequality was an ineluctable law of nature. When applying this to human society, they stressed the physical, mental, and even moral differences between individuals, calling some people superior and others inferior. While the purveyors of this ideology obviously belonged to the intellectually superior segment of society, the disabled, especially the mentally disabled, were called inferior and assigned a lower value. However, many Darwinists and eugenicists considered another category of allegedly inferior people just as dangerous—or perhaps more dangerous—than the disabled. We will now turn our attention to this next category: those of non-European races.

6. The Science of Racial Inequality ✍

The disabled and criminals were not the only ones whose lives were devalued by Darwinian-inspired social thought. Many social Darwinists and eugenicists consigned most of the world's population to the realm of the "inferior." They regarded non-European races as varieties of the human species—or sometimes even as completely separate species—that were not as advanced in their evolutionary development as Europeans. Of course, Darwinism was not the sole culprit in the rising tide of scientific racism in the late nineteenth century, but it played a crucial role nonetheless.[1]

Racism obviously predated Darwinism, but during the nineteenth century—in part through the influence of Darwinism—it would undergo significant transformations. Before the nineteenth century, the intellectual dominance of Christianity militated against some of the worst excesses of racism. Christian theology taught the universal brotherhood of all races, who descended from common ancestors—Adam and Eve. Most Christians believed that all humans, regardless of race, were created in the image of God and possessed eternal souls. This meant that all people are extremely valuable, and it motivated Europeans to send missionaries to convert natives of other regions to Christianity. As contact with other races increased during the nineteenth century, the Protestant missionary movement blossomed, sending out multitudes of missionaries to convert non-European peoples to Christianity, just as the Catholic Church had earlier done in Spanish and Portuguese colonies. Even though some Christian groups, especially in lands with race-based slavery, developed theological justifications for racial inequality, most Christian churches believed that people of other races were valuable and capable of adopting European religion and culture.

The Enlightenment, while eschewing many elements of Christianity, upheld most aspects of Judeo-Christian morality. In some cases,

Enlightenment thinkers were more consistent than the churches in applying Judeo-Christian morality to society. Many Enlightenment figures stressed cosmopolitanism and human equality, which militated against suppression of other races and led to greater opposition to slavery. The eighteenth-century German philosopher and pastor Johann Gottfried von Herder exemplified this attitude, when he wrote: "But thou, O Man, honour thyself: neither the pongo, nor the gibbon is thy brother: the american and the negro are: these therefore thou shouldst not oppress, or murder or rob; for they are men, like thee: with the ape thou canst not enter into fraternity."[2]

Drawing on Enlightenment ideals, many nineteenth-century liberals stressed human equality, including racial equality. This was especially true among those embracing an environmentalist view of human nature, which captivated many intellectuals in the nineteenth century. Believing that the human mind was a blank slate (a tabula rasa, to use Locke's terminology), they ascribed human disparities to differences in experiences, training, and education. Because of this, they thought "uncivilized" races could be elevated to the same level as Europeans through education. Many leaders of German anthropology in the late nineteenth century, especially the dominating figures of Rudolf Virchow, Adolf Bastian, and Johannes Ranke, reflected this liberal perspective and vigorously opposed incursions of biological racism (and Darwinism, too, for that matter) into their field.

However, there was also a dark side to the Enlightenment that would presage scientific racism. Polygenism—the belief that races did not descend from common ancestors—arose in the eighteenth century and clashed with monogenism, which had been dominant for centuries, because Christian teaching up to this time traced all human ancestry to a single pair created in the not-so-distant past. Voltaire and some other Enlightenment thinkers used polygenism as a weapon to attack Christianity's allegedly outmoded dogmas. Polygenism would continue to wield influence in the nineteenth century, until late in the century, when it was swamped by Darwinian explanations for the origin of races.

Even though Darwinism taught the common ancestry of all humans, at least if one looked back far enough in the past, this by no means implied racial equality. Far from it. Indeed many Darwinists claimed that Darwinism proved human inequality, including racial inequality. Darwin and most Darwinists, as we have already seen, emphasized biological variation within each species. When explaining human evolution, Darwin needed to respond to those who insisted that human rationality, speech, and morality were unique to humans and could not be the product of evolution. To overcome these objections, Darwin tried to show on the one hand that animals, especially primates, have primitive reasoning power, speech, and morality. On the other hand, he explained that some races have much lower intellectual and moral faculties than Europeans. Emphasizing racial inequality thus served an important function in Darwin's attempt to bridge the chasm between primates and humans. Even though he opposed slavery and sometimes expressed sympathy for non-European races, nonetheless he believed a wide gap separated the "highest races" from the "lowest savages," as he called them, who were inferior intellectually and morally akin to Europeans. This was not just a peripheral point of *Descent*, for in the introduction Darwin clearly stated that one of the three goals of his book was to consider "the value of the differences between the so-called races of man."[3]

Even so, didn't evolutionary theory imply that races could change, providing hope that lower races could be elevated? Yes, in one sense, but Darwinian gradualism would strip this objection of any significance. Darwin and most Darwinists believed that human races had formed over millennia. During the relatively short period of recorded human history, racial change had been unnoticeable. Ludwig Woltmann and other Darwinist racial thinkers, as well as later Nazi racial theorists, were thus not at all inconsistent when they insisted on both the evolution of human races (over vast stretches of time) and the basic stasis of racial types (during human history).

Darwin's emphasis on the power of heredity to conserve biological traits for long periods of time extended not only to physical, but also to mental and even moral traits. He regarded characteristics such as altruism, selfishness, bravery, cowardice, diligence, laziness, frugality, and many others as biological instincts that are hereditary. Thus, Darwin joined other biological racists in rejecting the influence of education, training, and the environment on shaping human nature. This had dire consequences for racial thought, since all attempts to bring European culture to the "uncivilized" peoples of the world would be futile, if it were true. Darwin was not original in formulating these ideas, to be sure, but he and many Darwinists vigorously promoted this kind of biological racism, and most biological racists after Darwin saw his theory as confirmation of their position.

Haeckel and many of his followers in Germany stressed racial inequality even more than Darwin did. Whenever Haeckel discussed human evolution in his works—and for him human evolution was of the utmost importance, so it was a frequent theme—he consistently stressed human inequality, especially racial inequality. Already in his first technical book on evolution in 1866 he stated, "*the differences between the highest and the lowest humans is greater than that between the lowest human and the highest animal.*"[4] He reiterated this point again and again during his career.[5] In analyzing the gradations between animals and humans, Haeckel asserted in his popular book of 1868, "If you want to draw a sharp boundary, you must draw it between the most highly developed civilized people on the one hand and the crudest primitive people on the other, and unite the latter with the animals."[6]

In order to make human evolution plausible, Haeckel tried to show that the gap between humans and their nearest animal relatives could be bridged by almost imperceptible gradations. The frontispiece of the first edition of his popular work, *Natürliche Schöpfungsgeschichte* (1868, *The Natural History of Creation*), was a series of twelve facial profiles, beginning with a European, then "descending" in order to an East Asian, a Fuegian, an Australian, a black African, and a Tasmanian. The pictures were arranged to show a gradual change in skull shape, and the profile of the last human, the Tasmanian, looked very similar to the profile of the gorilla in the seventh picture. After the gorilla came five other simian species (see figure 6.1). With six "steps" between the "highest" and "lowest" human races and only one "step" between the "lowest" human race and gorillas, the inference was obvious. But Haeckel made sure no one missed the point by commenting in his caption that his illustrations demonstrated graphically that "the differences between the lowest humans and the highest apes are smaller than the differences between the lowest and the highest humans."[7]

The proximity of "inferior" or "lower" races to simians is a frequent theme in Haeckel's writings. He referred to the Australian aborigines and the Bushmen of South Africa as similar to apes (*affenähnlich*). He further described some races in Africa and Asia as having no concept of marriage or the family; like apes, they live in herds, climb trees, and eat fruit. These races are not capable of learning European culture, for "it is impossible to want to plant human education (*Bildung*), where the necessary ground for it, human brain development, is lacking. . . . They have scarcely elevated themselves above that lowest stage of transition from anthropoid apes to apemen."[8]

Another way Haeckel emphasized the disparity between races was by insisting that different human races are distinct species. In 1868, he distinguished between ten "species" of humans, which he arranged in a racial hierarchy based on their alleged inferiority and superiority. He divided human "species" into two main groups, the "straight-haired" and the "wooly-haired." The latter included primitive extinct humans, Papuans, Hottentots, and black Africans, in order from lowest to highest. According to Haeckel's scheme of classifying races, the "wooly-haired species" were generally inferior to the "straight-haired," except the "straight-haired" Australian aborigines were the lowest extant human "species." The other

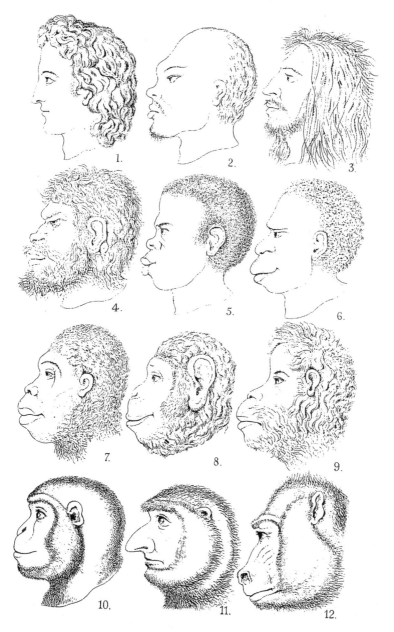

Figure 6.1 Frontispiece to Haeckel's *Natürliche Schöpfungsgeschichte* (1868), depicting six human races and six simian species, allegedly showing the proximity of the "lowest" humans to simians

"straight-haired species" (in ascending order) were: Polynesians, Eskimos, American Indians, Mongolians (Orientals), and Caucasians, whom he also called the Mediterranean "species." Later he fiddled with the details, increasing the numbers of "species" to 12, but the idea of a racial hierarchy—both between these "species" and between races within each "species"—remained constant.

Another line of evidence Haeckel used to distance human races was linguistics. Haeckel argued that languages do not appear to have a common ancestor, so they must have developed independently. Since the use of language is a defining characteristic for human beings, this means that human races must have evolved from separate ancestors. Of course, if one went back far enough, one would find a common ancestor, he admitted, but he didn't believe the common ancestor would be fully human.[9] Haeckel called this nonspeaking apeman Pithecanthropus alalus, and he thought it would look roughly like the painting Prof. Gabriel Max gave him in 1894 for his sixtieth birthday (see figure 6.2).[10] Thus, despite acknowledging monogenism formally, Haeckel tilted toward the polygenist side of the argument about human origins.

It may seem hard to imagine, but later in his career, Haeckel stressed racial differences even more, making statements even more provocative than his earlier ones. In his sensationally popular book, *The Riddle of the Universe* (1899), he stated,

> The difference between the reason of a *Goethe, Kant, Lamarck, Darwin* and that of the lowest primitive human, a Vedda, Akka, Australian Negro, and Patagonian, is much greater than the gradual difference between the reason of the latter and the 'most rational' mammal, the anthropoid apes and even [other] apes, dogs and elephants.[11]

In 1911, Haeckel further emphasized the distance between races in another way by placing the 12 human "species" in four separate genera![12] He also recommended that anyone wanting to understand racial inequality should read the writings of Count Arthur de Gobineau, the French racial thinker who captivated the attention of early twentieth-century German racial ideologues.[13]

In *The Wonders of Life* (1904), a sequel to *The Riddle of the Universe*, Haeckel devoted an entire chapter to racial inequality entitled "The Value of Life." Therein he argued forthrightly that not all people's lives have the same value. Though briefly mentioning the validity of this within every society, his main focus was on races: "Likewise also for world history the value of the various races and nations is very unequal." Haeckel's yardstick

Figure 6.2   *Pithecanthropus alalus*, an imaginary missing link between apes and humans, reproduced in Haeckel's *Natürliche Schöpfungsgeschichte* (1911 ed.)

for determining the value of life was the level of cultural achievements, but he believed that the development of culture was dependent on innate biological, intellectual, and moral traits. Thus "civilized" peoples with their allegedly higher intellectual capabilities have a higher value of life than the "primitive" peoples. According to Haeckel, "The value of life of these lower wild peoples is equal to that of the anthropoid apes or stands only slightly above them."[14] Haeckel's devaluing of "primitive" races, by placing them on par with animals, would be the first step toward a genocidal mentality, as we shall see in our chapter on "Racial Struggle and Extermination."

Inspired by Haeckel's vision of human evolution, Bernelot Moens, a teacher in the Netherlands, asked Haeckel for advice in 1905 about a scientific experiment he hoped to perform to demonstrate the close relationship between humans and apes. He hoped to use artificial insemination to cross an anthropoid ape with a "Negro" (it's unclear which black race he meant—Haeckel considered black Australians the lowest race, but more often than not "Negro" meant black Africans). Haeckel thought Moens' project was likely to succeed and encouraged Moens to proceed, though he refused to lend his name to the endeavor. In 1916, a German sexologist, H. Rohleder, approached Haeckel with a similar project, and Haeckel suggested trying to inseminate a chimpanzee with the sperm of a black African.[15]

Haeckel was not the only early Darwinist to promote the doctrine of racial inequality. In fact, he may very well have been influenced in his scientific racism by Karl Vogt, a zoologist whom Haeckel admired.[16] The scientific materialist Karl Vogt, in exile at the University of Geneva because of his participation in the Revolutions of 1848, was already famous before Darwin published his theory. He had resisted earlier evolutionary theories, but was one of the earliest German scientists to embrace Darwinism and apply it to humans. Vogt also became a leading anthropologist, helping found the German Anthropological Society in 1870.

Already four years after Darwin published *The Origin of Species*, Vogt was applying Darwinism to humans in his most important anthropological work, the two-volume *Lectures on Man*. Like Haeckel, Vogt continually stressed racial inequality and the mental inferiority of non-European races. He claimed that scientific evidence demonstrated "that the differences between particular human races is greater than between particular ape species, so that we must therefore recognize individual human races as species." Vogt provided a Darwinian explanation for the distance between the races, explaining that the three major human races had probably evolved from three different species of apes. The black Africans, however, were still closer to their ape origins than were the Europeans.[17] Vogt's position on human origins was thus even more polygenist than Haeckel's.

Haeckel's and Vogt's Darwinian racism was extremely influential in the late nineteenth and early twentieth century, appearing in recognizable form again and again in the writings of Darwinian ethnologists and other social Darwinists. Haeckel's friend Carneri was even more radical than Haeckel in claiming the close affinity of the "lower" human races with other animals. In his book *Morality and Darwinism* (1871) he stated that "entire human tribes stand lower than the animals. Not only is the mental activity of the elephant, horse, and dog significantly better developed than the lowest human species," but some human races possess no more ability to speak than do parrots. Such was the astonishingly myopic view of the leading Darwinian moral philosopher of the late nineteenth century.[18] Carneri, like Haeckel, believed that because of the mental inequality of races, only certain races were capable of developing higher culture. Specifically, he denied that Slavs were capable of creating higher culture; rather they depended entirely on German cultural developments.[19] Carneri's racial chauvinism seems especially inflammatory, since he was a liberal politician in the Austro-Hungarian Empire, which had a huge Slavic population.

No other leading Darwinist in Germany stooped quite to the level of ignorance displayed by Carneri in his racial views, but some came pretty close, and biological racism flourished in Darwinian circles. Krause and Büchner, the two most influential Darwinian publicists in nineteenth-century Germany (besides Haeckel himself) promoted scientific racism in roughly the same form as Haeckel's. In his popular work, *Man and His Position in Nature* (1870), Büchner claimed that black Africans, Asians, and Europeans (whom he sometimes called Aryans) were all distinct species.[20] Later, in his contribution to Hellwald's *History of Culture*, Büchner stated,

> History and ethnology show us that races are something so peculiar and fundamentally different physically and mentally, that one cannot on first view suspect their original unity, if one does not allow very long periods of time for gradual development.[21]

Despite his belief in the inheritance of acquired characteristics (Lamarckism), Büchner in his book, *The Power of Heredity* (1882), argued that physical, mental, and moral characteristics of races are basically constant, since they change so slowly. Thus education and training have little effect: "As little as one can make a real Christian from a Jew or a Caucasian from a Semite, so little can one make a civilized person out of one born to a wild [race]." Racial character is thus innate, and though Büchner never exhibited any anti-Semitism, he did think the Jews had a different racial character than Germans.[22]

The ethnologist Oscar Peschel, editor of *Das Ausland*, the leading German journal on geography and ethnology, was probably the first German scholar to publish a significant discussion of Darwin's new theory.[23] Peschel not only became one of the earliest proponents of Darwinian theory, but he was perhaps the first German to use Darwinian justifications for racial inequality. In 1861 and again in 1863 he wrote articles in

Das Ausland, both entitled, "Man and Ape." The point of both articles was to convince readers of the plausibility of human evolution by comparing humans and apes. Some races, according to Peschel, were closer to apes than others:

> The Negro is far removed from the European and close to the ape through its small build, through the relatively small breadth of its skull, through its relatively long upper limbs, and further the relatively short length of the thigh . . . Also the Negro is more animal, in that it gives off a disgusting odor, distorts its face in grimaces, and its voice has a harsh, grating tone.[24]

In Peschel's most important book, *Ethnology*, the theme of racial inequality surfaces repeatedly.[25] Peschel's work in ethnology was regarded so highly that he was named the first professor in geography at the University of Leipzig in 1871, which made him give up editing *Das Ausland* that year.

Peschel's successor as editor of *Das Ausland*, the Darwinian ethnologist Hellwald, continued to stress racial inequality, both in *Das Ausland* and in his major book, *The History of Culture* (1875). Like most scientific racists, he stressed the hereditary character of different races. The English have an innate disposition to be enterprising, according to Hellwald, while South Americans and South Sea Islanders have a hereditary tendency toward laziness. Though he considered all human races part of a single species, he not only stressed the mental inferiority of black Africans, but also claimed that the distance between races was ever widening. Further, even though he concentrated more on the racial inferiority of colored races, Hellwald was one of the few Darwinian ethnologists before the 1890s to exhibit overt anti-Semitism in his writings. The Jewish racial character, according to Hellwald, hindered the Jews from producing any great statesmen and artists (except musicians).[26]

In 1882 Hellwald passed on the editorship of *Das Ausland* to yet another prominent Darwinian ethnologist and geographer, Friedrich Ratzel, who later became professor of geography at the University of Leipzig. Before switching to geography, Ratzel received a doctorate in zoology and even wrote a popular exposition on Darwinism, *Being and Becoming in the Organic World* (1869), which remained relatively unknown, being eclipsed by Haeckel's popular exposition of Darwinism that appeared the preceding year. In preparing to write this work he relied heavily on Haeckel's *General Morphology*, and even though he later disagreed with some aspects of Haeckel's form of Darwinism, in 1872 he confessed that "in my scholarly work I live entirely in the theory of evolution." In the 1890s he still addressed Haeckel as his friend and teacher, and he hoped that everyone could clearly recognize by reading his works that he was a disciple of Haeckel.[27] Indeed, it is not difficult to discern the influence of Darwinian theory on Ratzel's ideas about biogeography, one of his favorite subjects.

Ratzel's treatment of race, especially in his later writings and lectures is considerably more nuanced than those of Peschel or Hellwald. He rejected the intensely chauvinistic biological racism that gained such prominence in the late nineteenth and early twentieth centuries, embracing instead the liberal view of anthropology championed by Adolf Bastian and Rudolf Virchow. He vigorously opposed Aryan racial theorists, calling them "race fanatics." In his university lectures on ethnology in the late 1890s, he specifically warned against the universal human tendency to create greater distinctions between races than are warranted by careful scientific investigation. He criticized those anthropologists who "are seeking people with tails" and who try to ascribe ape-like characteristics to black Africans.[28] By opposing those stressing the affinity of "lower" human races with apes, Ratzel was out of step with many leading Darwinists in Germany.

However, even though he was far less racist than many of his contemporaries and tried to minimize racial prejudice, Ratzel was not entirely persuaded of the biological equality of all human races. In *Being and Becoming* Ratzel reminded his contemporaries of the huge divide between humans and apes. Nonetheless, in the same sentence he admitted that the "lower" peoples are closer to apes than are their "higher-standing comrades."[29] In his later more influential works Ratzel often used the terms "higher" and "lower" to refer to different groups of people, but usually he was comparing them culturally rather than biologically, which is why he more often used the term "peoples" (*Völker*) than "races." Nonetheless, in some of his writings and university lectures he manifested elements of biological racism. In his major work on ethnology, *History of Mankind*, Ratzel argued for the substantial unity of the human races and stated that cultural achievements were not linked to biological differences. Nonetheless, when discussing racial interbreeding between Europeans and non-Europeans, he clearly presented the Europeans as biologically superior, the non-Europeans as biologically inferior, and the crossbreeds as intermediate. Furthermore, when discussing specific races, such as the Australian aborigines, he sometimes presented them as intellectually inferior to Europeans.[30] Ratzel once criticized a colleague for claiming that contemporary primitive peoples have "just as good human material" as civilized peoples had a couple of thousand years ago.[31] This indicates again that despite his own remonstrations to the contrary, Ratzel saw "primitive peoples" as biologically inferior, not just culturally inferior. In a university lecture on "Biogeography and Anthropogeography" he noted that the "race of people that we regard as the oldest, as the one furthest behind in evolutionary development, that is, the black, wooly-haired race," lives in the south, while the "youngest, most advanced" white race lives in the north.[32]

Ratzel also believed that racial differences should lead to practical policies of racial discrimination. While lamenting the extreme racial views of some racial theorists of his day, he nonetheless did not dismiss their views in toto. In a largely negative review of a racial theorists' book, he declared, "Even so we are in agreement with all these racial politicians, that the future of a people very much depends on breeding through the exclusion of one [racial] mixture and the promotion of the other; and we would be thankful to them [i.e., racial theorists] for thoughtful practical suggestions in the best interests of our own people." He continued by applauding the United States for restricting the rights of Indians and blacks and for introducing immigration restrictions for Asians. Germany would benefit, he thought, by introducing racial policies encouraging immigration from Scandinavia, while restricting it from Poland.[33] Ratzel certainly opposed the most egregious racial theories of his day, but he was not completely egalitarian, and his racism was linked closely with his understanding of Darwinian biology, which will become even more apparent when we examine his stance on racial extermination (in chapter 10).[34]

Before the 1890s, almost all the influential Darwinian anthropologists and ethnologists—along with most Darwinian biologists and popularizers—embraced scientific racism. Indeed, Darwinian materialists and monists were the leading apostles of scientific racism in Germany. While not formally depending on Darwinian theory, scientific racism appealed to Darwinists because of its stress on biological determinism and inequality. Scientific racism was not deduced from Darwinian theory, to be sure, but it served an important function in Darwinian discourse as evidence increasing the plausibility of human evolution. In short, racism helped them "prove" their theory. On the other hand, Darwinism lent greater plausibility to scientific racism, providing an explanation for the origins of races and their alleged inequality.

The connection between Darwinism and scientific racism appears all the more striking when we compare the Darwinian anthropologists and ethnologists with their non-Darwinian counterparts. The dominating figures in the German anthropological community in the late nineteenth century were the famous pathologist (and liberal political leader) Rudolf Virchow, Adolf Bastian, and Johannes Ranke, all of whom opposed Darwinian theory. Virchow aroused the ire of his erstwhile student, Haeckel, in 1877 by questioning the appropriateness of teaching evolution in schools, when the theory was so highly speculative. He also led the fight against viewing the Neanderthal Man as a primitive human ancestor, claiming instead that it was merely a pathological specimen. Ranke was also known for his anti-Darwinian stance, astonishing Hermann Klaatsch at the 1899 Anthropological Congress by calling his Darwinian views on human evolution fantasy, not science.[35] Along with rejecting Darwinism, Virchow, Bastian, and Ranke also rejected biological racism. Instead, they stressed racial equality, monogenism, and the influence of the environment and education on people. Scientific anthropology did not necessarily spawn theories of biological inequality, as German anthropology shows.

With the passing of the baton from Virchow, Bastian, and Ranke to a younger generation of anthropologists shortly after 1900, both Darwinism and scientific racism came to the fore. The anatomists Hermann Klaatsch, Gustav Schwalbe, and Gustav Fritsch, who had all embraced Darwinism in the late nineteenth century, emerged as leading figures in German anthropology, though none was well known outside the academy. All three stressed racial inequality and biological determinism. Klaatsch and Fritsch even proposed a Darwinian form of polygenism similar to Vogt's, except they derived present human races from two rather than three different ape ancestors.

Klaatsch's racial views are especially interesting, since he forthrightly applied his polygenism to ethical issues. In a speech to the International Congress on Criminal Anthropology in Cologne in 1911, he explained the significance of his studies on primitive races for understanding criminality. Klaatsch took an extended research trip in 1904–07 to study the Australian aborigines intensively, whom Haeckel and many others considered the lowest race in the world. In 1911, he still considered them an "isolated remnant of the oldest humanity," but after a close study, he concluded that they were physically and mentally akin to Europeans, especially to the fossilized human remains from Aurignac. He believed that the Aurignac ancestors of Europeans and the Australian aborigines were part of a common racial stock. On the other hand, he considered black Africans significantly inferior to both Europeans and Australian aborigines. The Africans, he thought, shared common traits with the Neanderthal race, which was largely supplanted by the Aurignac race in Europe.[36]

The Australian aborigines made a very good impression on Klaatsch, especially in regard to their moral character. He claimed they never stole, and loving their neighbors was a "fundamental characteristic of their being." This proved, according to Klaatsch, that the primitive state of humanity—at least the branch of humanity including them and the Europeans—was morally good. The Africans and Neanderthals, on the other hand, still had more "bestial" characteristics. Klaatsch surmised that European criminals probably had some remnant of Neanderthal traits overriding the good moral traits of their Aurignac ancestors.[37]

Klaatsch once stated that "modern science cannot confirm the exaggerated humanitarianism which sees brothers and sisters in all the lower races."[38] What practical conclusions did Klaatsch draw from this Darwinian racism? First, he deplored the racial mixture of whites and blacks. Second, he denied that blacks could be educated very much, so colonial powers should not expend too much effort in this regard. Finally, he rejected the notion that people of different races should have equal rights. "The humanitarian nonsense," he declared, "which grants equal rights to all on the premise of the unity of humanity, is to be condemned from the scientific standpoint." For Klaatsch, as for many other Darwinists of his time, science showed the folly of egalitarianism, especially racial equality. He also intimated that slavery was beneficial, not only for the slave-holding whites, but also for the black slaves.[39] Klaatsch and his cohort of Darwinian-inspired anthropologists thus overturned the liberal tradition of German anthropology.

Even some of the older generation of anthropologists, including Luschan and Waldeyer, converted to Darwinism around the turn of the twentieth century, simultaneously embracing scientific racism. The simultaneous shift toward Darwinism and biological racism was so pronounced that the historian Benoit Massin concludes:

> And for those embracing the new Darwinian approach in German anthropology, the implications of racial evolutionary hierarchies were even more radical: the replacement of the previous humanitarian ethics by a biological and selectionist materialism more concerned with the inequalities of evolution than the universal brotherhood or spiritual unity of humankind.[40]

The acceptance of Darwinism by the German anthropological community thus produced a shift from racial egalitarianism to inegalitarianism and a replacement of liberal, humanitarian ethics with evolutionary ethics.

To be sure, Darwinism does not necessarily imply scientific racism, and scientific racism did not necessarily depend on Darwinism, but the two shared affinities that made them not only compatible, but also alluring to each other. Historically Darwinism and biological racism are linked tightly together, as many historians have demonstrated. In the late nineteenth and early twentieth centuries, we almost always find them in tandem.

Indeed, as Darwinism captivated the thinking of German scholars by the end of the nineteenth century, biological racism escalated both in scope and in intensity. Before the 1890s, discussions of biological racism were usually confined to brief passages in longer books or articles, especially ones directly dealing with human evolution or ethnology. In the 1890s and early 1900s, however, race moved from the periphery to center stage with a veritable deluge of books and articles devoted to the topic. Also, before the 1890s biological racists usually focussed on the differences between Europeans on the one hand and black Africans, Asians, American Indians, and Australian aborigines on the other hand. Toward the end of the nineteenth century, quite a few biological racists radicalized racial distinctions by stressing the supremacy of the German or "Aryan" race and simultaneously denigrating the Jews. Many of these Aryan racists and anti-Semites—which included many scientists and physicians—appropriated the mantle of Darwinian science to enhance their legitimacy.

Darwinian racism received tremendous impetus through the emerging eugenics movement around 1900. Since eugenicists were avid Darwinists, many of them imbibed biological racism from Haeckel, Vogt, Büchner, and other leading Darwinists. Further, eugenics stressed the importance of heredity in determining physical, mental, and moral traits. It was only a short step from this general biological determinism to racial determinism. Further, eugenics was founded on the premise of human inequality, and even though racial inequality was not a necessary corollary of this, most eugenicists embraced the doctrine of racial inequality. Indeed, eugenicists would popularize one of their favorite terms—"*minderwertig*" (inferior)—in race discourse.

As the chief organizer of the German eugenics movement, Ploetz was wary about publicly emphasizing his racial views. He recognized that rabid racism might alienate some elements of the German scientific and medical community, and he wanted to gather as many scientists and physicians as possible into his big eugenics tent. Because he toned down his views a little in public, Ploetz's racism has often been underestimated. Nevertheless, Ploetz's racial views were an integral part of his program, providing the underlying motivation for his eugenics activities. In his youth Ploetz founded an organization with his friends Carl and Gerhart Hauptmann committed to preserving German racial purity. This was the start of a lifelong project, he explained in his unpublished memoirs:

> Through reading the works of Darwin, Haeckel and other biologists already at school, as well as through some novels by Felix Dahn and other glorifiers of German antiquity and medieval times I was permanently enthused for the Germanic race . . . and determined to make it my life's task . . . to help in Germany and other states with German-speaking populations to lead it upward again to purity and the height of the first millennia.[41]

This was not only the purpose of his youthful utopian club, but also of his later involvement in the eugenics movement.

His concern for Germanic supremacy continued unabated throughout his life. In 1890 he wrote from Paris to Carl Hauptmann that he was more convinced than ever of German racial superiority over the French.[42] Later, Ploetz's desire to name the German eugenics movement the race hygiene (*Rassenhygiene*) movement was motivated by racism, despite Ploetz's and other eugenicists asseverations to the contrary later, and its ambiguity abetted racist rhetoric. In coining the term race hygiene, he preferred it because

> the hygiene of the entire human species coincides with that of the Aryan race, which, except for a few smaller races, like the Jewish—which in any case is probably mostly Aryan—represents the civilized race par excellence; to further it [the Aryan race] is the same as furthering all of humanity.[43]

That Ploetz's race hygiene was not supposed to benefit all races was also apparent in his statement that the love of humanity "is nothing more than love for its Aryan part."[44] In reflecting back on the founding of the Society for Race Hygiene, he gave as one example of human inequality, the distance between the "low tribes" of Australian aborigines and the "most noble branches of the white race."[45]

Ploetz even founded a secret organization called the Nordic Ring in 1907 to promote a more racist form of eugenics. In 1911, Ploetz wrote a brief tract to recruit German youth into the Nordic Ring, in which he divulged his racial views more fully. After explaining why he believed that the Nordic race was the most advanced in the world, he stated that "for us the immediate foundation for the realization of our ideals can only be the Nordic race. . . . The object of our labor must be, in short, a Nordic–Germanic race hygiene."[46]

Despite his reticence at times, Ploetz divulged his racial views often enough. In his only book he not only stressed the racial inferiority of black Africans, but also claimed that the "Western Aryans" were the most advanced race in the world.[47] In a 1902 article he also emphasized the superiority of the Germanic race, whose members are "a totally different thing from a coolie or a Kaffir."[48] His article, "On the German–Polish Struggle" (1906), warned that the Poles were increasing at the expense of Germans in eastern Germany, which threatened the preservation of "the most valuable of the human races, the white race," because the Poles are racially inferior to the Germans.[49] At the First German Sociology Congress in 1911, Ploetz asserted that blacks in America were unintelligent and immoral.[50]

Not all eugenicists were as racist as Ploetz. Schallmayer rejected Nordic racism and extolled the Chinese for their advanced culture. Though less racist than Ploetz, he did not exactly believe in racial equality. In one article he stated, "Without a doubt the mental ability of the different human races is unequal, just as they are also different in their physical constitution."[51] In various writings he referred to races as higher and lower, specifically claiming that black Africans are mentally inferior to Europeans and unable to sustain a high culture.[52] In a letter to the American biologist and pacifist David Starr Jordan he stated that blacks and Eastern Europeans were racially inferior to the average American.[53] Also, in friendly correspondence with Ludwig Schemann, founder of the Gobineau Society, Schallmayer expressed sympathy for Schemann's writings, though he distanced himself from Woltmann and Gobineau.[54]

While Schallmayer and a few other eugenicists did not emphasize race as much as Ploetz, some eugenics advocates thought everything revolved around the race question. In his two-volume *Cultural History of Racial Instincts* Driesmans explained that race was the key factor determining historical developments, since only some races are capable of creating culture. Though stressing racial inequality, he parted company from Gobineau by rejecting the idea of a pure Aryan race and by advocating racial mixture, as long as it occurs between races that are equal. Driesmans integrated evolution into his racial ideology by claiming that races undergoing the stiffest struggle against the elements would advance faster than others. In general, then, races in cold climates are superior to those in the tropics, and the Germans had attained their racial superiority because they had endured the Ice Ages, while the Persians, Slavs, and Celts had escaped it.[55] This idea that the Ice Age had been instrumental in human evolution went back at least to Moritz Wagner, who in 1871 had suggested that the Ice Ages had been instrumental in human evolution, making the cradle of humanity Europe, not Africa or Asia.[56]

Even more prominent and influential than Driesmans was Ludwig Woltmann, a physician who founded the journal *Politisch–Anthropologische Revue* in 1902. Woltmann was a dynamic and articulate leader, who sought to synthesize major currents of European thought.[57] He began his scholarly career trying to harmonize Kant and Darwin, subsequently focussed on Marx and Darwin, and finally added Gobineau to the blend; with Gobineau's racism ultimately swamping his Marxism, despite his continued insistence that he was true to Marx's theory.[58] In 1901 Woltmann tried to establish a eugenics journal with Ploetz and Grotjahn as coeditors, but after establishing his own journal, he was disappointed when both men founded journals competing with his. His quarrel with Schallmayer over the Krupp Prize alienated him not only from Schallmayer, but also from other eugenicists who praised Schallmayer's work.[59]

In his Krupp Prize entry, *Political Anthropology* (1902), in which he applied Darwinian theory to social thought and legislation, Woltmann placed race at the center of his analysis. He believed that the blond-haired, blue-eyed, long-headed (or dolichocephalic, as anthropologists called it at the time) Nordic or Germanic race was "the highest product of organic evolution" and the "bearer of world civilization." Woltmann wasn't the originator of this idea, but he became one of the most influential popular-izers of it. Starting off with this assumption, he tried to fit all of history into a framework in which Nordic races produced all significant cultural advances. The Indians, Persians, Greeks, and Romans were all originally light-skinned and blond, according to Woltmann, and their downfall came from mixture with darker racial elements. In his zeal to find Aryan roots for all significant cultural achievements, he even claimed that Jesus probably descended from a blond Amorite (though why a materialist would want to claim Jesus as a fellow Aryan escapes me).[60] After publishing *Political Anthropology*, he scoured the museums of Italy and France, trying to prove that the Italian Renaissance and French cultural achievements were predominately the work of men with blond hair and blue eyes.[61]

Woltmann gathered around his journal a cadre of like-minded scholars, who called themselves the socio-anthropological school. Its most promi-nent members were the freelance anthropologist Ammon, the French social Darwinist Georges Vacher de Lapouge, and Ludwig Wilser. All three posed as data-driven, empirical scientists relying on Darwinian biology when they argued for the supremacy of the Nordic or Germanic race (which pre-dictably made Lapouge more popular in Germany than in his native coun-try). Ammon argued that the dolichocephalic Germans were born aristocrats, while the roundheaded (brachycephalic) elements were only fit for farming and manual labor.[62] In a published speech, *The Superiority of the Germanic Race* (1915) Wilser, who had imbibed Darwinian anthropol-ogy from Alexander Ecker at the University of Freiburg, divided humans into four races: light-haired northern Europeans, dark-haired southern Europeans, Africans, and Asians.[63] Woltmann and his socio-anthropological circle were extremely influential, especially among scientists, physicians, and anthropologists. Even though he contemptuously declined his third prize in the Krupp competition, his standing in the competition boosted his prestige in the scientific community. Woltmann's books received a boost from Ernst Rüdin, a psychiatrist who played a leading role in the Society for Race Hygiene, when he wrote positive reviews of them in Ploetz's eugenics journal.[64]

One prominent devotee of Woltmann's racist eugenics was Eugen Fischer, a young anatomist and anthropologist who became one of the most influential German anthropologists during the Weimar and Nazi periods. Fischer began lecturing on "social anthropology," his term for race hygiene, at the University of Freiburg in 1909. He considered Gobineau a forerunner of modern racial thought, which he promoted in his medical classes.[65] He praised Woltmann, Ammon, Wilser, and Lapouge, as well as Ploetz and Schallmayer, for their pathbreaking work on race and eugenics.[66] Because of his interest in the effects of racial mixture, Fischer was one of the first to apply Mendelian genetics to humans in his book on *The Rehoboth Bastards and the Bastardization Problem among Humans* (1913). The Rehoboth community in German Southwest Africa was composed of descendants of European men and African women. Fischer's careful firsthand study of the Rehoboth community lent scientific credibility to his racial prejudices. Though admitting that the people of Rehoboth were fairly intelligent and in his view superior to full-blooded black Africans, nevertheless he still considered them far inferior to Europeans in their creative abilities. Because of Mendelian inheritance, some people of mixed descent were valuable individuals, he admitted, "but just as many completely worthless individuals can be expected and the majority are inferior (*minderwertig*)." Fischer believed that racial crossing usually produces progeny approximately midway between the races of the parents. Thus he opposed racial mixture and supported racial segregation in German colonies.[67]

A medical student at the University of Freiburg, Fritz Lenz, who took courses from Fischer and served as secretary of the Freiburg branch of the Race Hygiene Society, also became an influential proponent of racist eugenics. In his student days before World War I, Lenz wrestled with philosophical questions, especially the vexing problem of how to ground ethics on a secular basis. He even wrote an unpublished book-length treatise at that time on "The Will to Value, Foundation of a Racial Ethics." In 1917 he published an article on his racially-based view of ethical values in *Deutschlands Erneuerung*, a journal coedited by Houston Stewart Chamberlain. He republished the same essay in 1933 with a preface explaining that his article contained the essentials of the Nazi worldview. Lenz argued in his article that "everything comes from the ideal of the race: culture, evolution, personality, happiness, redemption. . . . With every activity and with every inactivity we have to ask ourselves: Does it benefit our race? And to make our decision accordingly."[68] Lenz's prestige as a medical professor at the University of Munich—he was appointed to the first professorship in race hygiene in Germany in 1923—probably lent scientific respectability to his racial views, especially since he integrated his racist eugenics into the leading text on human genetics that he coauthored with Fischer and Erwin Baur. He also integrated racism into his university lectures. In a seminar he taught in 1919 on race hygiene, he devoted two full hours to Gobineau's racial theory.[69] Interestingly, however, Lenz argued forthrightly that his racial views were not based on science, but were ethical postulates ultimately incapable of proof.[70] Lenz's racism was thus based on faith, a faith akin to Hitler's views in *Mein Kampf*, as Lenz himself acknowledged. The Nazi regime rewarded Lenz by naming him director of the eugenics section of the Kaiser Wilhelm Institute and professor at the University of Berlin.

Another leading racist who forged close connections with Woltmann's socio-anthropological circle was Ludwig Schemann, founder of the Gobineau Society in 1894 and tireless promoter of Gobineau's writings. Schemann's organization was significant both for connecting leading scientific racists and for spreading racial propaganda. Most of the racial thinkers influencing Schemann's thought and most of his connections were social Darwinists and eugenicists. Schemann dedicated one of his books to Ammon, whom he considered one of the most important influences on him, but he also expressed great admiration for Woltmann, Wilser, Lapouge, Driesmans, and others of their ilk. He also promoted eugenics and cultivated relationships with leading eugenicists, especially Fischer, since both he and Fischer lived in Freiburg. The influential Pan-German League provided fertile soil for Schemann's propaganda efforts. The Pan-German League, as well as some of its leaders, like Heinrich Class and Adolf Fick, joined the Gobineau Society and contributed financially to it, while Schemann led the Freiburg branch of the Pan-German League.[71]

Another important promoter of social Darwinist racism in the early twentieth century was Julius Lehmann, a major publisher of medical books and a prominent medical journal in Munich. In the 1890s Lehmann joined the Pan-German League and began publishing books and pamphlets promoting their cause. He also tried to enlist Houston Stewart Chamberlain to write for his publishing house, but Chamberlain declined. After reading eugenics literature in 1908–09 he entered the eugenics movement with heart and soul, publishing thereafter numerous works by leading eugenicists, including the famous genetics text by Lenz, Fischer, and Baur. He also published works by the famous Nordic race theorist, Hans F. K. Günther, in the 1920s, as well other racist works.[72]

Many of the race theorists we have mentioned were anti-Semitic to some degree or other, and indeed the rising tide of anti-Semitism was decisively influenced by biological racism. Not all of this was directly tied to Darwinism, but much of it received some stimulation from Darwinian social thought. Willibald Hentschel, a leading anti-Semitic theorist, for example, had studied under Haeckel in the late 1870s and 1880s, though he later worked as a chemist rather than a biologist. In the early 1900s he successfully organized a drive to honor Haeckel by making a bronze statue of him.[73] The central points of his racist sociopolitical ideology were similar to Haeckel's views on race: humans are naturally separated into races and the Germanic Aryans have the highest "value of life" of all the races.[74] Hentschel believed that Darwinian theory had proven the subjection of humans to natural laws, and this should lead to new values: "One such newly-developed wonder is the biological valuation of life, which has led us to the concepts of race and selection." Thus, Hentschel determined the value of human life by biological characteristics, and for him one of the most important determinants of biological value was race. Indeed race—including anti-Semitism—and selection were foundational concepts in Hentschel's main theoretical work, *Varuna* (1901), which was reissued in 1925 in its fourth edition.[75] Racial concerns also dominated the utopian community he described in his popular book, *Mittgart*, which went through six editions by 1916.

Hentschel actively participated in anti-Semitic politics and propaganda, and his friend and publisher Theodor Fritsch, a leading anti-Semitic publicist, vigorously promoted Hentschel's *Varuna*, calling it the best theoretical work available on race. Fritsch insisted that Hentschel had elevated anti-Semitism to scientific status: "Knowledge about the distinctions between the human species and races belongs to the most recent results of scientific research. We have to admit that the slogan of the equality of all who bear a human face cannot withstand a rigorous examination."[76] Through his books, his journal, *Hammer*, and his organization named after his journal, Fritsch played a central role in the anti-Semitic movement in Germany. He promoted a form of anti-Semitism featuring the Darwinian elements of race struggle and eugenics.

Anti-Semitism was not specifically derived from Darwinian theory, of course. Nevertheless anti-Semitism found a ready ally in the biological racism circulating in the late nineteenth and early twentieth century, largely through the writings of avid Darwinists. The upsurge of racism around 1900 was not entirely of the scientific or Darwinian variety, either, though much of it was. Julius Langbehn's popular racist book, *Rembrandt as Educator* (1890) was even explicitly antiscientific and anti-Darwinian. Some racial thinkers considered Enlightenment rationalism shallow and railed at the rise of scientific materialism and positivism, which often denied free will and thus left no room for human agency. Instead they exalted intuition and volition.[77] Though some of these racial theorists, like Langbehn, dismissed scientific rationality for more irrational modes of thought, others combined elements of rationalism and irrationalism. Darwinism contributed to this synthesis of rationalism and irrationalism, because many Darwinists embraced a naturalistic account of human behavior, but founded morality and behavior more on instincts than on reason.

These ambiguities within Darwinism, as well as some racial theorists' ambiguous stance toward Darwinism, allowed even some of the more mystical race theorists to include Darwinian elements in their ideology. Houston Stewart Chamberlain, one of the most famous anti-Semitic writers around the turn of twentieth century, displayed considerable ambivalence toward Darwinism. While overtly rejecting Darwinian theory on philosophical grounds, calling it too materialistic, he nevertheless embraced key elements of it. In his memoirs he explained that, as a young man studying biology at the University of Geneva (Vogt was his zoology professor there), he had enthusiastically embraced Darwinian theory. Later, however, he turned against it. He came to embrace neo-Kantian idealism and vitalism, which he considered inconsistent with Darwinism, especially since he thought that organisms manifested certain forms (*Gestalt*) that were only malleable within strict limits.

Nonetheless, Chamberlain's racial theory retained significant elements of Darwinian thought, as he himself recognized. In his famous two-volume work, *Foundations of the Nineteenth Century* (1899), he admitted that Darwin had hit upon two key ideas important to racial theory: the struggle for existence and selection through breeding. Concerning struggle, in the closing paragraph of the first volume Chamberlain discussed racial competition, which may be fought with a variety of means, some of them peaceful. Nonetheless, Chamberlain averred, "No humanitarian chatter can side-step the fact that this [racial competition] means a struggle [*Kampf*]. . . . This silent struggle is even more than any other a struggle for life and death." Later in *Foundations* he called it irrational to try to end this racial struggle, as some social thinkers were proposing. Thus Chamberlain considered the racial struggle an inescapable process.[78] Racial struggle played a central role in Chamberlain's thought, since he believed that the Teutonic races were locked in mortal combat with the Jews.

Breeding was the other aspect of Darwinian theory that Chamberlain considered crucial to understanding racial theory. He highly recommended reading Darwin's *Animals and Plants under Domestication*, which demonstrated the "malleable possibilities of life." He compared the varieties produced through domestic breeding—such as various breeds of dogs or horses—to the human races. Just as dog breeds vary, not only in their physical characteristics, but also in their dispositions, so human races differ physically, mentally, and even morally. Chamberlain emphasized that these different breeds or races were unequal biologically. Even more importantly, in order to produce better breeds, they need to be treated unequally by selecting the best specimens to reproduce. Since he exalted the human

Figure 6.3   Cover of *Der Brummer* magazine in 1916 with caption: "Our captives in the West. 'Man, you're bringing him to Hagenbeck, aren't you? He can't catch better gorillas in Africa than we can on the Western Front!'"

mind and will above purely material processes, it should come as no surprise that Chamberlain favored purposeful selection or eugenics as the solution to preserving or improving racial character. He took a dim view of most racial crossing, however, though he did claim that some limited racial crossing could produce biologically improved varieties.[79]

While racism might have intensified in the late nineteenth century as a justification for European imperialism with or without Darwinism, it is clear that Darwinism played a key role in the rhetoric of racism (see figure 6.3). Scientific racism, which after the advent of Darwinism was synonymous with Darwinian racism, provided a powerful rationale for racial inequality, displacing the more egalitarian views of liberal German anthropologists around 1900. Darwinism also contributed to the radicalization of racism, especially among the scientific and medical elites in Germany. After examining proposals by leading Darwinists and eugenicists for dealing with the disabled in chapters 7 and 8, we will examine specific ways of dealing with other races in chapters 9 and 10 on Darwinian militarism and racial extermination.

3. Eliminating the "Inferior Ones"

7. Controlling Reproduction: Overturning Traditional Sexual Morality ✍

The Malthusian stress on mass death and destruction that Darwin adopted in his evolutionary theory presupposed high rates of reproduction. Thus, reproduction is central to the Darwinian vision of nature and humanity: organisms possessing traits that allow them to reproduce in greater numbers than their competitors survive and pass on those traits. Furthermore, sexual instincts and sexual selection played important roles in Darwin's explanation of human evolution. One wonders if Freud was guided (unconsciously perhaps?) by Darwin in formulating his theory of the twin drives of libido and thanatos, that is, sex and death (aggression).[1] In any case, sociobiologists and evolutionary psychologists today regularly appeal to Darwinism to make sex and reproduction the most important explanation for human behaviors. In his highly influential (but controversial) work, *Sociobiology: The New Synthesis*, Edward O. Wilson stated,

> In a Darwinist sense the organism does not live for itself. Its primary function is not even to reproduce other organisms; it reproduces genes, and it serves as their temporary carrier.... the organism is only DNA's way of making more DNA. More to the point, the hypothalamus and limbic system are engineered to perpetuate DNA.[2]

Not exactly racy stuff, but still the point is clear: Reproduction and sex are central to a Darwinian understanding of life and human behavior.

The penchant to move reproduction and sex to center stage in explaining many human behaviors, as some sociobiologists and evolutionary psychologists do today, is not a completely new phenomenon. Darwinian biologists and psychologists today are reviving ideas that flourished in the late nineteenth and early twentieth centuries. By pushing reproduction into the limelight, Darwinism helped spawn new ways of thinking about sexuality and sexual morality. As we have already seen, Darwinism spawned the eugenics movement, whose primary goal was to increase the biological quality of the human population by controlling reproduction. Since many Darwinists wanted to revise all human institutions to speed up evolution—or at least to hinder the biological degeneration many feared was occurring—evolution became the new arbiter for sexual morality: whatever improved biological quality was morally good, while whatever hindered it was evil. In 1911, Eduard David told a conference of sexual reformers, "In the realm of sexuality everything is moral which serves the upward evolution of the species."[3] Many Darwinists agreed with David and argued on this basis for serious revisions—or sometimes the complete overthrow—of traditional ideas about sex, marriage, and the family.

The devaluing of human life that we have already examined played a fundamental role in eugenics discourse on sexual morality. In trying to reformulate sexual mores, the key question eugenicists posed for themselves was: What is the best way to get rid of "inferior" or "unfit" people? Conversely, they wanted to find ways to promote the reproduction of the "fittest." As Gruber succinctly stated, "The fit must flourish, the unfit must disappear."[4] Some eugenicists focussed more on measures to eliminate those with "bad" heredity, while others tried to find ways to increase the numbers of those with "good" heredity. Most supported both approaches. While eugenicists did not always agree on which practical measures were best to achieve their goals, they were united in their zeal to subordinate sexual morality and behavior to the goal of evolutionary progress.

Though eugenicists generally agreed that traditional Christian sexual morality was detrimental to the welfare of the human species, they could not agree on what sexual mores would best promote human health and vitality. While agreeing on a common goal—the evolutionary advancement of humanity—and while agreeing that control of reproduction was the most important measure to accomplish this goal, they could not agree on concrete measures. Their proposals for sexual reform were, in fact, wildly divergent, ranging from free sex to monogamy (though often with loopholes) to polygamy. Some even called on evolutionary ethics to justify homosexuality. Evolutionary ethics thus seemed to offer proponents just about whatever sexual morality they preferred.

Sexual reforms were an important part of Forel's crusade for a new ethic based on science. He testified in his autobiography that when he became a psychiatrist, he

> seemed to hear a voice that called upon me . . . 'You must become the
> apostle of truth. What use to stay forever where you are, tending the lost
> victims of human stupidity, mere shattered remnants, behind the doors of
> your madhouses, and calmly allowing the cause of all this misery to persist?
> That is cowardice!' Social hygiene called for a complete revolution of opin-
> ion, if the evil was to be tackled at the root; above all, it called for a rational
> sexual selection of human beings.[5]

This "rational sexual selection" that Forel advocated was built on the
Darwinian premise of human biological inequality. Forel suggested that we
should conceptually divide society into two categories: "a superior, more
socially useful, sounder, or happier, and an inferior, less socially useful, less
sound and happy." Those in the "superior" half should reproduce bounti-
fully, while those in the "inferior" half should refrain from reproducing.[6]
Forel did not indicate how this should be enforced, nor did he advocate
government intervention; perhaps he hoped moral suasion would suffice.

Since "rational sexual selection" required a complete rethinking of sexual morality, Forel wrote an entire book on *The Sexual Question* (1905),
which became so popular that it went through sixteen editions in German
by 1931, as well as multiple English and French editions. In his book Forel
grounded sexual morality on evolution (the second chapter of the book is
devoted entirely to biological evolution), claiming that human happiness is
best served by improving human biological traits, including mental and
moral traits. He summed up his sexual morality in the precept: Do not
harm others through your sexual activity, but promote their happiness.
Morality must take into account not only those now living, according to
Forel, but also coming generations, since "the highest task of the ethical
deed is laboring for the welfare of the future generation."[7]

By judging sexual behavior solely according to whether it improves or
harms the biological traits of coming generations, Forel called for a radical
alteration of sexual mores. Though in general he favored monogamy and
opposed polygamy, he admitted that in some cases polygamy or at least
concubinage might be beneficial and thus morally acceptable. If, for example, a husband with "good" biological traits married an infertile woman, he
could pass on his "good" traits to the next generation only by taking
another wife or concubine. Forel also claimed that in some cases, such as
when a man with biologically "good" traits married a woman with biologically "bad" traits, adultery could be morally good. Since homosexuality and bestiality do not affect the biological well-being of humanity,

Forel regarded them as morally neutral, and he criticized present legal codes for punishing those indulging in these sexual activities. By reducing sexuality to its biological effects—especially on the offspring—Forel called for a radical overturning of Judeo-Christian sexual ethics.[8]

Not all eugenicists agreed with every detail of Forel's sexual morality, but almost all of them concurred with his main goals. Schallmayer stated, "The value of existing and proposed rules of social life and especially of sexual life may be measured primarily according to how it causes the race to flourish," and only secondarily according to social or individual utility or happiness.[9] Many eugenicists called the procreation of sick or weakly individuals a sin against humanity. The University of Basel physiologist Gustav von Bunge, an honorary member of the Society for Race Hygiene, predicted that it would take a long time to inculcate in people the truth "that the *procreation of sick, degenerate children is the most serious crime that a person could ever commit.*"[10] Eugenicists thus redefined sexual sin as any sexual relationship—including marriage—that produced "inferior" offspring, not as extramarital sexual liaisons.

Stöcker and the League for the Protection of Mothers agreed with Forel that eugenics and sexual liberation were part of a sweeping program of sexual reform. When the League first formed, it received considerable support from leading Darwinists and eugenicists, including Haeckel, Forel, Ploetz, Woltmann, Alfred Hegar, and Ehrenfels, among others, though a few eugenicists later distanced themselves from Stöcker's sexual radicalism. Stöcker used her journal as well as many speeches sponsored by the League to promote her "new ethics," which was built on a world view synthesizing Darwinian monism and Nietzscheanism. In this, she was following the lead of Tille, with whom she fell in love in the years 1897–99. When Tille's wife suddenly died, he proposed to Stöcker, but Tille had two children, and Stöcker did not want to be tied down to motherhood, so they went their separate ways.

The Darwinian influence on Stöcker's ethical views are readily apparent. The opening sentence of her article, "On a New Ethics," clearly displays the evolutionary thrust of her ideas: "If one believes in the eternal Becoming, in the flow of evolution, and holds struggle for the father of all things, then one can only see the moral task of humanity as seeking ever new, higher forms of morality."[11] Her affinity with Darwinian monism brought her and her League into a close relationship with Haeckel's Monist League. She not only spoke at some of their congresses, including the First International Monist Congress in Hamburg in 1911, but she also wrote articles regularly for their journal. She explained to fellow monists that her organization was attempting to tackle sexual and marriage reform "in the spirit of modern evolutionary theory."[12] In her 1913 speech to the Monist Congress, she insisted that reproduction must come under the sway of scientific rationality.[13]

Despite her inclination toward socialism, Stöcker stressed human biological inequality, just as most eugenicists did. She warned against biological "inferiority," stating, "If the mere existence of inferior people is a danger and a hindrance for the state, then to *hinder* this, with all the means of science, is not only our *right*, but our *duty*."[14] She hoped that her sexual reforms would ultimately elevate the human race. Though Stöcker stressed the influence of the environment on biological quality more than most eugenicists did, she still saw sexual reform as an important means to biological improvement.

One aspect of Stöcker's sexual reform that brought her into conflict with some leading eugenicists was her staunch support for single motherhood. Stöcker opposed the traditional stigma on women bearing children out of wedlock. She hoped that in the near future women would not only be permitted, but also financially supported to have children, whether they were married or not. She believed this would result overall in biological improvement, so she continually appealed to eugenicists to support her position. However, Ploetz objected to supporting all single mothers, and not just because he favored monogamy as the healthiest form of reproduction. Ploetz, in fact, had supported the League for the Protection of Mothers at first, since the founding document of the League had stipulated that only healthy mothers should receive support. When Stöcker and her supporters struck the word "healthy" from the League's platform and insisted that all single mothers should receive assistance, Ploetz protested that this was no longer consistent with eugenics ideals.[15] In fact, Stöcker continued to support eugenics, but her Nietzschean individualism caused her to value sexual liberation or "free love" more than Ploetz, and sometimes it overrode her otherwise strong eugenics concerns.

Stöcker and other feminists also butted heads with leaders of the eugenics movement over the issue of birth control. Some feminist eugenics proponents, including Stöcker, were the earliest leaders of the birth control or neo-Malthusian movement, when it began in the early twentieth century. Though their chief motivation in supporting birth control was sexual liberation, these feminists also argued that easy access to birth control would also advance eugenics goals. Stöcker hoped that widespread use of birth control would effectively separate sexual love and procreation, allowing people to make more rational choices about reproduction. Only then could they follow their moral duty not to reproduce if the interests of the coming generation would be harmed.[16] Also, these feminists, like most eugenicists, warned that the lower classes, whom they considered biologically "inferior" to the upper classes, would swamp society with their "bad" heredity if they continued reproducing at higher rates, as they were currently doing. Upper class women were already using birth control to restrict their reproduction, so easier access to birth control, these feminists argued, would primarily reduce reproduction among the biologically "inferior."

Ploetz, however, saw things differently. He, like all eugenicists, fully supported control of reproduction. That is what eugenics was all about. But Ploetz opposed the indiscriminate use of birth control by individuals based on their personal whim, which he thought would happen if the neo-Malthusian movement achieved its goals. This kind of birth control Ploetz considered counterproductive, since it was not based on eugenics criteria. Ploetz wanted birth control in the hands of the medical profession, not in the hands of individuals, who might pursue selfish goals, such as pleasure, wealth, or vanity, rather than the biological health of the nation and coming generations. He did not share Stöcker's optimism that most people would use birth control in a eugenically responsible manner.[17] Ploetz considered the threat of neo-Malthusian birth control so serious that he devoted his speech at the First International Eugenics Congress in London in 1912 to this topic, warning of the negative consequences of neo-Malthusianism for eugenics.[18] When some leading radical feminists, as well as Stöcker's League for the Protection of Mothers applied for membership to his Society for Race Hygiene, he vigorously and successfully opposed their membership.[19]

Another reason Ploetz opposed easy access to birth control was because—unlike Stöcker—he thought quantity was related to biological quality. Ploetz believed that a healthy population was an expanding population, since the greater the population expansion, the greater the pool of "good" variations from which to choose. He feared that limiting population expansion would reduce the numbers of biologically "good" people. He also feared that excessive use of birth control would reduce the ability of Europeans to compete against other races in the racial struggle for existence (see chapter 10).

Gruber and Rüdin, close colleagues of Ploetz in Munich, thoroughly agreed with him about the dangers of individualistic birth control. In a jointly-authored work they expressed grave concern over this issue: "The next and greatest worry of race hygiene—a much greater one than the relative increase of inferior people—is currently neo-Malthusianism, the purposeful limitation of the number of births, [sometimes] up to complete infertility." Of course, they favored regulating reproduction, but not based on individual egoism.[20] Gruber later wrote an entire book to combat the specter of population decline, which quickly went through three editions. He and other eugenicists were alarmed by population statistics in the early twentieth century showing a deceleration of German population expansion. They feared that Germans were beginning to adopt the French "two-child system," which had already put the brakes on French population expansion in the late nineteenth century. Gruber blamed individualism for the population decline and suggested measures to encourage population expansion.[21] However, despite his opposition to neo-Malthusianism, Gruber was not opposed to artificial birth control *per se*, as he explained to Grotjahn. In some cases he considered it beneficial to ensure the health and welfare of one's progeny. Nevertheless, Gruber added, while "regulation of bearing children is a command of the rational human order, conscious avoidance of bearing children by normal people is a crime."[22]

Many eugenicists agreed with Ploetz and Gruber that birth control might imperil the biological prowess of the nation and race. Luschan's warnings against the deliberate restriction of births sounded almost hysterical at times. He called it "the most imminent danger for the white man and for the civilised races" and "nothing less than *national suicide*."[23] Ribbert also feared the perils of birth control, cautioning that voluntarily limiting the size of families is "a grave danger to human evolution."[24]

Leading radical feminists found Ploetz's and other eugenicists' opposition to neo-Malthusian birth control incomprehensible. Adele Schreiber, a close colleague of Stöcker's in the League for the Protection of Mothers who later became a socialist (SPD) member of parliament, was fully committed to eugenics and highly recommended reading works by Ploetz and Rüdin. However, she was disappointed at the negative attitude many eugenicists displayed toward neo-Malthusianism. She considered the goals of eugenics unattainable without the ready availability of birth control, "with whose help alone upward breeding and selection, the elimination of worthless and bad elements, [and] the creation of a society free from the oppressive burden of the unfit of various kinds, can be achieved."[25] The socialist feminist Henriette Fürth agreed with Schreiber that birth control would biologically improve the human race.[26] As with so many other sexual reformers, Fürth made eugenics the arbiter of all sexual morality, stating, "Healthy procreation will have to be the sanctuary and the measuring-rod of the future sexual order."[27] This was not merely a German phenomenon, for feminist birth control advocates in other lands, such as Margaret Sanger in the United States and Marie Stopes in Britain, held remarkably similar views. In 1919 Sanger even stated, "More children from the fit; less from the unfit—that is the chief issue of birth control."[28] Feminist sexual reformers thus shared the goals of eugenicists, only disagreeing with them about the best means to achieve biological improvement.[29]

Not all eugenicists took as hard a line against neo-Malthusianism as Ploetz or Gruber. Grotjahn was much more sympathetic with feminism and criticized Gruber for his hard stance against birth control, but even he warned that using birth control without consideration for the biological value of the individual was an abuse.[30] In a book aimed at fighting the threat of population decline, Grotjahn declared that he was no neo-Malthusian, even though he opposed government restrictions on birth control. He ultimately agreed with Ploetz that an expanding population was necessary to ensure high biological quality, so he promoted high natality.[31]

Forel was not as convinced as Ploetz or Grotjahn that population expansion was necessary. He believed eugenicists should concentrate more on quality than quantity. However, he did not agree with neo-Malthusianism, either, for indiscriminate use of birth control did not give proper consideration to biological quality. Forel did not oppose artificial birth control *per se*, however, for he believed that "one must consistently teach neo-Malthusianism to the sick, the disabled, the imbecile, the bad, [and] the inferior races."[32] Forel, like almost all eugenicists, favored birth control for certain people, that is, those defined as "inferior." Ribbert, though rejecting limitations on family size through birth control, nevertheless stated that when considering the question of birth control, we must keep in mind "the comparative worthlessness of innumerable human beings," for whom birth control would be beneficial.[33] The conflict between many leading eugenicists and the neo-Malthusian movement was not over whether birth control was permissible, but rather it was over who should control it and on what basis. The radical feminists favored a more democratic control of reproduction, while most leading eugenicists wanted the state and/or the medical profession to play a decisive role. But all eugenics proponents agreed that humans differ in quality, that some are "inferior" and others "superior," and that decisions about sexual morality should subordinate individual interests to the interests of future generations or the species.

All eugenicists, including the feminists, supported some kind of birth control for the congenitally disabled or, as they termed it, the "inferior people." How to restrict the reproduction of the "inferior" was a thorny question, however. A few eugenics advocates considered moral suasion sufficient; simply convince the "inferior" of their moral obligation not to procreate and the problem would be solved. The sociologist Arthur Ruppin, a leading figure in the Zionist movement, argued strongly in his Krupp-Prize entry that those with mental and physical congenital illnesses should be hindered from reproducing. However, he opposed coercion by the state, instead advocating the incorporation of marriage restrictions into moral codes. State-sponsored marriage restrictions would be ineffectual, anyway, Ruppin reasoned, since those affected could simply reproduce outside the bonds of marriage.[34] Another intellectual who considered moral suasion the best path to success in eugenics was the leading socialist theoretician Karl Kautsky.[35]

Most eugenicists, however, viewed education and persuasion as only one prong of a multifaceted approach to eugenics that included compulsory as well as voluntary measures. One suggestion popular among eugenicists was to incarcerate the disabled, especially the mentally ill, in asylums to prevent procreation. They wanted to transform or at least supplement the intent of incarceration, which was originally therapeutic and/or for the protection of the patient and society, into a means of keeping the mentally ill from procreating. The prominent psychiatrist Kraepelin expressed the common view among his colleagues when he asserted in 1899 that heredity was "perhaps the strongest cause of mental illness," and therefore the state should institutionalize the mentally ill to prevent them from reproducing.[36] Kaup even suggested setting up work colonies for "inferior people," where men and women would be segregated to hinder them from having children.[37]

Since most eugenicists viewed criminality—or at least the tendency to criminality—as a hereditary condition, many eugenicists also suggested permanent incarceration for habitual criminals. Ribbert, for example, thought habitual criminals should be treated just like the mentally ill, permanently locking them away in asylums.[38] Luschan agreed, since "crime in the great majority of cases is a hereditary disease" and therefore "the Criminal [sic] should only be considered as a pure lunatic, who is not responsible for his ill doings." Not holding criminals responsible for their actions, however, did not imply that they would get off scot-free. In fact, the new eugenics conception of incarceration made their "punishment" even worse than under the old system. Luschan called for "*the complete and permanent isolation of the criminal*," rather than the current system of temporary incarceration and then release, which allowed criminals to have children in between stints in prison. Luschan expressed faith that vigorous eugenics measures would ultimately eradicate crime from society.[39]

Apart from erecting work colonies—which sounds ominously similar to labor camps—care for the disabled in asylums was expensive, so some eugenicists sought ways to hinder reproduction without internment. One measure that became increasingly popular among psychiatrists and physicians was sterilization, which was illegal then, even on a voluntary basis. The psychiatrist Paul Näcke led the campaign for compulsory sterilization of habitual criminals, alcoholics, and those with mental illnesses, all of whom he considered hereditarily degenerate. Sterilization, he argued, was more effective and cheaper than any other method of restricting reproduction. Marriage restrictions could be circumvented by reproducing outside wedlock, and incarceration is expensive. Since no safe method of sterilization for females yet existed, he thought sterilization should be restricted to males from ages 25 to 55.[40]

Näcke's proposal for compulsory sterilization was extremely controversial at the time, but it quickly won many adherents in the eugenics movement. Rüdin, a professor of psychiatry at the University of Munich and one of the leading figures in the early eugenics movement, became an early proponent of sterilization. In a 1903 speech on alcoholism, he suggested mandatory sterilization for all alcoholics desiring to marry. This would hinder the propagation of alcoholism, which he, like most of his contemporary eugenicists, considered a genetic condition.[41] An Austro-Hungarian diplomat to the United States, Geza von Hoffmann, reported favorably on sterilization laws in some states of the United States in his book on the American eugenics movement. He also advocated similar laws for Germany and Austria–Hungary to rid society of "bad heredity."[42] Other leading eugenicists began supporting compulsory sterilization before World War I, too, including Luschan, Max Hirsch, Ribbert, and Eugen Bleuler, Forel's successor as director of Burghölzli Clinic.[43]

Debates among eugenicists about marriage reform were just as contentious as the dispute over birth control. While some eugenicists wanted only modest changes to traditional conceptions of marriage and the family, others called for a complete revolution of these hallowed institutions. While some feminist advocates of eugenics favored complete sexual freedom, a few male eugenicists proposed replacing monogamy with polygamy. Most eugenicists, however, were more moderate in their proposals for marriage reform. They all agreed, however, that neither religion nor tradition should decide the matter; rather eugenics considerations should be paramount in marriage reform. Just about all eugenicists would have heartily agreed with Stöcker, when she stated, "One must take more seriously marriage in the interest of the race."[44] Stöcker, an advocate of free sex who never married nor had children (despite her rhetoric about motherhood being the highest calling for women), was not thereby encouraging all to marry, but she was charging those who do marry to subordinate their own interests to those of the community. In the view of most eugenicists, marriage was thus not merely an individual matter, but a matter of grave importance for the biological vitality of one's nation or race.

One of the more common complaints eugenicists raised about the institution of marriage was that the present system of choosing mates was counterproductive biologically and contributed to degeneration. Many eugenicists pointed out that economic or social factors often influenced the choice of marriage partners more than biological considerations. This gave an unfair advantage to those of higher socioeconomic status, who may or may not be biologically "superior." In order to ensure that the hereditarily "best" individuals married each other, they proposed freeing choices about marriage from any external pressures.[45] However, eugenicists were somewhat ambivalent about how people should choose their mates. Often they intimated that once freed from external considerations, sexual selection would take care of the matter by itself, that is, the "superior" individuals would surely "fall in love" with each other and pair off. Thus, they strongly supported the notion that marriages should be based on mutual attraction. However, eugenicists also counseled the use of reason in mate selection. One should carefully consider the health of the prospective mate and his or her family before deciding on marriage.

The physician and sociologist F. Müller-Lyer, a member of Haeckel's Monist League, made clear in his 1912 essay, "Marriage," that any reform of marriage must be based on eugenics considerations. His primary aim for the time being was to educate the public on the necessity of eugenics to prevent further biological degeneration. Some legislation might eventually be necessary, but only after the public had been won over to the cause. One of Müller-Lyer's goals was to disseminate the idea that begetting a child with a sick constitution is a crime almost as heinous as murder![46] Perhaps this was a bit of hyperbole, but even so, it manifests a twisted view of the value of human life that was characteristic of eugenicists.

One of the favorite suggestions of eugenicists for marriage reform was to require health examinations before marriage to ascertain the hereditary health of the prospective bride and groom. Members of the Society for Race Hygiene pledged to submit to a premarital health exam to determine their "fitness" for marriage. If they failed the exam, they were required to refrain from marriage and procreation.[47] The Monist League, following Haeckel's lead in supporting eugenics, also endorsed the idea of premarital health certificates. They passed a resolution at their 1908 congress (and again in 1912) to petition the government to require couples to undergo health exams before marrying and to exchange the health certificates with each other, so they could make an informed decision on marriage. Under this proposal, the state would not forbid marriage, but would leave the choice to the couple.[48] At the instigation of the physician Max Marcuse, a leading figure in the movement for sexual reform, the League for the Protection of Mothers at its 1907 congress passed a similar resolution supporting government-mandated premarital health certificates. Like the Monist League, it opposed any marriage prohibitions.[49] More restrictive measures would have to wait for greater public acceptance of eugenics.

Despite their efforts, eugenicists had little chance of legislating premarital health certificates at this time, so they had to devise other ways to convince the "inferior" not to marry. One method was to establish Marriage Counseling Centers to provide voluntary health examinations and information on heredity and eugenics to couples considering marriage. The local branches of the Monist League in Dresden and Königsberg, with assistance from the League for the Protection of Mothers, privately founded such centers in their cities before 1914.[50] Only in the 1920s, however, were eugenicists able to establish some government-funded Marriage Counseling Centers. All these centers were completely voluntary, but they disseminated eugenics advice and propaganda. The first one was established in Vienna in 1922 under the impetus of the medical professor Julius Tandler, who placed the socialist physician Karl Kautsky, Jr., at the helm. Many German states and cities followed suit in the 1920s, and by 1930, the state of Prussia alone had established 200 Marriage Counseling Centers.[51]

Many eugenicists—including Forel, Ploetz, Schallmayer, and Ribbert—did not have faith that purely voluntary measures would hinder reproduction enough among the "inferior elements" of the population, so they favored some kind of marriage prohibitions for the congenitally disabled. Eduard David unabashedly proposed that no one be allowed to marry without a physician's permission.[52] Alfred Hegar, a professor of gynecology at the University of Freiburg and an honorary member of the Society for Race Hygiene, was hesitant to call for much state intervention, since it would interfere with individual liberty and would arouse opposition. He preferred voluntary rather than compulsory measures, hoping that education and persuasion would suffice to achieve eugenics goals. Nonetheless, he thought that in some cases of mental illness and "instinctive criminality" the state should forbid marriage. Anyone who would favor a marriage between such "defective persons" is sinning against humanity, Hegar opined.[53] Some eugenicists, however, opposed marriage restrictions, not because they believed in individual liberty to marry, but because they feared it would backfire in practice. Merely forbidding "inferior" persons to marry would not necessarily stop them from procreating. "Inferior" males might even father more children out of wedlock than if they married.[54] David overcame this objection by advocating birth control for those not permitted to marry.

If marriage reforms to hinder "inferior" people from reproducing created minor disputes, even among eugenicists, proposals to replace monogamy with polygamy were even more controversial. The leading advocate of polygamy was Ehrenfels, who based his marriage reforms squarely on his interpretation of Darwinism. Since he believed that the "less valuable" could at times triumph in the Darwinian struggle for existence, he feared the contemporary system of monogamy would lead to biological degeneration. Monogamy allowed the "less valuable" to reproduce at the same rate—or even at higher rates, as many eugenicists feared was occurring in the early twentieth century—as the "more valuable." "As long as the custom of monogamy reigns," stated Ehrenfels, "nothing of importance can be altered in the devaluing tendency of selection."[55]

In his desire to sweep away monogamy, Ehrenfels found himself in agreement with some of the more radical sexual reformers of his day. Nordau, for example, forcefully argued against the conventional form of monogamous marriage in his book, *The Conventional Lies of Civilization*. Though Nordau's worldview centered on science, especially Darwinism, he did not use eugenics arguments to support polygamy. However, he did agree with Ehrenfels that human instincts—especially, but not exclusively in the male—were distinctly polygamous. Nordau argued that most organisms, including the higher mammals, only pair off for one breeding season or sometimes until offspring are born. He believed human instincts tended in the same direction, and thus marriage becomes an empty and burdensome institution after the honeymoon or perhaps the birth of the first child. Nordau thought almost all European males were polygamous in their sexual activity, anyway, despite moral prohibitions. Nordau proposed loosening the marriage bond to bring it into greater harmony with nature. He unabashedly argued that people should start acting more like animals in their sex life.[56]

Unlike Nordau and many other sexual reformers, however, Ehrenfels staunchly opposed replacing monogamy with greater sexual license. He considered this worse than the "decrepit" monogamous morality he opposed. Rather he called for a form of polygamy, in which the "more valuable" males could have multiple wives. He considered polygamy a more "natural" form of marriage than monogamy, since males can naturally reproduce more than females, and he thought that healthy male biological instincts demand more sex than females. Polygamy would thus allow males—but only those having "superior" biological traits—to pursue their natural instincts, which have adaptive value in the Darwinian struggle for existence.[57] For Ehrenfels polygamy was the panacea for biological degeneration, without which regeneration and ultimate victory over other races in the struggle for existence would be impossible.

The only other eugenicist who made polygamy the centerpiece of his eugenics reforms was the racial theorist Hentschel, whose marriage reforms were even more radical than those of Ehrenfels. Even Ehrenfels protested that Hentschel was breaking too sharply with tradition, but Hentschel dismissed all appeals to history and tradition with the remark, "We *are* history."[58] He hoped to establish utopian communities for the express purpose of breeding better people. He called his model community Mittgart, where the "best" racial elements could gather to practice "Aryan" eugenics. Candidates wanting membership in the Mittgart community would, of course, be carefully screened for biological quality. For Hentschel this included "Aryan" racial characteristics, for he included skin and eye color as factors to consider in judging one's biological "fitness" to join Mittgart. This selection process was crucial to ensure that progeny in Mittgart would be robust.[59]

The centerpiece of Hentschel's eugenics proposals was the Mittgart marriage, which had the same goal as Ehrenfels' polygamy: to maximize the reproductive potential of the "best" males. The Mittgart marriage was an exclusive relationship between one man and one woman, aimed solely at producing a healthy, vigorous child. The married couple did not live together, however, since all males, including married ones, lived in communal housing. Unlike traditional monogamy, the Mittgart marriage was temporary. It only lasted until the male impregnated his partner, after which the marriage would be dissolved and the male would remarry. The marriage would also be terminated if the woman did not become pregnant within a reasonable amount of time. If it turned out that a member was infertile, he or she would be banished from the community. Hentschel estimated that this system would require ten times as many women as men.[60] Though he claimed to have some fervent supporters, including a man ready to finance the project, his proposals found little resonance in Germany. Hentschel's five daughters were absolutely horrified by their father's ideas.[61]

Even his close friend, Theodor Fritsch, from whom Hentschel had imbibed the idea of establishing utopian communities, rejected his idea of the Mittgart marriage. Fritsch agreed with Hentschel that the present form of marriage was to blame for the ever-increasing problem of biological degeneration, since "the destiny of a person is decided in the hour of his conception." Monogamy, however, was not at fault. The problem, rather, was that too many people were choosing their mates for the wrong reasons—such as money or social standing—thus nullifying the allegedly beneficial effects of sexual selection. How could this problem be corrected? Fritsch considered monogamy ultimately the best form of marriage, but only if "full-valued" people can marry, and he did not think there were enough of them around. Therefore, he reasoned, since "the mass of half- and quarter-people running around is not fit for a regular marriage," polygamy may be necessary for a time to offset this imbalance. Fritsch then asked, Is this moral? Of course it is, he claimed, for "morality arises from the law of preservation of the species, of the race. Whatever ensures the future of the species, whatever is appropriate to elevate the race to a ever higher stage of physical and mental perfection, that is moral." For Fritsch, then, sexual morality and the institution of marriage were subordinate to the higher goal of evolutionary progress. Further, by calling some people only "half" or "quarter" people, Fritsch showed complete contempt for the value of human individuals.[62]

In eugenics circles Hentschel's marriage reforms found no support at all, but a few eugenicists were sympathetic with Ehrenfels' call for polygamy. None, however, pinned their hopes on it the way Ehrenfels did. Schallmayer was the most influential eugenicist approving of polygamy, but for him other reforms were more important. Nonetheless, in his 1903 book he mentioned that polygamy might be eugenically beneficial if restricted to "more valuable" men.[63] Whether Schallmayer already knew about Ehrenfels' proposals when he wrote this is unclear, since most of Ehrenfels' articles on polygamy appeared after Schallmayer's book was written. Later Schallmayer discussed polygamy in greater depth. Like Ehrenfels, he believed that the male sex drive is polygynous, and he concluded from this that polygamy probably is biologically advantageous. But he had to explain the anomaly that monogamy had triumphed in many societies, including the most advanced in the world (for him this meant Europe, of course). He surmised that it was not really monogamy that triumphed over polygamy, but rather it was democracy overcoming aristocracy. It may seem strange from a historical perspective (of course, Schallmayer was no historian), but he thought monogamy accompanied democracy, while aristocracy tended more to polygamy. Schallmayer believed that "in relation to the quality of a people's reproduction polygyny has without a doubt powerful advantages over monogamy," as long as only males with superior characteristics are allowed to practice polygyny. Schallmayer thus agreed with Ehrenfels that polygamy—if based on eugenics considerations—is a desirable reform, even though he did not seem particularly optimistic that his society would adopt it any time soon.[64]

Other eugenicists were less enamored of polygamy, though some admitted that in some circumstances it might be beneficial. Forel, as we have seen, opposed polygamy in most cases, but he did not reject it outright, since he thought it might be beneficial sometimes. In his 1912 doctoral dissertation, Lenz gave his stamp of approval to polygamy as a beneficial eugenics measure, but unlike Ehrenfels, he recognized that it did not stand any chance of implementation in the near future, so he concentrated on other measures instead.[65] Müller-Lyer admitted that Ehrenfels' proposals were, from a eugenics standpoint, eminently logical, but still he dismissed polygamy as too utopian.[66]

Though some eugenicists rejected polygamy as impractical or too radical, others considered it eugenically misguided. Gruber upheld traditional monogamy and vehemently opposed sexual reformers like Stöcker and polygamy advocates like Ehrenfels. He claimed that a nation's biological health could be promoted best by strengthening marriage and the family. For this reason, he proposed various reforms to encourage healthy married couples to have more children. One of Gruber's proposals that would later be implemented by the Nazi regime was to honor married mothers who bear many children with some sort of honorary title or award.[67] Ploetz agreed with Gruber that eugenics would be best served by promoting monogamy. Even though he published some of Ehrenfels' articles and carried on a friendly correspondence with him, Ploetz kept his distance from Ehrenfels and did not allow him to join the Society for Race Hygiene.[68]

Ploetz's views on marriage were not completely traditional, however, in that he favored easier divorce, especially for eugenics reasons. He divorced his first wife, because he thought she was trying to avoid having children. He also wrote to his friend Gerhart Hauptmann that modern people should not be bound to spouses against their will.[69] However, whatever his personal views on divorce were, as far as I know Ploetz never publicly encouraged divorce. On the other hand, some eugenicists did publicly call for easier divorce in cases where a marriage was eugenically unsound. Luschan proposed that wives of alcoholics should be allowed to divorce, since—like most eugenicists—he believed alcoholism was a hereditary condition.[70] Hirsch too advocated divorce as a eugenics measure.[71]

As we have seen, eugenicists represented widely divergent viewpoints on marriage and sexuality. However, perhaps more interesting than their differences were their commonalities. They all judged sexual morality by its effects on the hereditary health of future generations. This set them in opposition to traditional Judeo-Christian sexual morality and opened the door for radical sexual reforms. Even the eugenicists who remained committed to monogamy generally exalted eugenics considerations above traditional sexual mores. In their view, sexual morality—like all morality—was only valid inasmuch as it was useful to advance the evolutionary process. No matter how much Judeo-Christian baggage some of them still carried about, they had effectively replaced God with evolution as the source and arbiter of sexual morality.

8. Killing the "Unfit" ⌒

In their zeal to rid the world of "unfit" or "inferior people," some eugenicists were not content with half measures. Darwinism proclaimed the ultimate doom of the "unfit" anyway, who would inevitably perish in the struggle for existence. Some thought natural selection could be helped along, not only through sexual or marriage reforms, but also by killing those deemed "inferior," "unfit," "worthless," or "of negative value." Since many Darwinists and most early eugenicists were critical of the Christian virtues of compassion and pity for the weak and sick, they led the attack on Judeo-Christian prohibitions against killing innocent human life.

Before the advent of Darwinism in the mid-nineteenth century, there was no significant debate in Europe over the sanctity of human life, which was entrenched in European thought and law (though, as with all ethical principles, not always followed in practice). Judeo-Christian ethics proscribed the killing of innocent human life, and the Christian churches explicitly forbade murder, infanticide, abortion, and even suicide.[1] The sanctity of human life became enshrined in classical liberal human rights ideology as "the right to life," which according to John Locke, was one of the supreme rights of every individual. Until the second half of the nineteenth century, and to a large extent even on into the twentieth century, both the Christian churches and most anticlerical European liberals upheld the sanctity of human life. A rather uncontroversial part of the law code for the newly united Germany in 1871 was the prohibition against assisted suicide. Only in the late nineteenth and especially the early twentieth century did significant debate erupt over issues relating to the sanctity of human life, especially infanticide, euthanasia, abortion, and suicide.

Darwinism played an important role in the debate over the sanctity of human life, for it altered many people's conceptions about the value of human life, as well as the significance of death. Many Darwinists claimed that they were creating a whole new worldview with new ideas about the meaning and value of life based on Darwinian theory.[2] Darwinian monists and materialists initiated public debate and led the movements for abortion, infanticide, assisted suicide, and even involuntary euthanasia. Many of them also considered suicide a private matter beyond the scope of morality, and many favored capital punishment to rid society of "hereditary criminality."

The two leading experts on German euthanasia debates before World War I, Hans-Walther Schmuhl and Michael Schwartz, both recognized the Darwinian influence on euthanasia discourse. Schmuhl argued that eugenics constituted an attempt to promote a new ethics based on Darwinian science. He then perceptively explained, "By giving up the conception of the divine image of humans under the influence of the Darwinian theory, human life became a piece of property, which—in contrast to the idea of a natural right to life—could be weighed against other pieces of property."[3] Schwartz also mentioned the drive for a new Darwinized ethics as a significant factor in early euthanasia ideology.[4] Another leading scholar of the German euthanasia movement, Udo Benzenhöfer, devoted an entire chapter in his book on euthanasia to discussing the impact of social Darwinism and eugenics on the budding euthanasia movement in the late nineteenth century.[5] Recent scholarship on the history of the American and British euthanasia movements also emphasizes the pivotal role Darwinism played in devaluing human life and giving birth to the euthanasia movement.[6]

The earliest significant German advocate for killing the "unfit" was Haeckel, whose views on killing the weak and sick were, in his estimation, the logical consequence of his Darwinian monistic worldview. Already in the second edition of his work, *The Natural History of Creation* (1870), he lamented some of the dysgenic effects of modern civilization and expressed support for eugenics. In this context, he favorably mentioned the ancient Spartan practice of killing weak and sickly infants, implying that he advocated this practice. He wrote,

> If someone would dare to make the suggestion, according to the example of the Spartans and Redskins, to kill immediately after birth the miserable and infirm children, to whom can be prophesied with assurance a sickly life, instead of preserving them to their own harm and the detriment of the whole community, our whole so-called "humane civilization" would erupt in a cry of indignation.[7]

Was Haeckel espousing infanticide in this passage? Surely he was, even though his advocacy was thinly veiled. Haeckel evidently thought he was, in any case, since in 1904 and again in 1917, he confessed that he had indeed supported infanticide in his earlier book.[8]

In his later books Haeckel argued more explicitly in favor of infanticide for the congenitally disabled, which would, in his view, benefit both the individual being killed, as well as society in general. In legitimating his position, he used evolutionary scientific arguments. One of Haeckel's own contributions to evolutionary theory was the dubious claim that ontogeny (embryological development) recapitulates phylogeny (evolutionary ancestry). This means that as each individual organism develops from a single cell (such as a fertilized egg) to adulthood, it allegedly traverses the evolutionary stages of its ancestors. Based on this view, Haeckel argued that newborn infants were still in an evolutionary stage equivalent to our animal ancestors. The newborn child, he stated, "not only possesses no consciousness and no reason, but is also dumb and only gradually develops the activity of the senses and of the mind."[9] Newborn infants thus have no soul, so killing them is no different than killing other animals and cannot be equated with murder.

With respect to a physically or mentally handicapped infant, he wrote, "a small dose of morphine or cyanide would not only free this pitiable creature itself, but also its relatives from the burden of a long, worthless and painful existence."[10] The only reason we do not kill "defective" children at birth, according to Haeckel, is because we are following emotion rather than reason. "However, *emotion*," he emphasized, "should never abolish the grounds of pure reason in such important ethical questions."[11] In matters of life and death, then, Haeckel wanted reason to trump emotions, so sympathy and pity would have to take a backseat to cold, scientific calculation. In that impersonal equation, the value of human life varied according to the health and vitality of the individual.

Since Haeckel legitimated infanticide, it should come as no surprise that he used similar reasoning to justify abortion. However, his justification for abortion was quite different from the arguments of abortion rights advocates a century later. Haeckel considered it a scientific fact that human life begins at conception. So seriously did he hold this conviction that in 1917 he celebrated his eighty-fourth birthday with his family nine months ahead of time, explaining his former student, Richard Semon, that everyone is really nine months older than his official birthday.[12] However, since he believed that the human embryo recapitulates earlier stages of evolutionary development, it does not have the full value of adult humans. It is still on the level of other animals from which humans descended. He stated "that the developing embryo, just as the newborn child, is completely devoid of consciousness, is a pure 'reflex machine,' just like a lower vertebrate."[13] Abortion, in Haeckel's view, is thus no different from killing an animal.

Haeckel vigorously opposed the idea that humans have a soul or immortality, so for him suicide was also not immoral. He considered it perfectly reasonable for one to end his or her life, if it has become miserable or unbearable. For him this was not just a theoretical question, but had practical implications. On all his sea voyages, he carried a dose of cyanide in his pocket in case of shipwreck, for he would rather kill himself than wage a protracted struggle with death in the waves.[14] He objected to the German term for suicide—*Selbstmord* (self-murder)—for, he argued, suicide is not murder at all, but rather "self-redemption." Murder, Haeckel explained, is only the killing of human life against its will. This definition of murder excluded assisted suicide, too, which Haeckel likewise found unobjectionable. We should commiserate with people in distress, just as we do with animals, whom we rightly put out of their misery, he claimed. Indeed Haeckel believed we have a moral obligation to help people find "self-redemption," if that is what they desire. He stated, "Likewise we have the right—or if one prefers—the duty, to end the deep suffering of our fellow humans, if strong illness without hope of recovery makes their existence unbearable and if they themselves ask us for 'redemption from evil.' "[15]

Not only did Haeckel justify infanticide, abortion, and assisted suicide or voluntary euthanasia, but he also supported the involuntary killing of the mentally ill. He condemned the idea that all human life should be preserved, "even when it is totally worthless." He called cretinism and microcephaly "decisive proof" for the physical basis of the soul, since those suffering from these conditions "spend their entire life at a lower animal stage of development in their soul's activity." He complained that not only are many mentally ill people burdens to society, but so are lepers, cancer patients, and others with incurable illnesses. Why not just spare ourselves much pain and money, he asked, by just giving them a shot of morphine? To safeguard against abuse, Haeckel proposed that a commission of physicians make the final decision in each case, but the individual being reviewed would have no voice.[16] The leading Darwinist in Germany thus gave his scientific imprimatur to murdering the disabled, both in infancy and in adulthood.

That many Darwinists and eugenicists balked at following Haeckel's lead on suicide, abortion, infanticide, and euthanasia is not at all surprising. Prohibitions against killing innocent human life were not so easily cast off, even by those Darwinists devaluing human life. Even though they might reject Christianity and its call for compassion for the weak, and even though they might consider the disabled "inferior" and "worthless," still they were often loathe to suggest that we should kill some fellow human beings. Many, in fact, insisted that we should continue caring for the weak and sick, while taking measures to hinder their reproduction. Ribbert, for instance, despite his harsh rhetoric about the "inferior" and those "of negative value" being so burdensome, nonetheless admitted that society should continue supporting them. "Doubtless we cannot fail to know that it would be better if they did not exist," he wrote, "but once they do exist we must assume responsibility for them."[17] Grotjahn defined one-third of the population as inferior, but he not only opposed euthanasia and infanticide, but all killing, including war and capital punishment.[18]

But even though not all Darwinists and eugenicists went along with Haeckel's program of "rational" extermination of the disabled, it is striking that the vast majority of those who did press for abortion, infanticide, and euthanasia in the late nineteenth and early twentieth centuries were fervent proponents of a naturalistic Darwinian worldview. Some did not overtly link their views on killing the feeble to Darwinism, though many did. Even those not directly appealing to Darwinism were usually building their views on eugenics, which was founded on Darwinian principles.

Legitimizing suicide was the least controversial part of Haeckel's devaluing of human life. Just as with Haeckel, defending suicide was not merely theoretical for Ludwig Gumplowicz, an Austrian sociologist famous for his theory of racial struggle. Gumplowicz appealed to Darwinism as the basis for his sociology and indeed his entire worldview. In *Social Philosophy in Outline* (1910) he built upon these Darwinian foundations to argue for the propriety of suicide:

> To comply with the obvious will of nature is the highest morality: With a perceptible voice nature calls back into its bosom those who are sick and weary of life. To follow this call and to make space for healthy people filled with zeal for life is certainly no evil deed, but rather a good deed, for there are not too few people on the earth—rather too many.[19]

Central to Gumplowicz's justification of suicide was the Darwinian stress on overpopulation together with the notion that the sick should make way for the healthy. When Gumplowicz wrote this, he was nearing the end of his life, for he was diagnosed with cancer. Before these words were published, Gumplowicz and his blind wife had ended their lives together with cyanide.[20]

Other prominent Darwinian thinkers agreed with Haeckel's and Gumplowicz's view on suicide. Carneri thought that the right to suicide was self-evident, despite the fact that Judeo-Christian ethics—and European culture based on it—condemned it as immoral. "The right for a person to kill oneself," he wrote, "cannot be disputed. It is his life that he forfeits."[21] Schallmayer thought suicide served a beneficial purpose, since those who commit suicide are usually below average in mental ability.[22] Thus, suicide was good for eugenics selection. Ehrenfels agreed with this premise. Though he rejected infanticide and involuntary euthanasia as illegitimate, he believed that suicide should be made easier to allow the incurably sick and miserable to end their own lives.[23] Hentig believed that suicide is often linked to criminality, so he discussed at great length the significance of suicide for eugenics.[24]

By World War I, many German psychiatrists interpreted suicide as a biological rather than a sociological phenomenon. Instead of focusing on the social or economic causes of suicide, they blamed the "inferior" biological constitution of the victim. They did not necessarily claim the suicidal person's biological condition predisposed them to suicide in particular, but their "inferior" traits allegedly made them less able to cope with hardship. Despite the fact that they could demonstrate mental illness only in a minority of cases of suicide, still many insisted that the suicidal individual was "inferior." With many able-bodied young men dying in World War I, many Germans became less sympathetic with keeping alive those deemed inferior or "life unworthy of life." By the end of World War I, the attitude of many Germans about suicide had changed. Instead of striving to preserve the life of suicide victims, some now saw their demise as a "good riddance."[25]

If suicide is acceptable, then what about assisted suicide or voluntary euthanasia? Before the twentieth century, the word euthanasia (meaning "good death" in Greek; the German term is *Euthanasie*) referred only to measures that make death less painful and miserable, but not measures taken to shorten the patient's life. Only in the early twentieth century did the term take on its present meaning of purposeful intervention to hasten the death of sick people. Euthanasia can be either voluntary or involuntary, and voluntary euthanasia can be administered by the patient (then it is often called assisted suicide) or by a physician.

The first significant published debate in Germany over voluntary euthanasia occurred in the journal of the Monist League. Roland Gerkan, in the last throes of a terminal illness, wrote an article requesting that the Monist League support a euthanasia bill he had drafted, since the organization was destroying belief in an afterlife that would provide meaning for suffering. Gerkan's proposal supported only voluntary euthanasia after the petition of an individual to the proper legal authorities, and only after a panel of three physicians verified that the patient's illness was incurable. Gerkan's proposal included not only those with terminal illnesses, but also those with chronic diseases and the crippled, as eligible for voluntary euthanasia. However, his proposal specifically forbade involuntary euthanasia.[26]

Gerkan's article prompted responses by fellow Monists. Wilhelm Ostwald, a Nobel Prize–winning chemist whom Haeckel recruited as president of the Monist League and editor of its journal, vigorously defended Gerkan's proposal.[27] Eugen Wolfsdorf likewise supported euthanasia, which he considered a logical consequence of Ostwald's ethical postulate, the energetic imperative, which stated, "Waste no energy, but utilize it!" Caring for suffering family members is economically damaging and a waste of energy, in Wolfsdorf's view. Thus, he believed that Ostwald's energetic imperative superceded the principle, "God has given life, God must take it back."[28] Two other physicians in the Monist League, A. Braune and Friedrich Siebert, expressed their support for euthanasia, too.[29] The only Monist to publish a rebuttal to Gerkan's views was Wilhelm Börner, but even Börner was not arguing for the sanctity of human life. Börner merely argued that defining who should be eligible for euthanasia was too slippery, so his objection focused on the practical implementation of euthanasia, not on the worth or dignity of human life.[30] No Monist argued in this debate that the life of a suffering individual had intrinsic value.

Since voluntary euthanasia is generally less radical a position than involuntary euthanasia, one might think that support for voluntary euthanasia would precede advocacy for involuntary euthanasia. However, in Germany, this was not generally the case, largely because early discussions of euthanasia were intertwined with eugenics considerations. Voluntary euthanasia normally pertains to adults at the end of their life. However, eugenicists were concerned mostly about limiting reproduction, so those eugenicists who discussed killing "inferior" individuals generally wanted to do so as early in life as possible. Furthermore, since many of those defined as "inferior" were mentally ill, they could hardly give voluntary consent. Thus, eugenicists often showed little concern for voluntary euthanasia, while sometimes pressing for infanticide and involuntary euthanasia.

After Haeckel, one of the earliest Darwinian thinkers in Germany to discuss these matters was his friend Hellwald, who in 1875 discussed the Spartan practice of infanticide in his *History of Culture*. Hellwald carefully avoided explicitly endorsing infanticide, but he presented it in a generally favorable light, dwelling more on its beneficial effects in improving the race than on any objections to it.[31] Hellwald had already rejected any moral considerations from interfering with the right of the stronger in the Darwinian struggle for existence, so killing the disabled seems consistent with his amoral stance.

The biologist Dodel also did not explicitly endorse infanticide, but he concurred with Haeckel's views about the status of the infant's life.

He believed that the fetus and even the newborn infant is not fully human, since it is still repeating the developmental stages of its evolutionary ancestors. Furthermore, Dodel asserted, some babies never develop fully into humans. Though Dodel did not suggest killing them, he clearly stripped the mentally disabled of their status—and presumably their rights—as humans, which would place them in a precarious legal situation.[32]

Only after the triumph of Darwinism and along with the rise of the eugenics movement in the early twentieth century did German thinkers begin publicly espousing infanticide for disabled infants. In 1895, in the only major book Ploetz wrote, he broached the subject of infanticide in a rather oblique way. In the middle of the book, he described a eugenics utopia. In their zeal to rationally control reproduction, this society would not only kill all weak or deformed children immediately after birth, but they would also kill all twins, all children born after the sixth child, and all children born to a mother over 45 or a father over 50, because they (i.e., Ploetz) believed these were likely to be physically and mentally inferior.[33] Was Ploetz thereby advocating infanticide? Possibly, but probably not. For one thing, Ploetz painted a picture of his utopia that seems remarkably shocking and extremely inflammatory. For example, I doubt that he expected anyone to understand the killing of all twins as a serious proposal that he favored.

Furthermore, Ploetz never again defended infanticide, as far as I know, and he took a more ambiguous position on killing the sick in an essay he published the same year his book appeared. In his article in a philosophical journal, he first stated that

> in relation to the *persistently weak* [as opposed to those temporarily weak], i.e., the elderly, the incurable, and those who are otherwise defective, a society will be preserved all the better, the more these are disposed of. For their preservation requires sacrifice on the part of the strong and thereby reduces the ability of the whole to preserve itself. . . . Any faulty and defective individuals still produced [later] can only be disposed of by annihilation or expulsion.[34]

After making this provocative claim, Ploetz then argued that, despite the advantages, killing the weak ultimately would not be beneficial, because it would wreak havoc with our social instincts, which are essential for the healthy functioning of society. Ploetz did not think a society could have the necessary altruism to maintain itself, while at the same time killing the elderly and sick.

By making the killing of the disabled seem so rational, but then appealing to the social instincts to salvage prohibitions against murder, Ploetz took a rather perilous step. What if some—like Haeckel—argued that dispassionate reason should overrule our social instincts in this matter? Or what if some thought—as some eugenicists did—that our social instincts did not extend to the weak and sick? Ploetz's opposition to killing the weak and the sick has a certain logic to it, but in his explanation he showed absolutely no concern for individual lives. Neither the value nor the welfare of individuals enters his reasoning at all.

Even more provocative, though less influential, than Ploetz's book was another book published the same year, *The Right to Death* (1895) by Adolf Jost. I hesitated to include Jost in this study, since he did not appeal to Darwinism or eugenics directly. However, he admitted that his "natural world view" was in direct opposition to religion, and he also made clear that his view of the value of human life was shaped by his rejection of the supernatural. It is unclear to what extent Darwinism figured into Jost's "natural world view," but his view of human life was clearly naturalistic, just as the Darwinists we are examining in this study. In any case, Jost was the first to publish an entire book in German defending euthanasia. Jost's book was a plea not only for voluntary euthanasia, as one might expect from the title, but also for killing the mentally disabled. He insisted "that there really are cases, in which, mathematically considered, the value of a human life is negative." His entire book is an exercise in cool rationality to overcome objections to killing these "negative values."[35]

Jost's ideas appealed to the female socialist physician Oda Olberg, a staunch advocate of eugenics. In an article in a popular freethinkers' journal she agreed with Jost that individuals should have a "right to death," since this would spare them and society much suffering. Apparently she also approved of involuntary euthanasia, for " 'the right to death' that I advocate appears to me identical with the right of the individual to escape from useless suffering, and the right of society to expel physical or moral sources of infection or to hinder their development." Depicting some individuals as infectious diseases dangerous to society makes killing them seem necessary and even righteous. Killing becomes an act of self-defense rather than cold-blooded murder. In addition to advocating euthanasia—both voluntary and involuntary—Olberg used Jost's ideas as a springboard for promoting the right for women to get abortions.[36]

Though not as forthright as Jost, Forel—whose worldview clearly was built on Darwinism—largely agreed with him, approving of infanticide and euthanasia. Like most eugenicists, he continually warned against the reproduction of the "inferior." In his book, *The Sexual Question* (1905), in which he thoroughly discussed the role of Darwinism for sexuality, he implied that he supported infanticide for the mentally handicapped, whom he called "little apes."[37] He opened a 1908 published lecture on "Life and Death" with the question, "Is life worth living?" Not always, he replied, since those who are mentally abnormal or hereditary criminals, whose existence is "a plague for society," may not consider their lives worth living. Though he refrained from overtly passing judgment on them, Forel implied that he did not consider their lives worth living.[38] Two years later he made explicit, what was often implied in many of his earlier works on eugenics, the advocacy of killing the physically and mentally handicapped. He asked, "Is it really a duty of conscience to help with the birth and even the conception of every cripple, who descends from thoroughly degenerate parents? Is it really a duty to keep alive every idiot (even every blind idiot), every most wretched cripple with three-fourths of the brain damaged?" He answered with a resounding, No! The irrational idea of keeping such people alive derives from outmoded otherworldly ethics, according to Forel.[39] Apparently, by sweeping aside otherworldly (i.e., Christian) ethics, Darwinism opened the door for this-worldly killing.

A few other eugenicists also broached the subject of infanticide. Schallmayer did not explicitly advocate infanticide, but in his Krupp Prize–winning book he insisted that Europeans' criticism of Chinese infanticide was misguided. He asked for a rational, scientific investigation of Chinese infanticide to determine if it is any worse than European institutions, which contribute to high infant mortality. Since he was a critic of the capitalist economy, Schallmayer blamed it for the poverty and ill health of European children. Even though he had no data to compare infant mortality in the two societies, he surmised that the infant mortality in Europe was probably no more "moral" than infanticide in China.[40] Schallmayer thus stopped short of advocating infanticide, but he clearly did not think it was morally reprehensible, either.

In another article, Schallmayer explicitly criticized the Spartan form of infanticide, which Haeckel had extolled. However, he opposed it not because he saw it as murder and thus morally objectionable. Rather, he considered it eugenically ineffective, since the Spartans based decisions about killing infants on the physical appearance at infancy. Schallmayer believed that the appearance of the newborn was not a good measure of its "eugenical value." In order to make proper judgments in such matters, one would have to examine the hereditary traits of the baby's ancestors. He admitted that in cases of deformed infants, the matter was different, but even in this case, infanticide is not really necessary, and since it is so rare, most never reach maturity, and thus do not reproduce. Even in this passage, Schallmayer's opposition to infanticide is conditional, and he refrains from condemning it outright. He left open the possibility that under some circumstances—if it would be beneficial eugenically—it would be acceptable.[41]

Another eugenicist approving infanticide was the physician Agnes Bluhm, the leading woman in the German eugenics movement. At her speech to the First International Eugenics Congress she encouraged obstetricians to rethink their life-saving procedures in births involving "imbeciles," who represent, she averred, a "loss to the nation." Traditional medical ethics stressed saving all lives, regardless of the future health problems of the individual. In the case of the severely disabled, however, Bluhm hoped to transform medical ethics. In her view, an obstetrician "must no longer blindly seek to produce for the mother a living child, but must ask himself, in individual cases, whether he can take the responsibility as regards the [human] race."[42] She thus considered the killing of some infants a beneficent deed and hoped that physicians would stop trying to save the lives of handicapped infants.

In a 1911 article Hegar indicated his support for killing the mentally disabled, too. Before introducing this topic, though, he first discussed capital punishment for criminals. He criticized the present penal system that allows criminals to reproduce, passing on their "ethical inferiority" to future generations. "One cannot avoid the thought," he asserted, "that it might be better to quickly dispose of useless, corrupt and dangerous individuals, instead of supporting them till death in jail." Once he had (hopefully) gained the ear of the reader on this point, he moved on to the mentally ill. He declared that no really significant differences exist between the mentally disabled and criminals; both are dangerous and burdensome. One is mentally "inferior," the other morally "inferior." Why not kill people in both categories? Recognizing, however, that most people would oppose killing the mentally disabled, he hoped that his society would at least prohibit their marriages to help rid the world of these "inferior" people.[43]

Euthanasia became a hot topic of discussion in the 1920s in response to the provocative book, *Permitting the Destruction of Life Unworthy of Life* (1920), coauthored by the legal scholar Karl Binding and Alfred Hoche. Hoche was professor of psychiatry at the University of Freiburg, and he reflected the Darwinian devaluing of human life that we have already examined. In an autobiographical work, he set forth his view of life, explaining that to nature,

> the continued existence of the species is everything, the individual is nothing; she [nature] carries on an immense waste of seeds, but the individual, after she has given it—the mature one—opportunity to pass on its seed to the future, she heedlessly lets die; it is for her purposes without value.[44]

His view of the purposelessness of individual life was reflected in his controversial statement that "the dentist is worse than the guillotine."

Hoche naively called this statement scientifically unassailable, since the latter is painless.[45] Like many of his contemporaries who built their worldview on science, he failed to recognize that underlying his scientific worldview were philosophical presuppositions that were not derived from empirical science and about which science could not arbitrate.

Hoche's equanimity about death brought him face to face with ethical dilemmas. As a young physician in Heidelberg one of his patients was a nine-year old girl, who was dying of an unknown disease. In order to study the disease Hoche and his colleagues wanted to keep the child at the hospital, while the father demanded that they allow him to take her home to die. Hoche filled a syringe with morphine, struggling to decide whether he should kill the child before the father took her away, so they could perform an autopsy. He finally resolved not to kill her, because in weighing the interests, he did not think scientific curiosity was a strong enough justification for killing (and perhaps because he feared legal consequences, though he did not mention this as a consideration). Nonetheless, in relating this story in his memoirs Hoche argued that "there are circumstances, in which killing by a physician is no crime." One example where such killing would be permissible would be "if through the shortening of this one lost life immediate insights could be gained, which would save other better lives."[46] Hoche thus weighed the interests of the species against the welfare of the individual, and the former trumped the latter, even to the point of death.

In the book he coauthored with Binding, Hoche lamented that present moral codes prohibited the killing of those who are "completely worthless." He called for a revision of moral codes, which, under the influence of humanitarianism, overvalued human life. He continually stripped the mentally disabled of their dignity by referring to them as "mentally dead." Medical ethics, he asserted, change over time, so when society decides to permit the killing of incurable patients or the mentally ill, medical ethics cannot oppose it.[47] Psychiatry, which had originated to help the mentally disabled, was now being called on to help in their destruction. Hoche's views were more radical than most of his contemporary eugenicists, but it shows the peril of reducing humans to nature and mind to matter.

Abortion was another issue related to the value of human life that was influenced by the rise of the Darwinian worldview. The early debate over abortion in Germany was, of course, not always overtly linked to Darwinism, though sometimes it was, as we have already seen in Haeckel's case. In 1905, Siegfried Weinberg followed Haeckel's lead by advocating the legalization of abortion in an article published in Stöcker's new journal. He admitted that medical science had effectively undermined the old idea of the "quickening" of the embryo. This scientific advance militated against abortion. However, Weinberg believed that a more recent scientific discovery—Haeckel's theory that the human embryo is recapitulating the forms of its evolutionary forebears—showed that the embryo really is not fully human.[48] The jurist Otto Ehinger later used the same line of reasoning in Stöcker's journal to call for the legalization of abortion. He also used another argument related to Darwinism to support his position:

> Nature shows us with thousands of examples, even in the process of conception, that it wastes millions of seeds, in order to allow one of them to develop fully. Isn't a mother, especially under present economic relations, merely following its example, if she—perhaps with distress of conscience and pain—destroys her seed, because the children she already has must starve and waste away with an enlargement of the family?[49]

The biggest impact of Darwinism on the abortion debate came through eugenics discourse, which, as we have seen, was founded on Darwinian principles. Eugenics provided important impetus for those promoting the legalization of abortion. Most of the leading abortion advocates—Stöcker, Schreiber, Fürth, Olberg, and others—were avid Darwinian materialists who saw abortion not only as an opportunity to improve conditions for women, but also as a means to improve the human race and contribute to evolutionary progress. Stöcker and her League for the Protection of Mothers consistently used eugenics arguments to support the legalization of abortion, though ultimately they wanted to allow abortion for non-eugenics reasons as well.[50] Eduard David, in an essay on "Darwinism and Social Development," argued that eugenics was the proper social response to Darwinism, and he approved of abortion as one eugenics measure among others.[51] Lily Braun likewise became a strong advocate of both eugenics and abortion. She apparently favored infanticide, too, for once she scolded a doctor for not preventing the beating of a child with Downs syndrome: "If you physicians are not compassionate enough to free such children from the burden of life on this earth, you should at least protect them against cruelties."[52] Apparently for Braun, as for Hoche, killing someone was better than allowing them to live in pain.

Even those Darwinists who did not favor complete legalization of abortion often did not do so because of any regard for the life of the individual fetus. Rather many eugenicists opposed abortion because they thought easy access to abortion would lead to a population decline and would contribute to biological degeneration. Their concern was not with the individual life, but with improving the biological vitality of the German population as a whole. Ploetz was a forthright critic of Stöcker's and other sexual reformers' attempt to legalize abortion, because he feared it would result in a decline of the nation's biological quality. He proposed, however, that abortion be legalized only in cases in which the woman was raped or "the expected child with the greatest possibility would be grossly degenerate or infected with syphilis."[53]

Bluhm agreed with Ploetz, opposing abortion under most circumstances, since she believed that population growth was healthy for society. She further acknowledged that the fetus was not merely part of the mother, so it deserves protection as a unique individual. Just as no one would claim that an oak tree is just a part of the soil, even though it receives its nutrients from the soil, so no one should consider the fetus a part of the mother, she argued. However, she favored the legalization of abortion under circumscribed conditions, such as a mother's health problems, rape, or in case the offspring might be expected to be "inferior." Thus she saw abortions for the congenitally disabled as perfectly acceptable. She persuaded the League of German Women's Organizations, a large umbrella organization that coordinated efforts within the women's movement, to adopt her position on abortion.[54]

Eugenics considerations played a key role in making abortion more acceptable in the early twentieth century. No one illustrates this better than Anna Pappritz, a leading figure in the German women's movement. She agreed wholeheartedly with Bluhm that the fetus is a distinct individual, not merely a part of the mother. She argued forcefully against the legalization of abortion, since this would have a deleterious effect on the evolution of the race. Unlike most eugenicists, however, Pappritz embraced the Judeo-Christian conception of the sanctity of human life and opposed abortion on this basis. She agreed with eugenicists "that many healthy, strong, joyful people wear themselves out in service of the inferior, sick, cripples, and the elderly; in a word, the valuable are used up for the benefit of the worthless." But all these sacrifices and inconveniences are necessary, according to Pappritz, who vigorously opposed any suggestion that it would be better to let the disabled die. She presented abortion as a perilous step on the slippery slope toward murdering the disabled, stating, "If we infringe on the sanctity of life at one point, the danger presents itself, that gradually our very refined feeling in this area will become dull and we will sink down into savagery and barbarism." After making this impassioned plea against abortion and for helping the disabled, however, Pappritz ultimately undermined her own position by allowing abortions "if the well-founded prospect is present that the child is subject to degeneration through hereditary disease." Despite her rhetoric to the contrary, apparently eugenics ultimately prevailed over the sanctity of human life.[55]

Bluhm's and Pappritz's view that abortions should only be legalized in circumscribed situations was commonplace in the eugenics movement. Julius Tandler, a medical professor at the University of Vienna, also agreed with Bluhm that developing embryos are not part of the mother, but rather unique individuals. However, he also favored legalization of abortion, but only if a commission of physicians determined that the offspring is unworthy to live. Not the mother, but society should decide on the life or death of the fetus.[56] Rüdin was another leading figure in the eugenics movement to support abortions, but only for eugenics purposes.[57] Even opposition to abortion by many eugenicists was thus conditional, and women who might have "defective" offspring would be allowed—or even encouraged—to abort their children.

Tandler was not the only eugenicist to imply or sometimes explicitly argue that abortions for eugenics purposes should not merely be legalized, but mandated. Max Hirsch, who became editor of a leading eugenics journal in 1914, wrote a book that same year on *Abortion and Birth Control in Relation to the Declining Birth Rate*. First, he made a strong pitch for compulsory sterilization to prevent the birth of "defective offspring." A vigorous sterilization program would obviate the need for abortions in most cases. However, if a "defective" person became pregnant anyway, then abortion would be *"the only rational therapy, after prophylaxis failed."* Hirsch promised that a eugenics program featuring sterilization and abortion would secure Germany's happiness and prosperity, while limiting crime, poverty, and the economic "burden" of asylums.[58]

Another category of people some Darwinists considered biologically inferior and thus fit for destruction was habitual criminals, especially violent ones. This, of course, does not represent a major shift in European thought, since capital punishment had been sanctioned by most religious and secular authorities for millennia. However, Darwinists provided an entirely new rationale for capital punishment. Since Darwin and many Darwinists thought that moral characteristics, such as altruism or selfishness, diligence or laziness, honesty or mendacity, and the like were hereditary rather than the products of nurture, the only way to get rid of bad moral traits would be to keep the offender from reproducing. Lombroso's view of the "born criminal" thus led some Darwinists, including Lombroso, to advocate capital punishment as a way to prevent crime in future generations.

Haeckel was one of the earliest to advocate this new view of capital punishment. In the 1873 edition of his *Natural History of Creation*, immediately after discussing infanticide, he criticized those misguided humanitarians who wanted to abolish the death penalty. They did not recognize that killing criminals is "a good deed for the better part of humanity." It is just like pulling weeds from a garden, he alleged, since it hinders unreformable criminals from passing on their morally corrupt heredity.[59] Though many eugenicists preferred other methods for hindering the reproduction of criminals, such as permanent incarceration or sterilization, some upheld the necessity for the death penalty as well.

One of Haeckel's former students, Hentschel, promoted one of the more bizarre views on criminal justice and the death penalty. In his utopian Mittgart community there would be no property, so Hentschel thought this would minimize conflicts. If, however, conflicts developed nonetheless, they would not be settled by peaceful arbitration. Conflicts among boys would be settled by the "age-old" method of fisticuffs. Among adults, serious disputes would be settled by a duel between the two conflicting parties with a light saber. If the winner killed his opponent, he would not be punished, as long as he followed the rules. This form of contest corresponded to Hentschel's view that the stronger should triumph in the struggle for existence: "Giving the right to the stronger always supercedes all human rights."[60] Hentschel's view of justice would give free reign to bullies, showing utter contempt for the weaker members of society, who would have no protection from those with greater physical strength and agility. Of course, dueling was not unknown in Europe, but it was not considered the normal way to settle disputes and by the early twentieth century, it was almost universally condemned as a remnant of barbarism.[61]

Clearly not all Darwinists and not all eugenicists favored killing the "unfit." Most searched for other ways of ridding society of these "dangerous burdens," and some even vigorously protested against any form of killing. Nonetheless, it would be a mistake to underestimate the role that naturalistic Darwinism played in initiating and fueling the debate on suicide, euthanasia, and abortion. By reducing humans to mere animals, by stressing human inequality, and by viewing the death of many "unfit" organisms as a necessary—and even progressive—natural phenomenon, Darwinism made the death of the "inferior" seem inevitable and even beneficent. Some Darwinists concluded that helping the "unfit" die—which had for millennia been called murder—was not morally reprehensible, but was rather morally good.

Those skeptical about the role Darwinism played in the rise of advocacy for involuntary euthanasia, infanticide, and abortion should consider several points. First, before the rise of Darwinism, there was no debate on these issues, as there was almost universal agreement in Europe that human life is sacred and that all innocent human lives should be protected. Second, the earliest advocates of involuntary euthanasia, infanticide, and abortion in Germany were devoted to a Darwinian worldview. Third, Haeckel, the most famous Darwinist in Germany, promoted these ideas in some of his best-selling books, so these ideas reached a wide audience, especially among those receptive to Darwinism. Finally, Haeckel and other Darwinists and eugenicists grounded their views on death and killing on their naturalistic interpretation of Darwinism.

Those affected most by the newly emerging discussion on killing the "unfit" were the mentally disabled and criminals. Eugenicists continually labeled them "inferior" or "defective," emphasizing the burden—and even danger—they were for society. However, there was another major category of people affected by the growing discussion on killing the "unfit": those of "inferior" races. The Darwinian struggle for existence not only pitted members within a society against each other, but it also led simultaneously to competition between organized groups of people—tribes, nations, and races. Next, we turn to the implications of devaluing life for war and racial competition.

9. War and Peace ✍

Darwinism helped "lay the foundation for the bloodiest war in history," declared the pacifist William Jennings Bryan in his campaign to stir up Americans against evolutionary theory.[1] Bryan was, of course, referring to the horrors of World War I. He blamed Darwinism for creating a belligerent mentality among German intellectuals and political leaders. Bryan was not alone, for William Roscoe Thayer, in his presidential address to the American Historical Association in 1918 stated,

> I do not believe that the atrocious war into which the Germans plunged Europe in August, 1914, and which has subsequently involved all lands and all peoples, would ever have been fought, or at least would have attained its actual gigantic proportions, had the Germans not been made mad by the theory of the survival of the fittest.[2]

Many other Anglo-American scholars blamed social Darwinism for the outbreak of World War I, though—unlike Bryan—many hoped to rescue Darwinism from the taint of militarism. They considered Darwinian militarism an aberration from true Darwinism.

Vernon Kellogg, an entomologist at Stanford University, related a conversation he had with a German army captain, who had been a biology professor before the war. This captain believed the war was part of the universal struggle for existence that would bring human progress through natural selection. Kellogg did not believe this man's ideas were unique among Germans, for he asserted, "The creed of the omnipotence (*Allmacht*) of a natural selection based on violent and fatal competitive struggle is the gospel of the German intellectuals." Kellogg was so horrified by the "creed" of the Germans that he abandoned his pacifist stance to support the war effort against them.[3] Kellogg, like many other Darwinian biologists, rallied to the defense of Darwinism against this alleged aberration. Bryan and other creationists, on the other hand, saw the war as confirmation that Darwinism was morally bankrupt and dangerous.

In retrospect, both Bryan and Kellogg overemphasized the role of Darwinian militarism in fomenting World War I, though, to be fair, many scholars, both then and now (myself included), recognize that social Darwinism was a prominent ideology used to support militarism in Germany (and elsewhere) in the early twentieth century.[4] Only as an indirect, background cause could social Darwinism be implicated as a cause of the war, however. Most German and Austro-Hungarian politicians were not social Darwinists (some were probably even skeptical about Darwinism as a biological theory), and German militarists used many non-Darwinian arguments to justify war.[5] Bryan and Kellogg also completely overlooked the strain of Darwinian pacifism prominent in pre–World War I Germany and Austria. Some leading Darwinists and some leading pacifists in Germany argued that, far from supporting militarism, Darwinism, properly understood, actually favored the peace movement.[6]

Thus the relationship between Darwinism and militarism in German discourse was not at all straightforward. While some Darwinists and Darwinian-inspired social thinkers were thoroughgoing militarists, some of their pacifist opponents—including many leading figures in the eugenics movement—argued that, on the contrary, Darwinism opposed militarism. Superficially, the prominence of Darwinian pacifism in Germany seems to undermine my view that Darwinian naturalism tended to devalue human life. It is relatively obvious that militarism cheapens life, but one normally equates pacifism with a high regard for the value of human life.

But things are not always what they seem, since many Darwinian pacifists—especially those in the eugenics movement—did not base their opposition to war on universal human rights or the sanctity of human life. Nor did they object to wars out of psychological revulsion at the horror of human death and carnage (though this may have loomed in the background of their psyches). Rather, what they found objectionable about modern wars was that the *wrong* people were being killed—the strong and healthy rather than the weak and sickly. Modern war, in their view, helped spawn biological degeneration rather than biological progress. In short, it stymied evolutionary progress, and that was a cardinal sin in the eyes of many Darwinists.

The pacifism of many German Darwinists was paper-thin, and when push came to shove, many abandoned their pacifism. For some that shove came with the outbreak of World War I. Haeckel and others who had preached peace before the war, patriotically supported the war effort, as did most of their contemporaries (even the Social Democrats voted for war credits in 1914). However, until one understands the perilous foundations of Haeckel's pacifism, it may seem puzzling that he so staunchly upheld

German expansionism and remained committed to the war to the bitter end—despite the wanton carnage in the trenches. Even when many Germans were war weary and the German parliament passed a Peace Resolution in July 1917, Haeckel derided German politicians for wanting peace without annexations.

To be sure, some German Darwinists remained steadfast in their pacifist convictions during World War I, which was not easy, since pacifists were scorned and persecuted during the war. But even most of these were not absolute pacifists. With only a few exceptions, most Darwinists, even the most ardent pacifists, considered non-European races inferior and condoned warfare if it resulted in the destruction of "inferior" races. The problem with World War I, in their eyes, was that the Europeans—the highest race—were killing each other and weakening themselves vis-à-vis other peoples of the world. We will explore this topic of racial struggle and racial extermination at length in chapter 10, but for now, let us restrict our discussion to militarism and pacifism in general.

It is not surprising that many German Darwinists supported militarism. Prussian militarism had strong roots going back to the seventeenth and eighteenth centuries (and beyond). Most Germans, especially the educated elites, were intoxicated by Bismarck's success at militarily subduing Germany's neighbors in a series of three rapid wars between 1864 and 1871, creating in the process a powerful, united German nation. Germans did not need Darwinism to convince them that war is a noble enterprise. Frederick the Great and Bismarck were persuasive enough.

It would be a mistake, however, to conclude because of this that Darwinism had no impact on German militarism. Darwinian justifications for war would introduce new ways of thinking about international relations and military conflict. It would also provide scientific sanction for militarism in an age in which science was gaining prestige and—for some intellectuals—becoming the sole arbiter of truth.

Darwinian militarists claimed that universal biological laws decreed the inevitability of war. Humans could not, any more than any other animal, opt out of the struggle for existence, since—as Darwin had explained based on his reading of Malthus—population expands faster than the food supply. War was thus a natural and necessary element of human competition that selects the "most fit" and leads to biological adaptation or—as most preferred to think—to progress. Not only Germans, but many Anglo-American social Darwinists justified war as a natural and inevitable part of the universal struggle for existence.[7] The famous American sociologist William Graham Sumner, one of the most influential social Darwinists in the late nineteenth century, conceded, "It is the [Darwinian] competition for life . . . which makes war, and that is why war has always existed and always will." Nonetheless, while considering it ultimately inevitable, Sumner, just like some of the German Darwinists I discuss below, generally disapproved of war as policy.[8]

By claiming that war is biologically determined, Darwinian militarists denied that moral considerations could be applied to war. In their view wars were not caused by free human choices, but by biological processes. Blaming persons or nations for waging war is thus senseless, since they are merely blindly following natural laws. Further, opposition to war and militarism is futile, according to Darwinian militarists, who regularly scoffed at peace activists for simply not understanding scientific principles.

However, though denying that morality could be applied to war, they did believe that war fostered progress, including moral progress. Thus war not only was not immoral, it was the very source of morality. According to Darwin, tribes banding together with the greatest measure of selflessness and altruism would have a selective advantage in warfare against their more selfish neighbors. Since Darwin and many of his contemporaries considered altruism a biologically based instinct, the more altruistic tribe would be able to pass on its altruism to a greater number of offspring than the less altruistic tribe. But they would do this by killing as many members of their neighboring tribes as possible! In this way warfare not only selected the strongest and bravest, but also the "most moral."

Though admitting that war had been an important and even beneficial factor in human evolution in the past, Darwin shrank back from promoting war. In fact, he noted that nations and societies were becoming ever larger and predicted that one day a single nation might encompass the globe, rendering warfare obsolete. Thus Darwin himself was no militarist. However, he contributed to the rise of Darwinian militarism in several ways. First, he made liberal use of martial terminology to describe competition in the organic world. The struggle for existence became his most popular phrase to describe competition between organisms, but Darwin also referred to the struggle for life, the battle for life, and the war of nature. Germans added to this list by commonly referring to the struggle of all against all. Second, his reliance on the Malthusian population principle gave expansionists a powerful propaganda tool in a time when the European population was rapidly expanding and masses were emigrating. Finally, he construed human wars as a progressive force (at least in the past) in human evolution.

Because Darwin had been reluctant to discuss human evolution, he was not the first to suggest that wars are a form of the struggle for existence. Seven years after Darwin published his theory and five years before his *Descent of Man* appeared, the geologist Friedrich Rolle, one of the earliest disciples of

Darwin in Germany, wrote the first full-length book in German on human evolution. In his book he discussed at length the role of warfare as part of the human struggle for existence. He brushed aside any moral considerations, since the "right of the stronger" is not subject to morality. Rather, "In the contest between peoples the proverb was and always will be valid: 'Better that I smash you, than you smash me.' " Besides, he argued, war brings progress by favoring the fittest and ridding the world of the "less fit."[9]

In the same year that Rolle's book appeared, Otto von Bismarck, chancellor of Prussia, engineered a war with Austria, which was the second of three wars he used to unify Germany under the auspices of the Prussian crown. On the occasion of this 1866 war Peschel published an article using Darwinian arguments to justify Bismarck's militarism. He argued that the rise and fall of states is a phenomenon of nature under the sway of scientific laws, including the Darwinian struggle for existence. Moral concepts such as righteousness have no place in this natural struggle, he asserted. Rather, success of the strong and vigorous in the struggle among peoples is natural and needs no justification. He stated,

> Even we in Germany should view the most recent events [i.e., the war] as a lawful evolutionary process. . . . With such magnificent events it is no longer a matter of right or blame, but rather it is a Darwinian struggle for existence, where the modern triumphs and the obsolete descends into the paleontological graves.[10]

Peschel thus used Darwinism to justify warfare as a path to progress and improvement, as well as to dismiss all moral considerations.

Peschel and Rolle were only the first among many to interpret war as a form of the human struggle for existence. Rolle's friend, the zoologist Gustav Jaeger, followed the lead of Rolle and Peschel by interpreting both the 1866 war against Austria, as well as the 1870–71 war against France, as part of the universal struggle for existence. In an 1866 article in *Das Ausland* he claimed that since the advent of Darwinism, human history could now be "investigated with the same objectivity as every other natural event." In this brief article, he drew two conclusions from his biological examination of history: (1) wars, especially modern mass wars, are an important aspect of the struggle for existence; (2) wars occur approximately once every eighteen years, or the span of one generation.[11]

Jaeger's 1870 article was especially inflammatory, not only justifying war, but even wars of annihilation. From nature

> the scientist correctly draws the conclusion that war, and indeed the war of annihilation—for all the wars of nature are that—is a natural law, without

which the organic world would not only not be what it is, but could not continue to exist at all. Further this conviction must compel him to make the beneficial effects of this universal struggle of annihilation the object of his research.[12]

Jaeger claimed that war is necessary and beneficial, for without it, the world would be filled with unhealthy and disabled people. His concern about killing off the weak and sick would presage the concerns of later eugenicists. Jaeger was not the only Darwinist to see the Franco-Prussian War as an episode in the human struggle for existence. The University of Freiburg anatomist Alexander Ecker, a leading figure in the emerging field of anthropology, published a speech on *The Struggle for Existence in Nature and in the Life of Peoples*, portraying war and racial competition as a natural part of the human struggle for existence.[13]

Darwinian militarism reached an even wider audience in the 1870s through the popular writings of David Friedrich Strauss. Strauss began his career as a theologian, publishing his sensational *Life of Jesus* in 1835, which created a fire-storm of controversy because of its interpretation of the gospels as myth. Publishing this book effectively torpedoed his attempts to gain a professorship in theology. It is probably just as well that he did not become a theology professor, since by the 1860s Strauss no longer enjoyed reading theology, but rather science, especially Darwinism.[14] His immensely popular book, *The Old and the New Faith* (1872), heralded the replacement of Christianity—the old faith—with science. Darwinism, complete with natural selection and the struggle for existence, was a crucial component of Strauss's new faith. An inescapable part of the human struggle for existence, Strauss assured his readers, is war, which winnows peoples and nations according to their value. Though not a major theme in Strauss's book, his claim that different peoples have different values was a crucial component of later eugenics thinking. Strauss ridiculed those agitating for an end to wars, which he considered—on the authority of his scientific worldview—about as effective as trying to put an end to the weather.[15] Strauss's views resonated with many of his contemporaries, including leading Darwinists (like Haeckel) who praised his book.[16]

Even more blatant than Strauss in his justification of militarism was Hellwald, a prominent Darwinian ethnologist who edited *Das Ausland*, the main scholarly journal in Germany devoted to geography and ethnology, from 1872 to 1881. Hellwald was heavily influenced by Haeckel and Ecker and expressed complete sympathy with the views of Strauss, though he claimed he had developed his ideas before reading Strauss. In an 1872 article in *Das Ausland*, as well as in his widely read work, *The History of Culture in Its Natural Evolution* (1875), Hellwald explained human history as a process driven by the Darwinian struggle for existence. His allegedly scientific explanation for the development of human society replaced ethical considerations with violence, since

> science knows no "natural right." In nature only one right reigns, which is no right, the right of the stronger, violence. But violence is also the highest source of law [or right], since without it [i.e., violence] legislation is unthinkable. I will in the course of my presentation easily prove, that properly speaking the right of the stronger has also been valid at all times in human history.[17]

Again and again in his book Hellwald insisted that science had banished morality, since in the struggle for existence the ends justify the means—the winner of the struggle is automatically right.

When Hellwald brushed aside all moral objections to oppressive human institutions, barbarism, and even mass killing, he posed as a cool, objective scholar applying science to human affairs. Science has proven, he asserted,

> that just as in nature the struggle for existence is the moving principle of evolution and perfection, in that the weak are worn away and must make room for the strong, so also in world history the destruction of the weaker nations through the stronger is a postulate of progress.[18]

He showed not the slightest sympathy or emotion for the oppressed or dying, stating, "Whoever it may be, he must stride over the corpses of the vanquished, that is natural law."[19]

With his "scientific" justification for violent struggle among humans, it is not surprising that Hellwald construed war as a necessary part of the Darwinian struggle. He believed that warfare among primitive humans had indelibly imprinted a war-like character in the human species. He also considered war one of the most important factors promoting cultural progress, illustrating this point with the Spanish conquest of the Americas. Despite being accompanied by atrocities and incredible bloodshed, it was really, in his estimation, "an inexpressible blessing" to humanity.[20] Hellwald's utter disregard for morality and his glorification of violent struggle and bloodshed in human history make him one of the most radical early thinkers appealing to Darwinism to devalue human life.

Hellwald's work was held in high esteem by fellow Darwinists, including Haeckel, but not all Darwinian social thinkers were as brutal in their rhetoric as Hellwald. Nonetheless, as Darwinism became more entrenched among German intellectuals in the late nineteenth and early twentieth centuries, Darwinist militarism grew apace. Increasingly large numbers of scientists, physicians, and social thinkers from many different fields used Darwinian justifications for war and opposed the growing peace movement as unscientific. Though not quite as harsh as Hellwald, the end result of their views was pretty much the same—death through war for the "unfit" and, as a consequence, evolutionary progress.

Another scholar analyzing society through Darwinian lenses was the economist Albert E. F. Schäffle, professor at the University of Tübingen and later at the University of Vienna (and later yet a private scholar, after briefly taking on a position in the Austrian government). In several articles on the social implications of Darwinism and in his most important scholarly work, the four-volume *Structure and Life of the Social Body* (1875–78), he argued that the national economy should be organized in such a way that it minimizes conflict within society. He thus favored a moderate kind of state socialism. The purpose of avoiding friction within society, however, was not humanitarian, but rather to ensure success in the collective struggle, which Schäffle believed was even more important than individual competition. War, as part of the struggle for existence, was, according to Schäffle, "an elevating and stimulating force."[21]

Schäffle stressed that war is a manifestation of the struggle for existence among humans caused by the Malthusian population imbalance. However, he was not an avid proponent of militarism, for he believed that wars of destruction were characteristic of lower stages of human evolution, when humans were closer to animals. At higher stages of morality, humans condemn war as immoral. Also, higher civilizations could limit their births and escape population pressure without war. Thus, like Darwin, Schäffle believed that as humans evolved to higher stages of civilization, war would become less important. When civilized societies resort to warfare, Schäffle asserted, they are falling back into the behavior of beasts of prey. Despite his generally negative stance toward war, however, Schäffle denied that war could be universally condemned. He asserted that only specific wars could be justified or condemned, since "war is in certain cases justified as the most extreme means of self-preservation of independent peoples. . . . There is no unconditional right, except the right to self-preservation." With this kind of moral relativism the only yardstick for moral judgment is survival value in the struggle for existence. By Schäffle's standards, then, war is morally good if it promotes survival in the human struggle for existence, though in general he considered war an outmoded form of that struggle.[22]

Though many of the leading social Darwinist thinkers were scholars in some branch of the social sciences, biologists often provided them with their tools. Only rarely, however, did a biologist devote his career to studying the social implications of Darwinism. Such a biologist, however, was

Ziegler, who probably did as much as any other biologist to promote Darwinian militarism. Ziegler was horrified that the Social Democrats, especially their leading politician, August Bebel, were preaching socioeconomic equality, so he wrote a book-length refutation of socialism based on Darwinian theory. He specifically criticized Bebel for denying the necessity of war, which occurs because of population pressure and results in progress through natural selection:

> Bebel is also ignorant of the fact that according to Darwin's theory wars have always been of the greatest importance for the general progress of the human species, in that the physically weaker, the less intelligent, the morally lower or morally degenerate peoples must give place to the stronger and the better developed.[23]

Ziegler thus gives a very positive spin to war, which not only causes biological progress, but also moral progress by killing off those who are at a lower level of morality.

Many other prominent scholars upheld views similar to Ziegler's. One of the leading physical anthropologists in late nineteenth- and early twentieth-century Germany, Felix von Luschan, called on Darwinism to prove the inevitability of war. In a 1909 speech to the Society of German Scientists and Physicians he argued that one of the chief practical tasks of anthropology should be to help maintain "our" (presumably Germany's, though this is unclear) military defenses by keeping the nation biologically vigorous. He noted that wars will not end any time soon, since humans are subject to the struggle for existence.[24] Alfred Kirchhoff, geography professor at the University of Leipzig, agreed with Luschan in a posthumously published book. He also agreed with Ziegler that wars bring moral progress by ridding the world of "immoral hordes."[25] Similar sentiments are sprinkled throughout the writings of many prominent Darwinian scientists, popularizers, or social thinkers, especially when discussing warfare between different races (see chapter 10).

Justifying war from a Darwinian perspective became such an important topic in the early twentieth century that two German scholars, the sociologist Sebald Steinmetz and Klaus Wagner, devoted entire books to it. Steinmetz, like most Darwinian militarists, was strictly deterministic, portraying wars as unavoidable natural events in his work, *The Philosophy of War* (1907). He argued that war had been crucial in the early evolution of humans, stimulating the evolution of greater intelligence and higher levels of altruism. Without war, primitive people would probably have remained egoists, but war requires selfless devotion to the community.[26]

Steinmetz also stressed the role of war in natural selection. He called war a "world court," deciding "the entire value of a people." Underlying his analysis was the assumption that different peoples, nations, and races are unequal in value, and wars are the most reliable way to give the more valuable ones their rightful place. Of course, the flip side of this is that wars would bring "the extermination of the most worthless ones." Most of Steinmetz's discussion centered on the military contest between different races, which would pit Europeans against Asians and Africans to determine who would control the globe. But simultaneously, there would be wars within Europe and Asia that would also be real "wars of selection," killing off the least fit and elevating the surviving peoples.[27]

Even more rabidly militaristic than Steinmetz was Wagner, whose book, *War: A Political–Evolutionary Examination* (1906), appealed to all the predictable Darwinian themes to justify war as the universal "creative principle." Wagner, however, laid greater stress than some Darwinian militarists on patriotic love of fatherland and one's own people (*Volk*), which he claimed was part of the human social instincts. This instinctive feeling of unity with one's own people (*Volk*) was advantageous in the collective struggle for existence against other peoples. Thus Wagner portrayed patriotism as a force for evolutionary progress: "If our consciousness of our people (*Volk*) allows the fit to oust the unfit from this planet, so that high culture increases, then there is progress on earth."[28]

Wagner is as forthright as any of the Darwinian militarists about the connections between morality and war, so his views serve as a good summary of the ethics of Darwinian militarists. First, by construing war as a naturally determined event governed by Darwinian laws, he denied the relevance of moral judgments about war. War was now in the province of science, and any opposition to war on moral grounds was ruled out of bounds as unscientific. Wagner and other Darwinian militarists also turned the tables on moralistic peace advocates by claiming that morality was itself the product of wars. Killing in wars was not immoral, he thought, but rather on the average those who emerged victorious in wars were more altruistic than those who were vanquished.

Finally, because he believed war selected the physically strongest, the morally best, and the most highly cultured, Wagner and many Darwinian militarists depicted war as beneficial and morally good. It promoted evolutionary progress by ridding the world of the "inferior" elements. Wagner stated that it was a "shocking sin" if lands rich in natural resources were wasted by "inferior peoples." Rather the "most noble peoples must colonize the world."[29] Thus killing "inferior peoples" became almost a moral imperative for Darwinian militarists. They never seem to have considered that granting moral status to the evolutionary process itself was no more scientific than the anti-war morality they dismissed as unscientific.

Now that we have examined the key ideas of Darwinian militarism, we need to ask: What impact did it have, especially on the outbreak of World War I? As I already indicated earlier, Darwinian militarism was not a primary factor—nationalism and power politics were far more important and did not require Darwinian underpinnings. Nonetheless, it would be a mistake to dismiss the impact of social Darwinism altogether. One factor that contributed to the outbreak of the war in 1914 was the Austrian and German military staff's conviction that war was inevitable, so they might as well strike—or at least risk a war—while they held an advantage (they feared Russia would become stronger over time and make their military position ever more untenable). Social Darwinist militarism was one factor (among others) contributing to this skepticism about the long-term prospects of peace.

Gauging the influence of social Darwinist militarism within the German and Austrian military staff is difficult, since most military leaders did not publicly discuss their worldviews and philosophy of war. However, a growing number of German and Austrian military leaders did embrace Darwinian militarism before World War I. After retiring from the Austrian military, Gustav von Ratzenhofer wrote a systematic sociology based on Darwinism, featuring racial conflict (see chapter 10). The German officer Otto Schmidt-Gibichenfels became editor of *Politisch–Anthropologische Revue* in 1907 and used that platform to preach Darwinian militarism.[30] According to Istvan Deak, many Austrian officers at the beginning of the twentieth century were "voicing social Darwinist sentiments."[31]

One of those Austrian officers directly connected to the events of 1914 was Franz Conrad von Hötzendorf, the Austrian chief of the general staff, whom his biographer calls the "Architect of the Apocalypse" (i.e., World War I). His worldview was imbued with Darwinism, and he believed that the struggle for existence ensured that warfare between nations would never end.[32] In many of his writings he displayed his Darwinian vision of human struggle, but in his private journal he divulged his naturalistic philosophy even more forcefully.[33] As a young officer he had read Darwin and became convinced that the struggle for existence rules over human affairs. He stated, "The recognition of the struggle for existence as the fundamental principle of all earthly events is the only real and rational foundation of any policy." History, he thought, was a continual "rape of the weak by the strong," a violent contest decided by bloodshed. Like other Darwinian militarists, he dismissed all appeals to morality, asserting, " 'Right' is what the stronger wills."[34] Little wonder, then, that Conrad would want to put the Serbs in their place in 1914. And for him, their rightful place apparently was the graveyard.

The most influential propaganda for Darwinian militarism from a military leader was General Friedrich von Bernhardi's book, *Germany and the Next War* (1912), which stirred up so much controversy that it was republished in five German editions—and many English editions as well—before the outbreak of World War I. Bernhardi, enraged by Germany's diplomatic failure in the 1911 Moroccan Crisis, lashed out against peace activists and called on Germans to prepare for war. The first and most important argument he raised to buttress his militarism was that Darwinism proves that war is a "*biological necessity.*" He then argued—based on his understanding of Darwinian principles—that war was necessary to avoid biological and cultural degeneration. According to Bernhardi, Darwin demonstrated that "everywhere the right of the stronger reigns," and what seems brutal is really beneficial, since it "eliminates the weak and sick."[35] To spread his militaristic and expansionistic views more widely, he wrote a condensed version of his book, *Our Future*, the same year. Bernhardi's book alarmed foreign observers, and it played a key role in the perception of Anglo-American authors that social Darwinism was the driving force behind German militarism.

After World War I broke out, some scientists interpreted the war as a Darwinian struggle for existence. The famous hygienist Gruber, for example, published a speech in 1915 arguing that the war was caused, not by political decisions of individuals, but by German population expansion. Gruber, in true Darwinian fashion, claimed that resources simply were insufficient to sustain the population growth: "There simply is not enough for everyone!" Because each nation needs ever more living space (*Lebensraum*), "humans are necessarily the enemy of [other] humans."[36] Based on these views, Gruber helped organize the expansionist Fatherland Party in Munich during the war.

Not all Darwinists agreed that Darwinism supported militarism, and, in fact, many opposed militarism. Haeckel became an active participant in the German peace movement, and many German pacifists upheld a Darwinian worldview. However, Darwinian pacifism was not always as far removed from Darwinian militarism as one might think. Many Darwinian pacifists agreed with the militarists that the struggle for existence is universal among organisms, including humans, so there is no way of escape. Most of them also admitted that war was a manifestation of the human struggle for existence. However, they parted company with the militarists by maintaining that war was not a necessary element in the struggle for existence. Many animals, after all, do not fight against members of their own species, certainly not in large, organized groups (and what about plants?). They argued that humans were not violating scientific laws if they replaced violent struggle with peaceful competition. Darwinian pacifists usually tried to turn the tables on Darwinian militarists by arguing that, far from supporting militarism, Darwinism actually undermined militarism. However, when we closely examine the reasoning of most Darwinian pacifists, it becomes clear that they did not base their convictions on humanitarian concern for human life. Ironically, more often than not they helped contribute to the devaluing of human life.

Haeckel was one of the earliest Darwinian thinkers in Germany to express misgivings about war. Though insisting that humans are subject to the eternal struggle for existence, he often used peaceful economic competition as an example (though in the final analysis it was still neighbor against neighbor and the unfit ultimately perish).[37] Despite his belief in the human struggle for existence, Haeckel's opposition to militarism emerged in germ already in the second edition of *The Natural History of Creation* (1870). In a brief passage therein he warned about the deleterious effects of modern "military selection" on civilized nations: "The stronger, healthier, more normal the youth is, the greater is the prospect for him to be murdered by the needle gun, cannons, and other similar instruments of culture." However, the weak and sick were not allowed to fight, so they could have more children, leading the nation into biological decline.[38] Haeckel's concern, then, was based solely on eugenics considerations. It was not the killing that bothered Haeckel, but only the killing of the *wrong* people, which would stymie evolutionary progress and perhaps even cause degeneration. If war killed the weak and sick, while preserving the strong and healthy, apparently he would have no objections to it (this is confirmed by his stance on racial extermination, which we will examine at length in chapter 10). Nonetheless, Haeckel continued to oppose "military selection" throughout his life.

Based on his dim view of "military selection," Haeckel later joined the peace movement. In 1891 he wrote to Bertha von Suttner, one of the leading pacifists of Europe, expressing willingness to join the peace movement. He hoped the bloody struggle between nations and peoples would be replaced with peaceful competition.[39] Haeckel joined various peace organizations and continued to rail against war in his writings before World War I. The Monist League, founded by Haeckel in 1906, likewise promoted pacifism. One of the editors of the Monist League's journal wrote to Alfred Fried, "The idea of world peace naturally belongs to the cultural program of Monism." He later successfully recruited Fried to regularly write articles on pacifism for their journal, stating, "We consider pacifism one of the most important practical tasks of the German Monist League."[40] The first person to speak to the 1912 Monist League conference was Anna Eckstein, who urged the delegates to support her petition for world peace. Eckstein's speech was warmly received, and immediately thereafter Ostwald as president of the Monist League urged the audience to support her peace efforts.[41]

Haeckel's pacifism could not weather the outbreak of World War I, despite his claim in 1917 that he was still in principle a pacifist. Within two weeks of the outbreak of hostilities he wrote an article blaming England for the war (many other Germans blamed England, too), and three months later he argued that this war was a violent episode in the universal human struggle for existence.[42] In 1917–18 Haeckel opposed the German parliament's Peace Resolution, which called for peace without annexations, and he favored the newly created militaristic and annexationist Fatherland Party.[43] World War I revealed the fragility of the foundations of Haeckel's pacifism.

Haeckel's brief remarks on military selection exerted tremendous influence on Darwinian social thinkers. Already in 1872 Heinrich Fick, a law professor at the University of Zurich, suggested legal reforms to counteract the harmful effects of military selection. The key problem his reforms addressed was the reproductive advantage gained by those unfit for military duty in a nation with universal male conscription. Fick proposed that legislation prohibit those unfit for military duty from marrying, at least until they reach the age of those completing their military service. The first country to take measures to remedy the reproductive advantage of their "weaker elements" would, Fick assured his audience, achieve a decisive advantage in the struggle for existence with other nations.[44]

Alexander Tille, another Social Darwinist to grapple with "military selection" in his writings, also illustrates the ambiguities of Darwinian pacifism. In the process of blending together Darwinism and Nietzscheanism, Tille continually emphasized the right of the stronger in the struggle for existence and denied that the weaker had the right to continue living. Thus it may seem strange that he opposed militarism so strenuously. Nonetheless, his support for eugenics made him critical of the degenerative tendencies of modern war, and he even protested against the huge outlay of military expenditures, which ultimately hurt the nation economically. He expressed sympathy with Suttner for demonstrating against the insanity of war. Further, with a touch of utopian enthusiasm, he suggested that nations abolish national borders in favor of ethnic boundaries, so populations can expand or contract more naturally without causing armed conflict. This would allow peaceful economic competition to replace violent warfare.

Tille's pacifist stance, however, had nothing to do with any kind of concern for human rights, which he constantly attacked. Rather he supported pacifism, because he thought it would promote evolutionary progress better than militarism would. It would also make European nations stronger, so they could exterminate other races and conquer the globe (a position not so uncommon among Social Darwinists and eugenicists, as we shall see in greater detail in chapter 10).[45]

Opposition to military selection gained even greater currency with the emergence of the eugenics movement in the early twentieth century. Eugenicists shared Haeckel's concern with the degenerative tendencies of modern civilization, including modern warfare. They also shared Haeckel's ambivalence toward pacifism, as we can see quite clearly in the writings of Wilhelm Schallmayer. Schallmayer often criticized modern wars, and he bewailed the "racial damage" caused by World War I, but at times he admitted that wars are a selective agent in the human struggle for existence. In the past, he argued, wars were beneficial in elevating the human race, precisely because they resulted in the annihilation of "lower" peoples. Though Schallmayer usually stressed artificial selection within society and relegated racial and national conflicts to the background, the "national efficiency" for which Schallmayer strove ultimately aimed at making the body politic strong to engage competitors—other nations and races—victoriously in the human struggle for existence.[46]

Did Schallmayer think this national and racial competition would occur entirely peacefully? In many of his works it is hard to tell. However, Schallmayer clearly divulged his views in two 1908 articles. While opposing modern wars between European countries, because they are contraselective, he argued that wars between races that are unequal—such as between Europeans and black Africans—are beneficial, especially if they lead to the extermination of the "lower" races![47] His conclusion about war was that "on the whole the influence of war on human evolution should still be considered overwhelmingly favorable."[48] So much for Schallmayer's reputation as a pacifist and opponent of racism.[49]

Ploetz took essentially the same position as Schallmayer on militarism. He saw war as beneficial in the past, but detrimental in the present, at least among European nations. If war must come, however, Ploetz suggested drafting all young men, including the weak, into the army. Then, "during the campaign it would be good to bring the specially assembled [biologically] bad elements to the place where one needs primarily cannon fodder."[50] To be fair, Ploetz was probably not entirely serious about this proposal, embedded as it was in a discussion of a eugenics utopia, but it seems rather provocative nonetheless. But whether this specific proposal was serious or not, it is clear that Ploetz and most other eugenicists were opposed to war only if it produced biological degeneration, as they thought modern European wars did.

The vast majority of Darwinists—at least those who made their views about war known—either embraced militarism or an ambivalent pacifism. However, our discussion would not be complete without describing the more thorough pacifism of a small handful of Darwinist thinkers. Possibly the only German-speaking Darwinian biologist to embrace a more thorough form of pacifism in the late nineteenth century was the botanist Arnold Dodel at the University of Zurich. Dodel argued that even though humans cannot escape the struggle for existence, they can use reason to avoid the harsher forms of it. While agreeing with Haeckel's critique of "military selection," he also saw other Darwinian grounds for opposing war. As a moderate socialist, he had greater sympathy for egalitarianism and—unlike many social Darwinists—he did not spurn moral considerations when discussing war. On the contrary, he argued that human evolutionary progress had come through advances in human reason and morality, both of which militated against violence and bloodshed. Thus, humans had evolved to a point where wars were no longer necessary, and Dodel predicted that human reason and moral instincts would increase still more in the future.[51]

Two Nobel Prize–winning leaders in the German peace movement, Suttner and Fried (both from Austria), would rely on similar arguments to defend their pacifist convictions. Both were avid proponents of a Darwinian worldview, claiming Darwinian sanction for their pacifism. Suttner, in fact, was directly influenced by Dodel, not only reading his works, but also carrying on an extensive correspondence with him. Dodel, on his part, was excited about Suttner's books, especially her most famous work, *Lay Down Your Arms* (1889), which propelled her to a leading position in the nascent German peace movement.[52] This book was a sensational success, requiring 36 editions by 1905.

Suttner first read Darwinian literature in the late 1870s or early 1880s, and it decisively influenced the development of her materialist worldview. Over the next couple of decades she read all the more influential Darwinian writers—Darwin, Haeckel, Büchner, Sterne (Krause), and Bölsche—and frequently corresponded with Carneri (and I mean frequently!—they exchanged hundreds of letters), Büchner, and Dodel. Even more important in shaping her particular understanding of the evolution of human society, though, were the British thinkers Herbert Spencer and the historian Thomas Henry Buckle.[53] Spencer and Buckle both imbued Suttner with an optimism that evolutionary progress always leads upward to ever higher levels of perfection, including moral perfection. That optimism shone through in her acceptance speech for the Nobel Peace Prize in 1906, when she declared, "Those who have recognised the law of evolution and seek to advance its operation are convinced that what is to come is always a degree better, nobler, happier than what lies in the past." In that speech and many of her writings, Suttner stressed the role of evolution in producing humane ethics and human rights, especially the right to life, which formed the foundation of her peace efforts.[54]

Darwinism played a central role in many of Suttner's books, including her novel, *Daniela Dormes*, but even in *Lay Down Your Arms* the heroine read and agreed with Darwin's *Origin of Species*.[55] In two earlier books, Suttner had expounded her monistic worldview, which viewed evolution or change as the unifying principle of the cosmos. She believed that politics and society rested squarely on science:

> The extant institutions are determined by the current state of morality, morality is determined by the state of [one's] world view, and this finally is determined by the state of knowledge; in the final analysis therefore it is knowledge, or, in other words, the sciences, which provide the determining foundation of all social conditions.[56]

Thus Darwinian science was more than just peripheral to her pacifist concerns. She even agreed with Darwinists that one of the eternal laws of evolution is struggle, but she pointed out that struggle merely means competition, not war.[57]

For Suttner the triumph of pacifism was assured by the Darwinian struggle for existence through

> a gradual extermination of the belligerent tribes by peace-loving nations; an extinction of ethnic hatred through the increase of cosmopolitan ideas; a reduction of military honors in face of the growing glory of knowledge and the arts; an ever closer fraternizing league of world interests vis-à-vis the petty, vanishing special interests—and in this manner the prize of eternal peace can and will be achieved through the eternal struggle, which follows [natural] law.[58]

Evolutionary progress through the extermination of "primitive" peoples and races was a common theme among Darwinists of the late nineteenth and early twentieth centuries, but it comes as a shock to see it in the writings of such a devoted pacifist. Nonetheless, Suttner rejected a conscious policy of racial extermination, since she thought this would be stooping to the barbaric level of the "lower" races. She did, however, mention that Europeans had every right to defend themselves from these "wild people."[59]

Like Suttner, Fried was enthusiastic about the Darwinian revolution, calling it greater than the French Revolution in its effects.[60] He also had close connections with leading eugenicists, corresponding with Ploetz and Grotjahn. He even recommended Ploetz's eugenics journal to the American pacifist and eugenicist David Starr Jordan.[61] Eugenics was not only a theoretical interest for Fried, but a very personal issue, for in 1896 after a brief engagement he married, only to discover that his bride had severe mental problems. In 1904 he asked Ploetz to use his connections to help him secure a divorce, but Ploetz's help was in vain. Later he anonymously wrote a book describing his horrific experiences.[62] He also had an epileptic brother, whose physician, Grotjahn, promoted eugenics in letters to Fried.[63]

Despite his sympathy for eugenics, Fried rarely, if ever, used eugenics arguments to defend pacifism, which seems odd, since concern over "military selection" was a commonly used argument against war. Perhaps he recognized the superficiality of any pacifism founded on eugenics. In any case, he often used Darwinian rhetoric in other ways to advance his cause. He confessed that struggle is indeed universal, but—like Suttner—he denied that struggle means war. The lesson he drew from evolution was more Spencerian than Darwinian (again, following Suttner's lead): the ineluctable advance to ever higher levels of organization, which would culminate in a world government.[64] Fried remained true to his pacifist convictions, even under intense pressure during World War I.

However, despite his commitment to pacifism, Fried, like many of his contemporaries, was not quite sure how to apply pacifism to race relations. He clearly believed in the racial superiority of the European races, arguing that they had every right to colonize other parts of the earth. Before the Hague Peace Conference of 1899 he even suggested that Europeans should divide China among themselves, in order to avoid friction on this issue at the conference.[65] At times he insisted that colonization should occur peacefully. A united Europe could, he believed, colonize without bloodshed.[66] He also welcomed the Algeciras Conference in 1906, which would "regulate the civilizing exploitation of Morocco, of a people needing tutelage." He called the "subjugation of lower peoples under the leadership of the higher cultured peoples" a "task in the service of culture."[67]

However, some passages in his books betray less peaceful intentions toward "lower" races. In 1905 he reacted against the accusation that pacifists are pressing for an unrealistic world peace. He claimed that, on the contrary, pacifists recognize that pacifism is only possible among peoples who have reached a high level of culture. "The peace movement," he continued,

> by no means dreams of a possible ideal condition after centuries, where Germans and Botokunden, Frenchmen and Persians, Englishmen, Turks, and Bushmen will in peaceful harmony enjoy a time without wars based on legal principles. The peace movement supports itself consciously on the experiences of history, wherein war is a culture-promoting factor in the life of peoples up to a certain stage of cultural development.[68]

Fried shut out non-European races from the legal community of nations that he was advocating, which made their position rather perilous. To be sure, Fried favored peaceful exploitation, but he clearly asserted the right of Europeans to defend themselves against "uncivilized peoples standing outside the legal community."[69] That Fried, an ardent pacifist, supported European colonization and the subjugation of non-European peoples by Europeans, shows the power of racism—especially scientific racism—on the European mind in the late nineteenth and twentieth centuries.

Fried, Suttner, and Dodel represented a small fraction of thinkers trying to apply Darwinian science to social questions, and most Darwinists ultimately rejected their position. Some—those I call Darwinian militarists—opposed them outright, arguing that Darwinism proved the inevitability of violent conflict. Others—advocates of peace eugenics—sometimes expressed sympathy with the peace movement, but often their support for pacifism was rather shallow. They only opposed certain kinds of wars, particularly wars between European nations, where the brightest and best mowed each other down without regard to their biological traits. This kind of war, they claimed, would lead to biological degeneration.

But what about wars where "inferior" people die at higher rates than others? Even most proponents of peace eugenics had no objections to these. As long as "inferior" people—the weak and the sick—die in war, the war is allegedly beneficial. This alone is a shocking demonstration of the devaluing of human life by naturalistic Darwinists. But matters become even more serious once we turn our attention to the other category of people social Darwinists and eugenicists consistently labeled "inferior": those of non-European races. Many social Darwinists and eugenicists—even many calling themselves pacifists—justified or even advocated the extermination of "inferior" races, a topic we explore in depth in chapter 10.

10. Racial Struggle and Extermination ✍

S ince most Darwinists included war as one form of the human struggle for existence, the cross-pollination of scientific racism (which, as we have seen, had close links to Darwinism) with the Darwinian struggle for existence would bear bitter fruit in an era of renewed European imperialism. While many Christian missionaries and liberal humanitarians hoped to imbue the natives in other parts of the world with European culture, scientific racism proclaimed the futility of such endeavors. Instead, scientific racism suggested a different path to progress. In the 1896 edition of Friedrich Hellwald's magisterial four-volume *History of Culture*, Rudolf Cronau, relying on Social Darwinist arguments, dismissed the idea that the "lower races" could be elevated:

> The current inequality of the races is an indubitable fact. Under equally favorable climatic and land conditions the higher race always displaces the lower, i.e., contact with the culture of the higher race is a fatal poison for the lower race and kills them. . . . [American Indians] naturally succumb in the struggle, its race vanishes and civilization strides across their corpses. . . . Therein lies once again the great doctrine, that the evolution of humanity and of the individual nations progresses, not through moral principles, but rather by dint of the right of the stronger.[1]

Cronau—along with a host of leading scientists, physicians, and social thinkers who embraced Darwinian social explanations and eugenics in the late nineteenth and early twentieth centuries—thus argued that the key to progress was the annihilation of the "lower races," who stood in the way of advanced culture and civilization, because they were mentally incapable of creating and adopting "higher culture."

Furthermore, a number of social Darwinist thinkers argued that racial extermination, even if carried out by bloody means, would result in moral progress for humanity. Brutality would not necessarily triumph in the struggle for existence, since, as Darwin had argued in *The Descent of Man*, morality conferred a selective advantage.[2] Therefore, according to this twisted logic, since Europeans were morally superior to other peoples, the extermination of other races would rid the world of immorality. The University of Leipzig geographer, Alfred Kirchhoff, articulated this position as clearly as any in his posthumously published work, *Darwinism Applied to Peoples and States* (1910). He justified racial extermination as part of the Darwinian process, since "the righteousness of the struggle for existence, cool to the core, wills it." This statement conferred scientific sanctity to the harsh realities of

> the struggle for existence between the peoples, [which] causes the extermination of the crude, immoral hordes. . . . Not the physically strongest, but the [morally] best ones triumph. If there were not a diversity of peoples, if there were no international rivalries, where would be the guarantee for the preservation of the fitness of the peoples, not to mention for the progress of humanity?[3]

Thus, Kirchhoff—who was by no means alone in this matter, as we shall see—not only granted racial extermination the status of scientific inevitability, but he also made it seem righteous and noble.

The significance of discussions about racism and racial extermination in Germany in the pre–World War I era can hardly be overstated. Whether imperialism was a cause or an effect of racism (or both) can be debated, but racism undeniably served as a justification for German colonialism, especially its more oppressive features. Germany became infamous for crushing native uprisings, culminating in a controversial decree to annihilate all Hereros—men, women, and children—in German Southwest Africa (today Namibia) in 1904 during the Herero Revolt. Some scholars consider the German response to the Herero Revolt the first attempted genocide of the twentieth century. Also, Germany's foreign policy, especially the rhetoric surrounding Wilhelm II's controversial "world policy" (*Weltpolitik*) and concerns about the "Yellow Peril," cannot be understood outside the context of race discourse. Finally, Hitler's social Darwinist ideology—which developed in the years just around World War I—prominently featured racial conflict and racial extermination, and not only in relation to the Jews. To be sure, Jewish extermination was one of Hitler's highest priorities, but the extermination of other races was also part of his long-range agenda.

The late nineteenth century, especially the period 1890–1914, was an important turning point for discourse on racism in general and racial extermination in particular. Heinz Gollwitzer in his study on *The Yellow Peril* notes that in the final decade of the nineteenth century "the terminology of the racial struggle (*Rassenkampf*) was disseminated widely in the German press."[4] Also widespread by that time was the idea that races are locked in a Darwinian struggle for existence that determines the destiny of humanity. This struggle, whether carried on through ruthless bloodshed or by more peaceful means, would ultimately result in the annihilation of the "inferior" races. Even some German pacifists, while objecting to warfare among European nations, saw nothing wrong with exterminating those races they deemed inferior. In fact, some pacifists wanted to avoid wars in Europe for the express purpose of winning the racial struggle for world supremacy. In their view, Europeans should turn their attention away from killing each other and instead concentrate on annihilating the "inferior" races. Ideas about racial extermination were not unique to Germany, but became very influential elsewhere also. H. G. Wells epitomized an influential Anglo-American social Darwinist attitude when he stated that "there is only one sane and logical thing to be done with a really inferior race, and that is to exterminate it."[5]

Before the late nineteenth century, many atrocities occurred in the clash between Europeans and natives of other parts of the world. We all know that Christianity failed in practice to stop the extermination of many tribes and peoples (though, to be fair, some Christian leaders, especially missionaries, protested against the oppression of natives by Europeans). However, under the dominance of Christian thinking, racial extermination could never develop into a coherent ideology.[6] The increasing secularization of European thought in the nineteenth century, in tandem with Darwinian ideas, opened the door for racial extermination by sweeping aside traditional (Christian) moral objections and by reducing humans to soulless organisms. Darwinism also contributed to the growth of scientific racism, as we have already seen.

Discourse about racial extermination drew heavily on Darwinian concepts, especially the struggle for existence. Darwin's idea of natural selection through the struggle for existence was derived from Malthus's view that organisms have a biological tendency toward overpopulation, causing most organisms to perish before reproducing. Thus, the mass destruction of organisms, including humans, was, according to Malthus and Darwin, inevitable. Social Darwinists consistently stressed population pressure as a continual source of human conflict, including racial conflict. The population expansion in Germany and most of Europe in the nineteenth century lent plausibility to Malthus's idea about the tendency of humans to reproduce faster than their food supply. As the population expanded, Germans were emigrating in large numbers, leading some German scholars and politicians to begin pressing for colonies as an outlet for German emigration. This migrationist colonialism with its call for German colonial settlement would necessarily bring Germans into conflict with native peoples currently inhabiting those lands.[7]

The Darwinian struggle for existence between humans, resulting from the Malthusian imbalance between reproduction and the food supply, could take place on an individual level, but many nineteenth-century thinkers stressed national or racial competition as much or more than individual competition. Darwin himself stressed individual competition more, but he resorted to group competition to explain the origin of social instincts, altruism and morality in social animals and humans. Darwin clearly believed that the struggle for existence among humans would result in racial extermination. In *Descent of Man* he asserted, "At some future period, not very distant as measured by centuries, the civilised races of man will almost certainly exterminate and replace throughout the world the savage races."[8] In 1881 Darwin warned W. Graham against underestimating the efficacy of natural selection in advancing human civilization, since "The more civilised so-called Caucasian races have beaten the Turkish hollow in the struggle for existence. Looking to the world at no very distant date, what an endless number of the lower races will have been eliminated by the higher civilised races throughout the world."[9] Darwin once wrote a note that "natural selection is now acting on the inferior races when put into competition with the New Zealanders."[10]

As the leading Darwinist in Germany, Haeckel not only became a leading spokesman for scientific racism, but he also stressed the human struggle for existence and interpreted racial struggle as a significant part of it. Like Darwin, Haeckel believed that Malthus's population principle was essentially correct, meaning that many humans necessarily die without reproducing. Already in 1863, in his first public declaration of support for Darwinism, Haeckel insisted that humans, just like all other organisms, are engaged in a struggle for existence.[11] In *The Natural History of Creation* he clarified that the struggle for existence is most intense among members of the same species, and he explicitly included humans. Examples Haeckel provided for the human struggle for existence were normally peaceful. He stressed economic competition rather than violence, but in the final analysis it was still neighbor against neighbor and the unfit must ultimately perish.[12]

What were the implications of Haeckel's view of the human struggle for existence for racial competition? Based on his stance on racial inequality, the Malthusian population principle, and the human struggle for existence, it would seem to follow that one aspect of the human struggle for existence would be racial competition that would lead to the extermination of "inferior" races. Indeed this is precisely the position Haeckel articulated. In *The Natural History of Creation* he wrote that Europeans were taking over the whole world, driving other races, such as the American Indians and Australian aborigines, to extinction. "Even if these races were to propagate more abundantly than the white Europeans," he stated, "yet sooner or later they would succumb to the latter in the struggle for existence."[13] Later in the same book he explained that racial competition is part of the struggle for existence, this time including African races among those succumbing to Europeans.[14]

With these views of racial struggle, it should come as no surprise that Haeckel supported German colonialism. He was a founding member of the Pan-German League in 1890–91, and many of his writings—including those written during his participation in pacifist activities—are sprinkled with comments supportive of German colonialism.[15] In his 1917 book, *Eternity*, he expressed horror that Germany's enemies in World War I were using non-European troops against them, calling this an "underhanded *betrayal of the white race*." These members of "wild" races simply did not have the same value as Europeans, according to Haeckel: "A single well-educated German warrior, though unfortunately they are now falling in droves, has a higher intellectual and moral value of life than hundreds of the raw primitive peoples, which England and France, Russia and Italy set against us." He predicted this fraternizing with other races would damage European authority in their colonies. Haeckel also made it clear that he supported German annexations in Europe and colonial acquisitions in Africa.[16]

Haeckel was one of the earliest and certainly the most influential Darwinist to argue that racial extermination was a natural, unavoidable consequence of the human struggle for existence. In the 1860s to the 1890s a number of prominent Darwinian biologists, ethnologists, and other social thinkers took up the same theme in their writings. Though some stressed the peaceful nature of this racial competition-to-the-death, some admitted that the struggle often produced violent warfare. A few even seemed to glory in the brutality of the racial conflict, though many expressed regret at the inevitable extermination of the "inferior" races. These Darwinists all agreed, however, that the extermination of the "inferior" races was on the whole, a positive development leading to progress.

The earliest example I have found in German sources of Darwinian justifications for racial extermination, appeared less than a year after Darwin published his theory. Oscar Peschel, editor of *Das Ausland*, the only scholarly journal on geography and ethnology in mid-nineteenth century Germany, began promoting Darwinian theory immediately after Darwin's *Origin of Species* appeared. In 1860 he wrote an article on races, in which he argued that the two highest races are Caucasians, because they are the most intelligent, and black Africans, because they are the best-adapted race for the tropics. In the face of competition from whites and blacks, other races were dying out. The American Indians and Polynesians "could not be saved, their time had come, as soon as a white face appeared." Peschel exonerated Spaniards for slaughtering Indians, claiming that it was not brutality—they were merely following natural law. He stated,

> This is the historical course. If we view it with the eye of a geologist, and indeed a geologist which accepts the Darwinian theory, we must say that this extinction [of human races] is a natural process, like the extinction of secondary animal and plant forms [in earlier geological eras].[17]

He thus accepted all the human death and misery of racial extermination with equanimity, and legitimated his stance with an appeal to nature and science.

Later, in his major work on ethnology, Peschel discussed the reasons for racial extinction, which, he believed, would inevitably befall all hunting and gathering peoples as well as some nomadic peoples. "This paleontological occurrence should not be mysterious to us," he stated. But his own explanation for racial extinction still seems rather mysterious, for Peschel suggested that it was not the result of European brutality, nor disease nor alcohol. Rather the natives had simply lost the desire to live any longer. Faced with contact with "higher" races, the "lower" races simply stopped reproducing, sometimes even killing their offspring. Thus they brought about their own death.[18]

Peschel's position resonated with Hellwald, Peschel's successor as editor of *Das Ausland* and, like Peschel, a Darwinian ethnologist. In his 1880 book, *The Natural History of Humans*, Hellwald called the decline of the Australian aborigines, which he and many contemporaries considered the lowest race in the world, "race suicide," since allegedly they had simply stopped reproducing. Hellwald showed no regret for the Australian aborigines, Tasmanians, or other races he consigned to extinction. Instead, he quoted with approval from an 1870 article by Peschel:

> Everything that we acknowledge as the right of the individual will have to yield to the urgent demands of human society, if it is not in accord with the latter. The decline of the Tasmanians therefore should be viewed as a geological or

paleontological fate: the stronger variety supplants the weaker. This extinction is sad in itself, but sadder still is the knowledge, that in this world the physical order treads down the moral order with every confrontation.[19]

Nature thus trumps ethics every time, so there is no point in decrying natural processes like racial extinction.

In his earlier major work, *The History of Culture* (1875), which he dedicated to Haeckel, Hellwald was even more callous in his treatment of racial extermination. He derided all ethical considerations, maintaining that the ends justify the means and that "in nature only one law rules, which is no law, the law of the stronger, of violence."[20] He thoroughly agreed with the social Darwinist writer, Robert Byr, whom he quoted: "Whoever it may be, he must stride over the corpses of the vanquished; that is natural law. Whoever shrinks back in hesitation from this, deprives himself of the chance for existence."[21] Like most Darwinists, Hellwald considered the extermination of "inferior" races a natural development necessary to bring progress, but unlike many of his contemporaries, he seemed to revel in the brutality of that struggle.

While denying that humans have any innate conception of laws or rights, Hellwald insisted that they have aggressive instincts driving them—together with the Malthusian population imbalance—inevitably toward war and racial conflict. The Spaniards' brutal massacre of American Indians was an example of this racial struggle for existence, and "this conquest itself has been an inexpressible blessing" (see figure 10.1). He also asserted that the British were foolish to end slavery, which would only result in "the extinction of the Negroes," who can survive in slavery, but cannot compete with the white races in free competition.[22] In a later book he blamed the American Indians—whom he called beasts of prey—for fighting a war of destruction against the "more civilized" white race. Hellwald made it sound as though the whites were merely defending themselves against wild beasts when they annihilated the Indians.[23]

Another early disciple of Darwin, the geologist Friedrich Rolle, who founded a short-lived journal in 1867 dedicated to promoting Darwinism, upheld views of racial extermination similar to those of Peschel and Hellwald. In his book on human evolution he discussed at length warfare and racial competition as part of the human struggle for existence, in which the inequality of races plays a prominent role. He considered the black Africans much closer than white Europeans to apes, both physically and mentally. Rolle specifically mentioned the American Indians, Australian aborigines, and black Africans as lower races dying out in competition from white Europeans. He brushed aside any moral considerations, since the

Figure 10.1   Illustration in Friedrich Hellwald's *Kulturgeschichte* (4th ed., 1896), with caption: "Spaniards allow their dogs to rip up Indians"

"right of the stronger" is not subject to morality. Rather, "In the contest between peoples the proverb was and always will be valid: 'Better that I smash you, than you me.'" Besides, even if the destruction of the lower races is sad in some ways, he argued, it is advantageous for the survivors of the struggle and thus results in progress.[24]

Another Darwinian biologist to endorse racial extermination was Oscar Schmidt, professor of zoology at the University of Strassburg. In the 1870s Schmidt wrote and lectured about the human applications of Darwinism, claiming scientific validity for human inequality and the struggle for existence. He asserted that natural selection "is a pure question of might," not right.[25] Because he viewed some races as mentally inferior, he concluded:

> If we contemplate the ethnology and anthropology of savages, not from the standpoint of philanthropists and missionaries, but as cool and sober naturalists, destruction in the struggle for existence as a consequence of their retardation (itself regulated by the universal conditions of development), is the natural course of things.[26]

Schmidt thus used science to displace ethics and sympathy, transforming racial extermination from an atrocity into an inevitable—and by implication, beneficial—phenomenon.

Much more influential than Rolle or Schmidt was Ludwig Büchner, whose views on racial struggle reached a larger audience. Despite his opposition to war, Büchner continued to believe that races were locked in a Darwinian struggle for existence that would ultimately result in the annihilation of "inferior" races. His position on racial inequality and racial competition was similar to Haeckel's early views. In the early 1870s he wrote, "The white or Caucasian human species [!] is ordained to take dominion of the earth, while the lowest human races, like Americans, Australians, Alfuren, Hottentots, and such others, are proceeding toward their destruction with huge steps." Büchner was not as explicit as Rolle, but he implied that racial extermination is a positive development.[27]

So how did Büchner square his opposition to war with his view on racial extermination? I do not know for sure, for despite often discussing both issues separately, he never explained the relationship between war and racial competition. There are two ways that Büchner could have answered this question, however, both of which were upheld by some of his contemporaries. He could have held that: (1) "inferior" races would die out through peaceful competition; or (2) warfare was justified against "inferior" races, but not against other "civilized," that is, European, peoples.

Opting for the former explanation was Ernst Krause (usually writing under the pseudonym Carus Sterne), who behind Haeckel and Büchner was probably the most influential popularizer of Darwinism in Germany. Krause earned a doctorate in botany in 1866, but academic appointments were scarce in late nineteenth-century Germany, and he turned to writing about science, especially Darwinism, for a popular audience. He also edited *Kosmos*, a scientific journal devoted to Darwinism that enjoyed the support of leading Darwinian biologists, including Haeckel and Darwin himself. Krause took the same view of racial inequality as Haeckel, admitting that the "inferior" races were dying out as a result of contact with the more "civilized" races. However, this was not so much the result of bloody conflict as peaceful competition, as the example of the Tasmanians proved. The British even tried to save the Tasmanians, according to Krause, but in vain. "For the scientist this highly painful drama had great interest, inasmuch as it showed him how the struggle for existence in some circumstances entirely loses the character of violence and yet just as infallibly favors the rise of the more capable race."[28] Krause remained rather vague about the cause of the Tasmanians' deaths, calling it mysterious, but somehow ascribing it to the struggle for existence made it seem more explicable.

Thus, according to Krause, Europeans were only indirectly responsible for the death of "lower" races, whose extinction would widen the gap between humans and apes.

By the end of the 1870s quite a few social Darwinists had already discussed racial extermination in their writings, but in the 1880s two new currents of thought would emerge to foster social Darwinist racism even more in the German academy. First, law professor and sociologist Ludwig Gumplowicz developed his Darwinian-inspired sociology of group conflict, focussing on *Racial Struggle*, the title of his 1884 book. Second, the geographer Ratzel began publishing his theory of *Lebensraum* (living space), which, though more subtle in its treatment of racial struggle than Gumplowicz's theory, nonetheless served imperialists well as a justification for racial conquest.

Gumplowicz's ideas were not radically new. As we have seen, many social Darwinists had already presented racial conflict as a part of the Darwinian struggle for existence, just as Gumplowicz did. But Gumplowicz systematized and elaborated on many of the ideas we have already examined, placing racial struggle at the center of his analysis. He stated, "The *racial struggle for dominion* in all its forms, whether open and violent, or latent and peaceful, is therefore the actual *driving principle, the moving force of history.*" Gumplowicz believed that racial hatred was ingrained in the human breast, manifesting itself in racial conflict, including violence and bloodshed; violent enslavement and extermination of races is simply a part of the natural order.[29] Where Gumplowicz differed from most social Darwinists of his time—and even more so from Gobineau and his German adherents—was by defining race as a sociological category, not a biological category.[30] As a Polish Jew teaching at the German-speaking University of Graz in the multiethnic Austro-Hungarian Empire, Gumplowicz had firsthand knowledge of ethnic rivalries. By presenting ethnic hostility as inevitable and "scientific," Gumplowicz provided further grist for the mill of racist imperialists. Gumplowicz's term "racial struggle" became firmly entrenched in race discourse by the early twentieth century.

Unlike Gumplowicz, Ratzel rarely used the term "racial struggle" in his writings, though in his university lectures he told his students that since races are no longer physically separated by natural boundaries in the modern world, "the world is everywhere filled with racial struggles."[31] The concept of racial struggle, however, resonated quite well with his concept of *Lebensraum* (living space), which became a powerful intellectual construct supporting imperialism and the extermination of "primitive peoples," as Ratzel and many of his contemporaries called them. In *Being and Becoming* Ratzel argued that the extermination of "primitive peoples" by Europeans was a powerful example of Darwinian natural selection in operation:

> And can we any longer doubt the existence of natural selection, when we read, how the last remnant of primitive peoples melt like snow in sunshine, as soon as they come into contact with the European, and how the European peoples for three centuries have populated entire large continents.[32]

Ratzel's geographical theory, which he hoped would become "nothing less than the foundation of a new theory of humanity," focussed on human migrations and the "struggle for space" (*Kampf um Raum*), or as he later termed it, *Lebensraum*.[33] Ratzel derived this idea from his friend, Moritz Wagner, whom he considered second only to Haeckel in contributing to evolutionary theory among German scientists.[34] Wagner was a geographer and ethnologist who devoted much of his career to formulating a theory of biological evolution, featuring migration as the key mechanism behind speciation. Wagner had already used the phrase "struggle for space" in 1868 (actually, Rolle had used it two years earlier, but it is unclear if Wagner derived it from him).[35] Though Wagner saw his migration theory as an alternative to Darwin's theory of natural selection, most Darwinists incorporated his insights on migration into Darwinian theory. There was no reason, after all, why migration and the struggle for existence could not both have validity (we must remember that other mechanisms, such as Lamarckism, were often synthesized with Darwinian selection in the late nineteenth century, too).

Ratzel claimed that his concept of the struggle for *Lebensraum* was simply more accurate terminology to describe the Darwinian struggle for existence. In his book entitled *Lebensraum* he overtly applied the concept to humans, noting on the first page that biogeography must include anthropo-geography. He gave examples of struggles between human groups, such as the extermination of the American Indians by Europeans. Thus Ratzel's concept of *Lebensraum* clearly included the extermination of "less civilized" peoples or non-European races.[36] With Ratzel's stress on population expansion forcing people to migrate, together with his emphasis on controlling land, it is no surprise that he advocated German colonialism, both in his writing and in his participation in the Pan-German League.[37] Ratzel was purposely subtle in his writing, confessing in 1897 to a colleague,

> The main idea in this political geography cannot be expressed, i.e., not in print. Briefly I can formulate it thus: If we Germans are to take the great position in humanity determined for us, then we must become clear about

what a people (*Volk*) on this earth *can* and *should* do by virtue of its situation.[38]

Ratzel thus intended his book, *Political Geography*, to stir up his fellow Germans to colonize other parts of the globe, even though he rarely mentioned a specific role for Germany.

He did, however, stress the need for European colonization: "For all rapidly expanding peoples, finding a place for their emigrants is a matter of life and death and will become the occasion for political difficulties for those who have no large colonies at their disposal." But how should this colonization proceed? Ratzel discussed three main ways to colonize other lands: (1) by allowing the indigenous populations to control the land and establishing trade; (2) by subduing the native populations and exploiting their labor; or (3) by Europeans taking over the land and farming it. Since the whole purpose of colonization was to find new living space for an expanding population, he favored the third form as the most effective. He also left no doubt about the need for wars of conquest, which "quickly and completely displace the inhabitants, for which North America, southern Brazil, Tasmania, and New Zealand provide the best examples." While Ratzel never relied on biological racism to support the extermination of other peoples, he did present the conflict between different peoples as part of a Darwinian "struggle for space."[39] Thus both Ratzel and Gumplowicz, though treating race as a cultural or sociological category more than a biological one, both presented the extermination of "less civilized" peoples as a natural consequence of the Darwinian struggle for existence.

Gumplowicz's idea of racial struggle resonated with Gustav Ratzenhofer, a retired Austrian military officer with a naturalistic worldview. He became a leading disciple of Gumplowicz and developed a systematic sociology based on Darwinism and racial struggle. Gumplowicz called Ratzenhofer a genius, whose work was even more complete than his own.[40] Ratzenhofer, drawing on both Gumplowicz's and Ratzel's social theories, believed politics was driven by population pressure, causing a Darwinian struggle for expansion: "Gaining territory (*Raum*) is fundamentally the purpose of every political struggle."[41] Since he believed that races varied considerably in their physical and especially mental abilities, in the racial struggle for existence some would triumph and exterminate their competitors, thereby increasing the distance between humans and apes.[42] He prophesied that "the civilized races will annihilate the less valued ones in the former colonial territories." Ratzenhofer dismissed any attempt to apply ethical standards to politics or the racial struggle. Ethics and politics, he said, simply don't mix. In fact, like many fellow Darwinists, he considered ethics the product of the human struggle for existence, especially the racial struggle.[43] After 1900 Ratzenhofer also came under the influence of Chamberlain, espousing Aryan racism and anti-Semitism. He believed that ultimately most races on the globe would be exterminated, leaving only the "superior" Aryan race to compete with the Jews.[44]

Some imperialists took Darwinian rhetoric to heart and used it to support ruthless colonial policies. Karl Peters, famous for bringing German dominion to Southeast Africa (presently Tanzania), believed that human history was subject to natural laws, including Darwin's law of natural selection through the struggle for existence. He argued that "if Darwin is right, then perhaps a time would be conceivable, in which they [Germans and British] would be the sole lords of this earth." Though Peters did not explicitly endorse racial extermination, his thinking seems to push in that direction. For him colonial policy was aimed not at elevating indigenous peoples, as some imperialists claimed, but rather "the ruthless and resolute enrichment of one's own people at the expense of the other weaker peoples." Further, he saw colonies as outlets for German population expansion, and claimed that in the future only the Indo-Germanic and the East Asian races would survive.[45] Taken as a whole, his views imply the ultimate extermination of Africans and other "inferior" races by Europeans.

By the 1890s and early 1900s Darwinism had become well-entrenched in Germany. Racial theorizing, most of which was laced with Darwinian rhetoric, was heating up, capturing the imagination of ever wider audiences. Earlier most discussions of racial struggle and extermination were tucked away in brief passages in longer articles or books on various topics (Gumplowicz was an exception), but in the 1890s and especially after 1900 there was a proliferation of books and articles discussing racial struggle. For some thinkers race became the universal key to interpreting history, society, and culture.

For many of those applying race struggle to history, racial degeneration became the favorite explanation for the decline of past civilizations. The historian Otto Seeck in his book, *The History of the Downfall of the Ancient World*, explained that the Greeks and Romans had declined because they had eliminated the biologically best people through civil wars, internal violence, and corruption. Biologically inherited cowardice flourished late in the Roman Empire, according to Seeck, leading it to ruin.[46] While Seeck focussed on internal reasons for racial decline, many race theorists believed that the downfall of Greek and Roman civilization occurred because of racial mixture with "inferior" races.

Eugenics discourse would also at times help in the dissemination of ideas about racial extermination. We have already seen that eugenicists preached biological inequality and campaigned to rid the world of "inferior" (*minderwertig*) or "degenerate" people. Most eugenicists concentrated on regenerating their own nation through measures either encouraging, discouraging, or even prohibiting reproduction of certain individuals, depending on their perceived biological value. However, even when race was not a primary consideration, it often lurked in the background. Biological improvement of Europeans would give them a greater advantage in the struggle against other races, while biological degeneration—which many eugenicists feared was occurring—might lead to disaster for Europeans in the global struggle for existence.

Woltmann, like other social Darwinists we have examined, believed not only in racial inequality, but also that races were locked in an eternal struggle for existence that led to the extinction of "lower" races. Woltmann admitted that this process occurred first and foremost through "the war of extermination." He believed that the Germanic race, as the only culture-producing race, would eventually conquer the globe: "The Germanic race is called to encompass the earth with its dominion, to exploit the treasures of nature and the labor forces, and to make the passive races serving members of their cultural development." Unlike some of his contemporaries, however, Woltmann did not call on his fellow Germans to unite to fight the struggle for existence against other races. On the contrary, he considered the racial struggle among different branches of the German race beneficial, too, for

> the German is the greatest and most dangerous enemy to the German. To banish this enmity from the world would mean to abolish the development of culture at its foundation; it would be a childish endeavor to break the laws of nature through dreaming.[47]

For Woltmann, then, unlike many earlier social Darwinists, racial struggle even included in-fighting within the Germanic race.

Lapouge, a member of Woltmann's circle, also viewed history as a process of Darwinian natural selection and forthrightly stated that the destruction of millions of people of "inferior" races was beneficial for the future of humanity. In an 1887 essay Lapouge stated,

> In the next century people will be slaughtered by the millions for the sake of one or two degrees on the cephalic index [i.e., cranial measurements popular with physical anthropologists]. . . . the superior races will substitute themselves by force for the human groups retarded in evolution, and the last sentimentalists will witness the copious extermination of entire peoples.[48]

Jennifer Michael Hecht argues that Lapouge was not actually advocating mass extermination in this passage, but rather calling for measures to avert future extermination. However, on the whole, it seems Lapouge did favor racial extermination. Hecht implies this, in any case, for she presents him as a forerunner of the Holocaust and stresses not only his Aryan racism, but his "anti-morality" and his attack on the prohibition against killing.

Another member of Woltmann's circle, Ammon, also regarded the struggle between races a crucial and inescapable part of the struggle for existence. He exulted in war as a part of the Darwinian struggle for existence:

> In its complete effect war is a *good deed* for humanity, since it offers the only means to measure the powers of nations and to grant the victory to the fittest. War is the highest and *most majestic* form of the struggle for existence and cannot be dispensed with, and thus also cannot be abolished.[49]

Ammon not only theorized about war; he also joined the Pan-German League and sponsored meetings promoting German navalism (and thus imperialism).[50] In 1900 he wrote an article arguing that because of population expansion, the white races must gain new space: "The inferior races (blacks, Indians) would thereby succumb in the struggle."[51] He also considered mixed races inferior and thereby doomed to failure and "destruction through the struggle for existence, for they were only produced as an unavoidable byproduct of the process producing better [races]."[52] Not only did Ammon promote the idea of racial struggle in his writings, but he also tried to recruit the Austrian racial theorist Matthäus Much to write articles for a Baden newspaper on the racial struggles in the Austro-Hungarian Empire.[53]

Though Woltmann and his circle were outsiders to academic circles, they exerted considerable influence, both inside and outside the universities, despite the opposition they encountered. Many anthropologists in the academic community in the first two decades of the twentieth century, especially the younger generation, adopted social Darwinist racism and considered racial struggle and racial extermination inevitable. The prominent anthropologist Eugen Fischer, a follower of Woltmann's racial theories, in his 1908 study of the so-called "bastards" of Rehoboth (descendants of European fathers and black African mothers in German Southwest Africa) maintained that in free competition with Europeans the blacks and half-blacks would perish. How then should Europeans treat these "inferior" people? Fischer recommended strict segregation from the natives and treating natives in whatever way is most beneficial to Europeans. For the time being, it may be wise to allow them to increase their population, so they can provide labor for Europeans, he thought. But once they are no longer useful, he asserted, they must be done away with. He stated,

> Therefore one should guarantee to them only the measure of protection that they *need* as a race inferior to us, in order to survive, but no more, and only so long as they are useful to us—otherwise [allow] free competition, which in my opinion means [their] demise! This viewpoint sounds almost brutally egoistic—but whoever thinks through the racial concept and the points portrayed in the above section on "Psychology," cannot take any other view.[54]

Thus Fischer claimed the authority of his scientific investigations for a brutal policy of colonial exploitation and even racial extermination.

Even those anthropologists who resisted Woltmann's race theories did not always escape the lure of social Darwinist ideology. The first professor of anthropology at the University of Berlin, Felix von Luschan, was liberal enough in his racial views that he was invited to address the Universal Race Congress in 1911, a meeting committed to fostering reconciliation among races and peoples.[55] They were in for a surprise. Luschan began by admitting that no race is inferior to another, but later warned against allowing "coarser or less refined elements," such as blacks, Asians, and even Eastern Europeans, to immigrate into "civilized" nations. Near the end of his speech he opposed the very purposes of the congress he was addressing, stating,

> The brotherhood of man is a good thing, but the struggle for life is a far better one. Athens would never have become what it was, without Sparta, and national jealousies and differences, and even the most cruel wars, have ever been the real causes of progress and mental freedom. As long as man is not born with wings, like the angels, he will remain subject to the eternal laws of Nature, and therefore he will always have to struggle for life and existence. No Hague Conferences, no International Tribunals, no international papers and peace societies, and no Esperanto or other international language, will ever be able to abolish war. . . . natural law will never allow racial barriers to fall . . . Nations will come and go, but racial and national antagonism will remain.[56]

So much for racial reconciliation.

Luschan was not alone, for the two leading theorists of social Darwinist militarism, Steinmetz and Klaus Wagner, upheld similar views. Steinmetz argued that wars of all kinds, whether between "civilized" peoples or between different races, are selective and thus bring progress. He explicitly discussed the contest between Europeans, Asians and Africans, which will be decided not by peaceful competition, but by war. He welcomed racial conflict, calling for "Selection and extermination in Asia and Africa and between both of them and the old, tired, peace-loving Europe!"[57] Wagner's stance was even more extreme, for as a follower of Woltmann and his circle, race was his primary concern. In the contest between Europeans, Asians, and Africans, he asserted,

> Only one group can remain as ruler. The two others will be destroyed, where they are in the way of the stronger race, and enslaved, where they can serve them. . . . We Germans have the power to destroy and smash the might and future of the two other groups, if we clearly see this necessity, vigorously arm ourselves, and keep our blood pure . . .[58]

As we can see with Wagner, Steinmetz, and Fischer, rhetoric about racial extermination—horrible as it was already in the 1860s to the 1870s—was becoming increasingly inflammatory.

It was also disseminating widely. Prominent professors, like Max von Gruber, a leading hygienist and eugenicist, were advocating German population expansion and imperialism to wrest land away from indigenous peoples. Gruber dismissed racial equality as a fiction, and explained racial antipathy as a natural consequence of racial differences. Each ethnic group, such as the French or Slavs, sees nothing wrong with "tromping on such low worms" (i.e., other ethnic groups), if it is advantageous to them. Gruber strongly implied that he saw nothing wrong with "tromping on" other peoples and races, since in his world of Malthusian scarcity, one could only choose between trouncing others or being trounced.[59] Other figures on the fringe of the eugenics movement, like Hentschel and Driesmans, were intensely racist and did not hesitate to discuss the extermination of the "inferior" races.[60] Viennese occultist race theorists, led by Guido von List and Jörg Lanz von Liebenfels, likewise warned about the threat of other races and the blessings of their elimination.[61] All these figures used social Darwinist arguments to justify or even advocate racial extermination.

It is not surprising that Darwinian militarists justified racial extermination. What seems incongruous, however, is that some pacifists and proponents of peace eugenics, who often criticized the views of social Darwinist militarists, also advocated racial extermination. This position is only intelligible once we understand that most German pacifists were not absolute pacifists and opposed war only in its modern European form. Many pacifists considered population expansion a legitimate cause for colonization, but some, like Otto Umfrid, a leading figure in the German Peace Society, believed colonization could be a peaceful process—Europeans should only move to areas of low population density.[62] But not all German pacifists were committed to peaceful expansion, especially in areas occupied by races they deemed inferior. What is striking about much pacifist rhetoric is the frequent appeal to peace among the "civilized" nations and peoples, implying that the "uncivilized" may not be included.[63]

Ploetz, as we saw in the last chapter, was ambivalent about war, but he was not at all ambivalent about racial struggle. In his 1911 tract written to recruit German youth into the Nordic Ring he warned that Germans and their fellow Nordic types were engaged in a "struggle for existence against the other races." In order to win this struggle, he suggested increasing reproduction and then providing economically for the growing population, if necessary by warfare, since, "Only the race that consistently has the greatest excess natality will, in the end of the struggles, have conquered the world."[64]

Another indication that Ploetz was not committed to peace in racial relations was the Archery Club he founded in Munich in 1911, which included in its membership some leading Munich eugenicists. Except for a brief mention of "cultivating a *völkisch* (ethnic) mentality," the official statutes seem innocuous.[65] However, Lenz, a member of the Archery Club, informed Schemann that the goal of the club was "the cultivation of racial thought," so he donated Schemann's book on Gobineau to the club's library.[66] Ploetz confided to Gerhart Hauptmann that the real purpose of the Archery Club was to cultivate the "primitive instinct of preservation" and to strengthen the German race by recalling the old tribal solidarity.[67] Using bows and arrows to emphasize racial unity and strength is suggestive of martial intent. Ploetz made this point even more explicitly, when he congratulated Luschan for upholding the necessity of war among races and nations in his 1911 speech at the Universal Race Congress. Ploetz opposed the congress because of its emphasis on racial reconciliation.[68]

Ploetz's concern over the racial struggle for existence emerges clearly enough in his writings and speeches, though often he toned it down a bit. In his speech to the First International Eugenics Congress in 1912 on "Neo-Malthusianism and Race Hygiene," Ploetz opposed birth control as a danger to Europeans, especially the "highly endowed Nordic (Teutonic or Germanic) race" in its struggle against African and Asian races.[69] Thus for Ploetz, eugenics was a means to win the Darwinian racial struggle.

Though less racist than Ploetz, Schallmayer likewise saw eugenics as a means to revitalize the white race in its competition with other races.[70] Like other eugenicists, he interpreted the decline of "primitive" races through the lenses of Darwinian struggle. The less fit races—here Schallmayer listed the Australian and Pacific natives, the Hottentots, and the American

Indians—"*were only capable of existing, as long as they were protected from competition with the superior European race.*"[71] On the other hand, unlike many of his contemporaries, Schallmayer spoke highly of the Chinese and noted that the black Africans were not dying out in the face of European colonization.

Some eugenicists were even more actively involved in the pacifist movement than Ploetz or Schallmayer, but even so, most of them continued to believe that certain races would ultimately perish in the Darwinian struggle for existence. August Forel, a prominent Swiss psychiatrist who probably deserves to be called the grandfather of the German eugenics movement, was well known for his participation in the peace movement (and the socialist and women's movements as well). He remained true to his pacifist convictions during World War I—the carnage only made him more anxious to promote peace. Ironically, however, he was even more racist than Schallmayer, whom he criticized for not placing enough emphasis on racial inequality.[72] In his memoirs he confessed that the race issue was an important concern to him. Based on his view of racial inequality and his desire to find ways of promoting biological progress, he wrestled with the following questions: "Which races can be of service in the further evolution of mankind, and which are useless? And if the lowest races are useless, how can they be gradually extinguished?"[73] Though he never provided a clear answer to the latter question, he did on occasion suggest that "inferior" races need to be eliminated. In an 1899 article, "On Ethics," he argued that ethics needs to focus on two issues: suppressing races "dangerous to culture" and improving "our own race" through eugenics.[74] To be sure, Forel never advocated slaughtering those of other races, but he clearly wanted their eventual elimination. He also advised against extending sympathy to those of all races, since some are "simultaneously so fertile and of so inferior quality" that if we treated them nicely and allowed them to reproduce among us, "they would soon have eliminated us." If this happened, he warned, "the most cruel barbarism, which rests in their instincts, would gain the upper hand."[75] Forel, like so many other social Darwinists, thus thought that eliminating "inferior" races will rid the world of immorality.

Another advocate of peace eugenics who remained true to pacifist convictions during World War I was Helene Stöcker, the leader of the League for the Protection of Mothers. Stöcker tirelessly promoted her "new ethics," which drew heavily on Darwinism, eugenics, and Nietzsche. Concerning war, she declared, "It is clear that from this cultural stance we come to condemn war, this mass murder, which is opposed to all eugenics ideals."[76] During World War I Stöcker remained firm in her opposition to war, warning about its deleterious effects on the health and vitality of society. However, Stöcker's concern was embedded in larger concerns about racial competition. She believed that

> this war of the white races among themselves immensely threatens the domination of the white race in relation to the yellow and black races. If before the war we could hear so often about the dangers of the yellow race, since then this danger has multiplied tremendously.[77]

In this same article she called the goal of eugenics "the dominion of the most valuable race," that is, the white race. Unity among the white races would facilitate this goal. Thus Stöcker saw pacifism as a means to facilitate European domination of other races.[78] It is hard to understand how Europeans could dominate other races without engaging in colonial warfare, but perhaps she believed that peaceful competition would decide the racial struggle. In any case, after World War I Stöcker embraced an absolute pacifism that rejected all wars, including wars of imperialism.[79]

Stöcker was not the only eugenics enthusiast concerned about the "Yellow Peril." Ehrenfels was so convinced of the contraselective effects of modern warfare that he suffered a mental breakdown in the middle of World War I, depressed by the slaughter of those he deemed biologically most valuable.[80] Despite his antipathy for war, however, he continually warned about the "Yellow Peril," since he saw Asians as the chief threat to Europeans in the racial struggle. His fear of Chinese racial prowess led him to the conclusion that "if there is no change in current practices, this will lead to the annihilation of the white race by the yellow race."[81] Ehrenfels' interpretation of racial struggle and racial annihilation was clearly Darwinian. He claimed that he had "biologically proven" that monogamous Westerners could not survive in the struggle for existence with the polygamous Mongolian race [i.e., East Asians], unless Europeans took measures to regenerate themselves biologically.[82] Ehrenfels called for polygamy as the key to biological improvement. Though Ehrenfels was not explicit on this point, presumably the end result of his program to elevate Europeans would be the annihilation of East Asians, as well as other races, whom Ehrenfels considered far inferior to Europeans and Asians.

G. F. Nicolai, a physiologist at the University of Berlin, was so committed to pacifism that he was imprisoned during World War I for opposing the war in his lectures. While in prison, he wrote an anti-war book, *The Biology of War*. In his book Nicolai was consistent in opposing all wars, including wars against "inferior" races. However, his reasoning is interesting. There is no need to fight wars against the "inferior" races,

Nicolai argued, since

> Practically these peoples will undoubtedly be exterminated gradually by the white race, but it has been apparent for a long time that it would be highly foolish to wage war against these peoples. They disappear on their own, when they come into contact with whites, for the bloodless war is always more effective than the bloody one.[83]

He admitted that in the racial struggle East Asians were the most dangerous competitors to Europeans, but instead of fighting them militarily, he suggested using eugenics to improve the Europeans biologically. This would lead to ultimate victory, while military conflict is more hazardous.[84]

It should be clear that among the German educated elites the notion was widespread that racial extermination is an inevitable process that may be lamentable, but is ultimately beneficial for humanity. Only through racial extermination could humanity improve biologically and advance to higher cultural levels, since the "lower" races are not mentally capable of producing culture. I have demonstrated the prevalence of such ideas among social Darwinists and eugenicists, including some who called themselves pacifists. Racism existed long before Darwinism, of course, but Darwinism did radicalize racism in the late nineteenth century, providing scientific justification for racial inequality, racial competition, and even racial extermination.[85] In harmony with the rising tide of positivism and materialism, Darwinism also helped sweep aside ethical considerations, which had been a restraint on racial extermination (albeit not always very effective).

Many Germans still opposed racial extermination, to be sure. Racial egalitarianism was still strong in some circles, especially among socialists and Catholics. Both these groups exerted political influence, and they used their representation in German parliament to oppose policies directed at exterminating indigenous races in German colonies. Another reason some Germans, including many colonial administrators, opposed racial extermination was because of the widespread belief that tropical regions were unsuitable for European settlement. Those upholding this view—including many colonial physicians who supported eugenics policies—favored increasing indigenous populations, so Germans could exploit their labor. Thus some German eugenics proponents sought ways to improve the non-European races, both quantitatively and qualitatively. Their ultimate goal, however, was to benefit Germany, not the indigenous peoples.[86] Despite these contrary currents, racial extermination was becoming more and more intellectually acceptable by the early twentieth century.

Though anti-Semitism has dominated the attention of many scholars examining German racism—for obvious reasons—we can see that ideas about racial extermination were not necessarily connected to anti-Semitism. More often than not when social Darwinists and eugenicists spoke about inferior races, they meant non-European races, especially American Indians, Australian aborigines, blacks, and East Asians. Though some of the social Darwinists and eugenicists I have discussed were anti-Semitic in varying degrees—some rabidly so—very few of them ever referred to the extermination of Jews. On the other hand, some social Darwinists even opposed anti-Semitism, and some German and Austrian Jews (Gumplowicz, for example) justified racial struggle and racial extermination, just as other German thinkers did. Of course, there were some anti-Semitic thinkers advocating elimination of the Jews, and even a few radical anti-Semites advocating extermination, as Daniel Goldhagen has reminded us (while overstating his case).[87] But the notion of racial extermination was much more widespread in forms not associated with anti-Semitism, especially among the educated elites.

In fact, not all leading anti-Semites, even those embracing the idea of Darwinian racial struggle and extermination, thought the Jews should be eliminated by violent means. Viewing the Jews as "subhuman," Theodor Fritsch believed that "the struggle against Jews is a question of progress and spiritual elevation" and "a mighty lever of moral purification." He rejected violence against the Jews, however, since boycotts and ostracism would suffice.[88] He vehemently denied reports that he and his associates wanted to exterminate Jews. On the contrary, he declared, he wanted them to have the opportunity to develop their own culture—but only in some far-away land, not in Germany.[89]

Adolf Harpf (pseudonym for Adolf Hagen) proposed an even more peaceful means of eliminating the Jews—mixed marriages between Germans and Jews. Harpf was a journalist who wrote many articles and pamphlets on biological racism, including several numbers of Lanz von Liebenfels' *Ostara* journal. Deeply influenced by Darwinian theory, he wrote a *Festschrift* on Darwinian ethics for Carneri on his eightieth birthday. Harpf saw a Darwinian racial struggle leading to extermination everywhere that races came into contact: "History gives us examples in every epoch of the unavoidable struggles of annihilation between human races. . . . In short everywhere we find and have found racial hatred and, necessarily arising from it, racial war, where racial differences are present." Even though he viewed Jews as unequal and irreconcilable to Germans, he thought the only humane way to escape the struggle of annihilation was by mixing the races.[90] This solution did not find much resonance in anti-Semitic circles, largely because most biological racists considered racial mixing harmful, leading to degeneration. Harpf and Fritsch may have eschewed violent solutions to the "Jewish Question," but their doctrine of racial differences and racial struggle were pernicious nonetheless.

So what practical effect did these ideas about racial extermination have? First of all, the most immediate impact was on German imperialism and colonial policy. German colonialist propaganda lacked the religious and humanitarian rhetoric so characteristic of British imperialist discourse, relying instead on social Darwinist and biological racist arguments. No wonder, then, that German colonial officials often denied that natives had a right to live, unless they were serving German interests.[91]

Both Gollwitzer and Ute Mehnert in their analyses of German discourse about the "Yellow Peril" agree that social Darwinism was a fundamental feature of rhetoric about racial struggle against East Asians. Wilhelm II made liberal use of this rhetoric, informing Theodore Roosevelt in 1905,

> I foresee in the future a fight for life and death between the "White" and the "Yellow" for their sheer existence. The sooner therefore the Nations belonging to the "White Race" understand this and join in common defense against the coming danger, the better.[92]

Thus underlying Wilhelm II's "world policy" was a view in which racial struggle and racial extermination were inevitable.

The dehumanizing tendencies of racism also played a key role in the genocide of the Hereros during the Herero Revolt (1904–06), since most Germans considered black Africans little more than animals.[93] One German missionary lamented that

> the average German here looks upon and *treats* the natives as creatures being more or less on the same level as baboons (their favourite word to describe the natives) and deserving to exist only inasfar as they are of some benefit to the white man. . . . Such a mentality breeds harshness, deceit, exploitation, injustice, rape and, not infrequently, murder as well.[94]

The German crushing of the Herero Revolt was the first attempt at genocide of the twentieth century. General Trotha, the German military commander in Southwest Africa, ordered the annihilation of all Hereros—men, women, and children—in what he called a "racial war." He explicitly justified racial annihilation using Darwinian concepts. In a newspaper article he stated that Germans should not resist his order of extermination on the basis of the economic value of the Hereros, which was a common argument against Trotha. He stated, "We cannot dispense with them [the natives] at first, but finally they must retreat. Where the labor of the white man is climatically possible, . . . the philanthropic disposition will not rid the world of the above-mentioned law of Darwins, the 'struggle of the fittest [*sic*].' "[95] Thus Trotha tried to counter the moral indignation of his opponents by claiming scientific sanction for racial struggles.

Trotha's decree of annihilation aroused considerable controversy in Germany. Many German officials were horrified by Trotha's policy, including the governor of the colony and the German chancellor Bernhard von Bülow, who criticized it as "contradictory to all Christian and humane principles." In the Reichstag debates over Trotha's decree, however, only the Social Democrats and the Catholic Center Party took a principled stance against racial extermination; most liberals, on the other hand, manifested virulent racism.[96] General von Schlieffen, Trotha's superior, privately supported Trotha, believing that "Racial war, once it has broken out, can only be ended by the destruction of one of the parties."[97] For practical reasons, however, Schlieffen eventually convinced a reluctant Emperor Wilhelm II to annul Trotha's decree. It was much too late for most of the Hereros, however, since an estimated 81 percent of the Hereros actually perished through the brutal German policies. The attitudes behind Trotha's decree were clearly shaped by the rhetoric of racial struggle and racial extermination.[98]

Finally, an ideology of racial extermination would captivate the minds not only of Hitler, but also of many other Germans of his time, who would support him and cooperate with his attempts to create a racial utopia. Hitler's ideas were by no means idiosyncratic, even though they were not dominant in German thought, either. In chapter 11 we examine the way that Hitler understood and applied many of the ideas we have examined in this book, especially the ideology of racial extermination, which bore wicked fruit when applied with his pitiless logic.

4. Impacts ∽

11. Hitler's Ethic ⌒

D id Hitler even have an ethic? Since Hitler is the epitome of evil, some will think it absurd even to consider the possibility that morality played a significant role in his worldview. In order to perpetrate such radical evil, many assume, he must have been either an immoral opportunist or else as an amoral nihilist. Gisela Bock, for instance, in her works on eugenics and sterilization properly stresses the Nazi devaluing of human life and rejection of humane ethics.[1] However, by accenting only what the Nazis opposed morally, she only gives one side of the story. Did they replace humane ethics with an inhumane ethic, or did they abandon morality altogether?

Undoubtedly Hitler was immoral, but Eberhard Jäckel has convinced most historians that Hitler was not a mere opportunist. Rather he was a principled politician with a well-defined worldview that he pursued relentlessly.[2] Jäckel does not specifically analyze Hitler's views on ethics and morality, but it is clear from Hitler's writings and speeches that he was not amoral at all. On the contrary, he was highly moralistic and consistently applied his vision of morality to policy decisions, including waging war and genocide. It may be difficult for us to grasp this, but in Hitler's worldview war and genocide were not only morally justifiable, but morally praiseworthy. Hitler was ultimately so dangerous, then, precisely because his policies and decisions were based on coherent, but pernicious, ethical ideas.[3]

One also cannot comprehend Hitler's immense popularity in Germany without understanding the ethical dimension to his worldview and his political policies. Hitler not only promised to bring prosperity, health, and power to the German people, but he also promised moral improvement. Many scholars have noted the utopian appeal of Nazism, which aimed at creating a higher and better person. The police state he erected not only persecuted political enemies, but also tried to eliminate criminality and social deviance.[4] Nazi propaganda continually portrayed the Nazis as decent, clean-cut, upstanding members of society. Hitler continually criticized modern, urban society for its rampant immorality, especially the proliferation of sexual immorality and prostitution.

Because of Hitler's support for "family values," some mistakenly assume that Hitler was a moral conservative. If we examine some specific moral issues—abortion, the role of women, or homosexuality—then Hitler's views do reflect a conservative position. However, taken as a whole, Hitler's ethical views do not comport well with traditional morality, since he based his morality on an entirely different foundation than did most conservatives. Hitler's morality was not based on traditional Judeo-Christian ethics nor Kant's categorical imperative, but was rather a complete repudiation of them.[5] Instead, Hitler embraced an evolutionary ethic that made Darwinian fitness and health the only criteria for moral standards. The Darwinian struggle for existence, especially the struggle between different races, became the sole arbiter for morality.[6] Thus, whether or not we view Hitler as a conservative or reactionary politically, he surely was not a conservative morally.

In his early speeches and writings Hitler often appealed to nature to justify his vision of struggle.[7] He denied that anyone could escape the universal struggle for existence. All efforts to avoid struggle would be useless in the final analysis, for nature would assert itself and destroy those unwilling to fight for their existence. The only really rational policy, then, would be to conform to the laws of nature. Hitler repeatedly asserted that the only rights that any people had are rights won through victory in the struggle for existence. He scorned humaneness and Christian morality, which would promote weakness, thereby producing decline, degradation, and ultimately the demise of the human species. In a 1923 speech Hitler explained the relationship between struggle and right:

> Decisive [in history] is the power that the peoples (*Völker*) have within them; it turns out that the stronger before God and the world has the right to impose its will. From history one sees that the *right* by itself is completely useless, if a mighty power does not stand behind it. Right alone is of no use to whomever does not have the power to impose his right. The strong has always triumphed . . . All of nature is a constant struggle between power and weakness, a constant triumph of the strong over the weak.[8]

Taken in isolation, this sounds thoroughly amoral. If victory through struggle is the only arbiter of what is right, as Hitler clearly stated, then how can there be any objectivity or criteria for morality?

Indeed, elsewhere in *Mein Kampf* Hitler clearly denied that morality has any objective, permanent existence. He argued that all ethical and aesthetic ideas—indeed all ideas except those that are purely logical deductions—are dependent on the human mind and have no existence apart from humans, who have not always existed.[9] Hitler's view that morality is purely a human construction undermines any system of ethics claiming transcendence, such as Judeo-Christian ethics or Kantian ethics. Hitler clearly did not believe in the existence of immutable, universal moral standards. This aspect of Hitler's thought seems to reinforce the portrait of him as an amoral Nietzschean nihilist, especially since he so often spoke of the need for overcoming obstacles through the exercise of a strong will, thus sounding much like Nietzsche with his concept of the will to power.

However, Hitler, like many other evolutionists we have examined in this work, evaded complete nihilism and moral relativism by adopting a form of evolutionary ethics. He believed that human morality was a product of evolutionary development, representing the highest stage of evolution ever yet reached. Like so many of his contemporaries, his view of evolution was imbued with the notion of progress, and he considered morality one of the greatest evolutionary advances. Of course, given his own belief that morality does not exist apart from the human psyche, Hitler had no objective, scientific basis for decreeing that human morality represented a "higher" or "better" stage of evolution, for he had no objective criteria to determine what is "higher" or "better." Hitler was not alone in this error, for many of his predecessors and contemporaries upheld the same view of evolutionary progress, while denying any basis for judging developments as better or worse.

Since Hitler viewed evolutionary progress as essentially good, he believed that the highest good is to cooperate with the evolutionary process. The struggle for existence has produced humans, and if we try to oppose the laws of nature by abandoning struggle, we will hinder progress and promote degeneration. In a passage from *Mein Kampf* warning about the deleterious effects of artificially limiting population expansion, Hitler cautioned about the perils of trying to terminate the human struggle for existence:

> If reproduction as such is limited and the number of births decreased, then the natural struggle for existence, which only allows the strongest and healthiest to survive, will be replaced by the obvious desire to save at any cost even the weakest and sickest; thereby a progeny is produced, which must become ever more miserable, the longer this mocking of nature and its will persists. . . . A stronger race (*Geschlecht*) will supplant the weaker, since the drive for life in its final form will decimate every ridiculous fetter of the so-called humaneness of individuals, in order to make place for the humaneness of nature, which destroys the weak to make place for the strong.[10]

Hitler thereby redefined humaneness by stripping individuals of any rights and by arguing that the destruction of the weak by the strong is humane. He thus turned traditional morality on its head.

But how should humans fight this struggle for existence? When applying Darwinian principles to human societies, Darwinists had to wrestle with the relative weight to give to competition between individuals within a society versus competition between different groups or societies. As we have seen earlier, many Darwinists—including Darwin himself—believed that group competition, such as war and racial antagonisms, played a crucial role in the development of human societies and even in the evolution of morality. Hitler aligned himself with this approach, believing that racial competition was the primary factor driving human evolution and history. Thus, in order to promote evolutionary progress, Hitler opted for a two-pronged strategy involving both artificial and natural selection: eugenics within German society to improve the health and vitality of the "Aryan race," and racial struggle and warfare toward those outside the German racial community. Hitler's eugenics and racial policies were linked as part of a larger program to preserve and improve the human species, since he considered the Germans the highest race, not only physically, but especially intellectually. Even more important for our purposes, Hitler believed that Germans were morally superior to all other races. Thus, eliminating other races and replacing them with Germans would bring moral improvement to the entire world.[11]

Thus, those passages in Hitler's speeches and writings stressing struggle are not really as amoral as some suppose. For instance, Hitler proclaimed in *Mein Kampf* that

> this preservation [of culture and culture-producing races] is tied to the iron law of necessity and the right of victory of the best and the strongest. . . . Whoever wants to live, must struggle, and whoever will not fight in this world of eternal struggle, does not deserve to live. Even if this is harsh—it is simply the way it is![12]

Notice first that Hitler referred not only to the victory of the strongest, but also "the best," which implies those who are morally superior. If one examines the context of this passage, one sees that Hitler was asserting that the "Aryan" race alone is responsible for building human culture. When discussing the racial superiority of the "Aryans," Hitler constantly emphasized their contributions to human culture, which he believed were based on their biological characteristics. A few pages after this statement Hitler explained that one aspect of "Aryan" superiority was their moral character. He claimed that the reason "Aryans" can create culture—and indeed are the only race that can do so—was because they have the most developed sense of sacrifice for the sake of others. Thus "Aryans," in Hitler's view, are less egoistic and more altruistic.

Hitler contrasted his vision of altruistic Germans, the highest exemplars of human morality, with the Jew as the epitome of immorality.[13] He continually accused the Jews of greed, deceit, sexual deviance, and other immoral deeds, thereby justifying his view of them as an inferior race. Thus inferiority did not just mean physical and mental inferiority, but moral inferiority (remember that all three of these were biologically based, in Hitler's view). Indeed, when discussing the racial inferiority of the Jews, Hitler stressed their alleged moral defects far more than their physical or mental traits. In fact, the only reason the Jews were a threat to Germany (in Hitler's view) was because of the immoral methods they allegedly used to cheat Germans out of their rightful heritage. Because he viewed them as immoral, Hitler often referred to the Jews as parasites or bacillus, who were infecting and destroying the health and vitality of the German nation. Hitler took this metaphor dead seriously, remarking in February 1945—after millions of Jews had perished in the Holocaust—that Germany's actions against the parasitic Jews "has been an essential process of disinfection, which we have prosecuted to its ultimate limit and without which we should ourselves have been asphyxiated and destroyed."[14]

Hitler's race-based view of morality sometimes led him to embrace moral tenets in harmony with traditional Christian morality. For example, by favoring monogamy and opposing abortion and homosexuality, Hitler's morality came in line with traditional conservatives. However, Hitler was not at all concerned with upholding traditional morality for the sake of tradition or religion, but rather he embraced these moral positions because he believed they were the best ways to promote biological improvement for Germans. The difference between these two approaches becomes evident when we examine Hitler's view of abortion. He opposed abortion only for healthy German women, since he believed that an expanding population is the key to biological vitality for the nation. However, he approved abortion for those of "inferior" races or in cases where the infant would likely have a congenital illness.

While agreeing with some traditional values, Hitler's evolutionary ethic turned other values on their head. He was able to justify just about anything morally, if it contributed to the evolutionary improvement of humanity. Since he believed the "Aryan" race was the highest form of humanity, this meant practically that anything that promoted the victory of the "Aryan" race was morally right. Thus Hitler could state in a 1923 speech,

But we National Socialists stand here [on the Jewish Question] at an extreme position. We know only one people (*Volk*), for whom we fight, and that is our own. Perhaps we are inhumane! But if we save Germany, we have accomplished the greatest deed in the world. Perhaps we perpetrate injustice! But if we save Germany, we have abolished the greatest injustice of the world. Perhaps we are immoral! But if our people (*Volk*) is saved, we have paved the way again for morality.[15]

Thus Hitler justified any immorality committed to those outside the racial community, as long as it contributed to the welfare of the "Aryan" race. By advancing the cause of the "Aryan" race, morality would ultimately triumph.

Hitler also explained this same basic idea in *Mein Kampf*, when he explained that his worldview

by no means believes in the equality of races, but recognizes along with their differences their *higher or lower value*, and through this knowledge feels obliged, according to the eternal will that rules this universe, to promote the victory of *the better, the stronger*, and to demand the submission of *the worse and weaker*. It embraces thereby in principle the aristocratic law of nature and believes in the validity of this law down to the last individual being. It recognizes not only *the different value of races, but also the different value of individuals.* . . . But by no means can it approve of the right of an ethical idea existing, if this idea is a danger for the racial life of the bearer of a higher ethic.[16]

Hitler's sole criteria for morality was its effect in promoting or hindering the interests of the highest race, which was highest, not only in physical and mental prowess, but also in its morality. If any moral code whatever posed "a danger for the racial life" of Germans, according to Hitler, then so much the worse for that moral code. Thus it is clear that Hitler exalted racial considerations over any religious, humanitarian, or traditional ethical standards.

Hitler derided any morality inimical to the increased vitality of the "Aryan" race, especially traditional Christian values of humility, pity, and sympathy. He considered these unnatural, contrary to reason, and thus detrimental and destructive for the healthy progress of the human species. He spurned the idea of human rights, calling it a product of weaklings. "No," he explained, "there is only one most holy human right, and this right is at the same time the most holy duty, namely, to take care to keep one's blood pure," in order to promote "a more noble evolution" of humanity.[17]

Another way that Hitler believed he could advance the interests of the "Aryan" race was by encouraging the reproduction of the "highest" individuals within that race and getting rid of "inferior" individuals. Thus eugenics, infanticide, and euthanasia played a central role in his vision for social improvement. These views also show how far removed his evolutionary ethic was from more traditional morality and certainly from the Judeo-Christian proscription of killing innocent human life. He often discussed eugenics and population policy in his speeches, sometimes intimating that the "unfit" were dispensable. However, only rarely did he explicitly discuss killing those he deemed inferior. In an unpublished 1928 manuscript, however, Hitler clearly showed how eugenics and infanticide fit into his Darwinian vision of nature:

> While nature only allows the few most healthy and resistant out of a large number of living organisms to survive in the struggle for life, people restrict the number of births and then try to keep alive what has been born, without consideration of its real value and its inner merit. Humaneness is therefore only the slave of weakness and thereby in truth the most cruel destroyer of human existence.

Hitler then resorted to Haeckel's favorite example—the ancient Spartans—to defend the killing of disabled infants, since, in his view, their lives have little or no value. According to Hitler, killing such individuals was actually more humane than allowing them to live. He thus construed killing the disabled as morally good, overturning the Judeo-Christian moral code, which protected the disabled.[18]

In Hitler's mind Darwinism provided the moral justification for infanticide, euthanasia, genocide, and other policies that had been (and thankfully still are) considered immoral by more conventional moral standards. Evolution provided the ultimate goals of his policy: the biological improvement of the human species. But this is a very slippery goal, for it depends on preconceived ideas about what constitutes improvement. Hitler's own conceptions of improvement drew upon his own values, so he thereby smuggled his own cultural presuppositions and moral opinions into an ostensibly scientific goal. He also could blend racial prejudices into this scheme by defining some races as inferior and others as superior. Of course, Darwinian racism contributed significantly to this enterprise.

If evolution provided the ends, the Darwinian mechanism suggested the means: increase the population of the "most fit" people to displace others in the struggle for existence. In his zeal to speed up the evolutionary process, he promoted both artificial selection (eugenics) and policies to intensify natural selection. Competition and conflict would advance the cause of the stronger, more fit individuals and races. But so would killing off those individuals and races deemed inferior. Morality could not be determined by any codes of the past, but only by the effects it has on evolutionary progress.

So where did Hitler get these ideas about ethics and morality? Despite the incredible amount of research that has been done on Hitler and Nazism, this is still a difficult question, for while we have a mountain of information about Hitler's own ideas—based on both public and private utterances—we have only scant evidence about the formative influences on his worldview. Much of the available evidence about intellectual influences on Hitler as a young man is sketchy, unreliable, or circumstantial. He rarely mentioned any important authors or books that helped shape his thought. His roommate for a short time in Vienna, August Kubizek, claimed that Hitler often frequented the library in Vienna and read voluminously, including works by Dante, Schiller, Herder, Schopenhauer, and Nietzsche.[19] Hitler probably did read some serious, scholarly works as a young man, but judging from what we know of his later life, these were likely not the most important formative influences on his worldview. Possibly even more significant were Viennese newspapers, political tracts, and pamphlets, which we know he followed with intense interest. If this is the case, then Hitler probably imbibed much of his evolutionary ethics and racist social Darwinism second-hand. This would not be at all surprising, since the Viennese press was saturated with racist social Darwinism during Hitler's time there.[20]

Some scholars, however, have pressed the case for the overwhelming influence of one or another particular thinker on the formation of Hitler's worldview. Hardly anyone has pursued the thesis of a single dominating influence on Hitler more relentlessly than Daniel Gasman with his Haeckel-to-Hitler hypothesis. Gasman's arguments may seem plausible at first, because there are undeniable parallels between some parts of Haeckel's and Hitler's ideologies. We have already seen that Haeckel was one of the earliest German thinkers to discuss the extermination of "inferior" human races by the "superior" Europeans and the killing of the disabled. Hitler's views on the human struggle for existence and the relationship of nature and society are similar to Haeckel's, and some of Hitler's statements about evolution could easily have been cribbed from Haeckel. One example among many is a 1933 statement of Hitler that clearly had its origin in Haeckel (but maybe indirectly): "The gulf between the lowest creature which can still be styled man and our highest races is greater than that between the lowest type of man and the highest ape."[21]

However, Gasman's approach is too blinkered, ignoring the huge disparities between Haeckel and Hitler. Haeckel was a nineteenth-century liberal nationalist who supported Bismarck for unifying the German nation, but who never espoused dictatorship and one-party rule. In many respects Haeckel lined up with liberal progressives of his time, promoting the peace movement and homosexual rights, among other liberal causes. Gasman makes altogether too much ado about Haeckel's anti-Semitism, which, though misguided, was not likely a source for Hitler's anti-Semitism. For one thing, Haeckel's anti-Semitic utterances are extremely rare, and they are much milder than Hitler's. Also, there were many anti-Semitic thinkers in the early twentieth century whose views are much closer to Hitler's. Eugenics and euthanasia, the very aspects of Haeckel's thought that Hitler probably did imbibe, were not the sole property of Nazis, but were embraced by a diverse crowd of secular social reformers. Even many socialists, including some Marxists, jumped on the eugenics bandwagon.

Also, Gasman cannot prove that Hitler ever actually read any of Haeckel's works, so whatever influence Haeckel allegedly exerted on Hitler may have been mediated by others. Indeed Haeckel's works were widely read in the early twentieth century, and it would not be surprising if Hitler read one or more of them. However, many eugenicists, racists, and anti-Semites peddled Haeckel's ideas, too, and they were widely discussed in the popular press, so it is not at all unlikely that Hitler imbibed them through others. Thus, Gasman is right to point out that Haeckel's ideas were an important influence on Hitler, but they by no means provided the ideological foundation for fascism, and Haeckel was by no means a proto-fascist.[22]

Another individual credited as *The Man Who Gave Hitler His Ideas* is Jörg Lanz von Liebenfels, a Viennese occult Aryan racist who—along with Guido von List—formulated a doctrine known as Ariosophy (derived from root words meaning "Aryan wisdom"). The Austrian psychologist Wilfried Daim uses two lines of evidence to link Lanz and Hitler. First, he tries to demonstrate that Hitler read Lanz's racist periodical, *Ostara*, which was readily available at a kiosk near Hitler's dwelling in Vienna. After World War II, Daim interviewed Lanz, who claimed that Hitler had once visited him, requesting a couple of missing copies from his collection of *Ostara*. Lanz's testimony is not entirely credible, however, since he also claimed that Lenin was one of his disciples. Thus Lanz's story about Hitler may have been self-serving, inflating his real importance. However, many scholars, myself included, find it entirely plausible that Hitler read Lanz's writings in Vienna.

Daim's second line of argument is to show the many parallels between the views of Lanz and Hitler. Indeed, Lanz's crass Aryan racism did have many points of agreement with Hitler's racial ideology. First and foremost, Lanz was committed to Darwinism and interpreted racial struggle as a necessary part of the evolutionary process. Like many other eugenicists, he was concerned about biological degeneration, which he thought was a product of racial mixing. He continually stressed the need for increased competition, so that natural selection could operate. The "Aryan" race would triumph in unfettered competition, according to Lanz, since they were naturally an aristocratic race with greater talent and ability than any other race. In addition to sharpening the competitive struggle, however, he also advocated eugenics (artificial selection) to help the evolutionary process along.[23]

Lanz lauded the racist publicist Adolf Harpf, who contributed some articles to Lanz's journal, for revealing the centrality of race in determining the destiny of peoples and nations. Lanz remarked that "race is the driving force behind all deeds."[24] Later Lanz discussed how his racial views affected his view of morality: "For us the human, the fully human, heroic Aryan, is the measure of all things, and thus whatever is advantageous to the Aryan, the heroic human, must be moral, and whatever is disadvantageous to him is immoral."[25] He specifically denied that the biblical command to love one's neighbor included those of other races. He claimed that the universal application of love was a perversion of Christian teaching being foisted on the Aryans by inferior "ape-men."[26] Like Hitler after him, Lanz would measure morality by its impact on the "superior" race, and love was reserved only for one's racial comrades.

Lanz's race-based morality could justify all manner of immoral deeds, as long as they were perpetrated against "lower" races. He continually remonstrated against charity given indiscriminately, since this would benefit those of "inferior" racial stock at the expense of the "Aryans." He compared people of "inferior" races to weeds needing to be pulled out, because they were a financial burden on the "better" racial elements.[27] Lanz recognized that this weeding out was not necessarily going to be peaceful. In one article he praised the racial policies of the Bible, which is a "hard, racially proud and racially conscious book, which proclaims death and extermination to the inferior and world dominion to those of high value."[28] Thus, Lanz's ideas did presage in some way Hitler's harsh racial policies.

However, one problem with the attempt to portray Hitler as a disciple of Lanz is that the ideas they shared in common were widespread in Vienna and elsewhere in Austria and Germany in the early twentieth century. We have encountered many of these same ideas about Darwinism, eugenics, and morality elsewhere in this book, often in works predating Lanz's. Second, though Lanz exalted the blond, blue-eyed Aryan as the highest race, he rarely discussed the Jews. The racial enemies that Lanz mentioned most often were the Negro and Mongolian races, as well as those Europeans with an admixture of the blood of these other races. Undoubtedly most Jews would fit Lanz's description of the inferior races, but anti-Semitism was not central to his racial ideology. Also, even if Hitler was influenced by

Lanz, he appropriated the ideas rather selectively. For example, Hitler never embraced the mystical interpretations of the Bible, which were a central feature of Lanz's ideology, nor did he ever use Lanz's esoteric terminology. While Lanz reveled in the occult nature of his racism, Hitler usually presented his racial ideology as rational and scientific.

Another Viennese occult racial theorist who may have exerted influence on Hitler was Guido von List, an associate of Lanz's who was the first Aryan racist to use the swastika prominently (Lanz adopted it after him). Hamann suggests that Hitler's worldview paralleled List's even more closely than it did Lanz's. However, she also notes that List's ideas circulated widely in the Viennese press, so Hitler could easily have imbibed List's ideas second or third-hand.[29] Also, many of List's ideas were culled from other racial theorists, so Hitler could have derived them from many different sources. Nicholas Goodricke-Clarke is probably right in asserting, "Ariosophy is a symptom rather than an influence in the way that it anticipated Nazism."[30] Lanz and List may have influenced Hitler—directly or indirectly—but Hitler was certainly not one of their disciples.

Peter Emil Becker suggests another Viennese racial theorist as a candidate for "the man who gave Hitler his ideas": Josef Reimer.[31] Indeed, many of the ideas Reimer circulated in his book, *A Pan-German Germany* (1905), are contained in Hitler's worldview. Reimer promoted Pan-German nationalism and the primacy of race in human affairs, stating, "*Our race with its culture is of higher value* than the other peoples and races of the earth."[32] He used Darwinian theory to justify both eugenics and racial struggle. Indeed the similarities with Hitler's worldview are striking. However, Becker admits there is a problem: We have no evidence that Hitler ever read Reimer's book. Also, Reimer admitted that his book was heavily influenced by other racial thinkers more prominent than himself. First and foremost he admired the work of Woltmann, but he also credited Lapouge, Wilser, Gobineau, and Chamberlain with influencing his thinking.[33] Reimer, like Lanz and List, is more likely a symptom of the prevailing racist ideologies in Vienna at the time Hitler lived there. It is always possible that his book influenced Hitler, but even if it did, it was only one among many other writings by racial theorists influencing Hitler's ideas.

When he wrote *Mein Kampf*, Hitler never even mentioned Haeckel, Lanz, List, nor Reimer as influences on the development of his worldview in Vienna. However, one influential political figure in Vienna that Hitler did mention in *Mein Kampf* is Georg von Schönerer, the leader of the Austrian Pan-German Party. Hitler was enamored of Schönerer's Pan-German nationalism and—at least by 1919—with his biological anti-Semitic racism. In *Mein Kampf* Hitler stated, "Theoretically speaking, all the Pan-German's [i.e., Schönerer's] thoughts were correct."[34] Many of Hitler's ideas about biological racism, ethics, eugenics, and anti-Semitism were already presaged by Schönerer's speeches in the 1880s—before Hitler was born. Schönerer portrayed the competition between the noble "Aryans" and the ignoble Jews as a racial struggle for existence. He believed that moral traits were hereditary, and Jews were using their immoral biological character to try to destroy the more morally upright Germans. His biological determinism led him to conclude that the only solution to this ethical dilemma was to destroy the immoral Jews.[35] He justified this position on moral grounds: "Love everything noble, good, and beautiful, just as it is embodied in Germanness, and hate everything ignoble, bad, and vulgar, just as it is manifested in Semites! In the arena of the struggle for national existence it must be: 'Eye for eye and tooth for tooth!' "[36] Thus Schönerer thought that eliminating Jews would improve the moral climate.[37]

Another Viennese figure briefly mentioned in *Mein Kampf* is Houston Stewart Chamberlain, one of the most influential anti-Semitic writers of the early twentieth century. William Shirer and Roderick Stackelberg both present Chamberlain as one of the most important intellectual shapers of the Nazi worldview.[38] Shirer remembers that Nazi writers in the 1930s regularly feted Chamberlain as a "spiritual founder" of Nazism. Indeed, as influential as Chamberlain was among early twentieth-century anti-Semites, it would be surprising if Hitler did not either read Chamberlain or at least read discussions of Chamberlain's ideas in the popular press. Hans Frank later claimed that Hitler read Chamberlain while he was incarcerated in Landsberg Prison in 1923–24.[39] In any case, Chamberlain recognized the affinity of his ideas with those of Hitler about racial struggle, eugenics, and anti-Semitism (see chapter 6). Right after Hitler visited him in Bayreuth in October 1923, the ill Chamberlain exulted that Germany had finally found a savior. However, one indication that Chamberlain was not the sole—and maybe not the primary—influence on Hitler's racial thought is Chamberlain's consistent use of the word Teutons when referring to Germans. Hitler preferred the term Aryans. Whatever influence Chamberlain exerted on Hitler was probably diffuse and synthesized with the ideas of other racial theorists.

One problem with tracing Hitler's worldview back to the direct influence of any of these Viennese thinkers was that, so far as we know, Hitler's intense anti-Semitism probably did not emerge until after World War I.[40] This suggests that some of his central ideas were still being shaped after he left Vienna. In his study of *Hitler's World View* Eberhard Jäckel argues that Hitler's ideas were still in flux after 1918, and even altered in some points (especially in his foreign policy) after he wrote *Mein Kampf*.[41] Our knowledge about Hitler's ideas before 1918 are sketchy, though we do know that he was already a Pan-German nationalist during his time in Vienna. I consider it likely that he already believed in the primacy of racial struggle in human history.

The intellectual milieu in Munich may have contributed as much to Hitler's worldview as did his experience in Vienna. At the very least it reinforced ideas he had already imbibed in Vienna. The historian Hellmuth Auerbach has shown that the ideas Hitler promoted were widely circulating in Munich in the years immediately after World War I.[42] Indeed Munich was an important center for the eugenics movement, and Ploetz even organized a secret Nordic racist branch of his Society for Race Hygiene in Munich. While there is no evidence that Hitler ever met Ploetz or attended any of his meetings in Munich, Hitler did have close contact with Julius Lehmann, a major publisher of medical and scientific books who promoted racist eugenics and was a leading member of Ploetz's organization. Lehmann was able to spread his ideas of Pan-German nationalism, eugenics, and racism widely through the journal, *Deutschlands Erneuerung*, which he published.

Already in the 1890s Lehmann embraced Pan-German nationalism, rising to prominence in the leadership of the Pan-German League. According to Gary Stark his nationalism "reflected the Darwinistic spirit which permeated so much of the age."[43] Lehmann was enthusiastic upon reading the anti-Semitic racist ideas of H. S. Chamberlain and Paul Lagarde. Then, in 1908–09 he read several works on race hygiene and devoted himself to that cause, placing his publishing business at its disposal. He soon came into contact with leading eugenicists in Munich, including Ploetz, Gruber, and later Rüdin and Lenz.[44] Before World War I he joined both the Society for Race Hygiene in Munich and Ploetz's Archery Club (see chapter 10).[45] In the 1920s he became the leading publisher of works on eugenics and racial theory, including the famous Baur–Fischer–Lenz text on human genetics and eugenics, as well as numerous works by the notorious racial theorists, Hans F. K. Günther and Ludwig Schemann.

Lehmann not only had considerable contact with Hitler from 1920 on, but he was a leading figure in nationalist and racist circles that overlapped with the Nazis. Lehmann was a leading figure in the Thule Society, a nationalist organization founded in Munich in the aftermath of World War I. The Thule Society's membership list "reads like a *Who's Who* of early Nazi sympathizers," reports Ian Kershaw.[46] Leading Nazis connected with the Thule Society included the co-founders of the party, Anton Drexler and Karl Harrer, as well as Gottfried Feder, Dietrich Eckart, Hans Frank,

Rudolf Hess, and Alfred Rosenberg. The Nazis first newspaper, the *Völkischer Beobachter*, was originally owned by the Thule Society, which sold it to the Nazi Party in December 1920. In early 1920 Lehmann joined the fledgling Nazi Party after hearing Hitler speak several times, and he also published some Nazi propaganda.[47] In the 1920s Lehmann sent Hitler various books on eugenics and racial theory, including the Baur–Fischer–Lenz text and Günther's works.[48] This does not prove that Lehmann had a significant influence on the formation of Hitler's worldview, since many of the books Lehmann sent him were published after Hitler wrote *Mein Kampf*, and we cannot be sure Hitler read all of them. However, they certainly confirmed him in his ideas (if he indeed read them).

However, despite my inability to conclusively prove it, Lehmann probably did have a significant influence on Hitler's thinking—either directly or indirectly. Certainly the parallels between the views of Lehmann and Hitler are conspicuous, and Lehmann was a dynamic leader in nationalist circles in Munich. He zealously circulated his views, often publishing racist and eugenics literature at a loss to his publishing firm, covering the cost from profits gained through publishing medical texts. Also, considering Hitler's political milieu and his friendship with Lehmann, it would be surprising if Hitler did not read *Deutschlands Erneuerung* in the early 1920s. From 1919 to 1923 this journal contained numerous articles on nationalism, racism, and eugenics, including several by Gruber, Lenz, Günther, and Chamberlain. In one of Günther's articles in 1921—entitled "Hatred"—he criticized the ethic of love toward all humanity (*Menschenliebe*). Instead, he preached love only to those inside the racial community, but hatred toward those outside.[49] *Deutschlands Erneuerung* even carried an article in 1923 by the American eugenicist Lothrop Stoddard on "The Growing Flood of Colored Races against White World Domination."

Not only did the Nazi ideologist Rosenberg contribute an article in 1922 on "Anti-Semitism," but the famous letter in which Chamberlain endorsed Hitler's leadership appeared in the January 1924 issue of *Deutschlands Erneuerung*. Three months later Hitler himself published an article therein attacking Marxism as a tool of the Jews to undermine German culture. One of his primary criticisms of Marxist ideology was that it "replaces the aristocratic principle of nature with the masses," and it denies nationality and race. While using Nietzschean terminology, such as the will to power, and calling for the "conversion or destruction of the last Marxist" in Germany, he also insisted that this destruction was ethically motivated: "*The Marxist internationalism will only be broken through a fanatically extreme nationalism of the highest social ethic and morality.*"[50]

Thus Hitler posed as a moral crusader crushing the forces of immorality, represented by Marxists and Jews.

In addition to *Deutschlands Erneuerung*, Lehmann also published many books on racism and eugenics in the early 1920s, including most prominently the two-volume Baur–Fischer–Lenz work on human genetics and eugenics and Hans F. K. Günther's *Racial Science of the German People*. Hitler's personal library contained a copy of the third edition of the Baur–Fischer–Lenz text (1927–31), as well as Lenz's earlier pamphlet, *Race as a Principle of Value: The Renovation of Ethics* (1917). We do not know if Hitler owned earlier editions of Baur–Fischer–Lenz, but he probably did, since Lenz and Günther both report that Hitler read the second edition of Baur–Fischer–Lenz while in Landsberg Prison in 1923–24. Lenz was not exaggerating when he stated in 1931 that "many passages in it [Baur–Fischer–Lenz] are mirrored in Hitler's expressions."[51] For example, in his section on racial anthropology Fischer stressed the inequality of races. He also explained that race is "one of the most decisive factors in the entire course of the history of a people," largely because of mental differences between races. Lenz, who wrote over half the first volume and all of the second volume, promoted a Nordic racist form of eugenics. He presented the Nordic race as the most creative and intellectually superior race on earth, and though he toned down his anti-Semitism to make his book more marketable in medical schools, he nonetheless depicted the Jews negatively. He warned against measures protecting "inferior" individuals and encouraged restrictions on reproduction. Further, he supported infanticide for disabled infants, calling such killing a humanitarian measure.[52] The Baur–Fischer–Lenz text, which went through five editions by 1940, was highly respected and received overwhelmingly positive reviews in medical and scientific journals, not only in Germany, but also in other lands. Most scientists at the time accepted the racist eugenics of Lenz as mainstream science, not pseudo-science, as it is usually considered today.[53]

Whatever racial ideas Hitler did not pick up from Fischer or Lenz, he could easily have imbibed from Günther, whose ideas were quite similar (Günther had recommended the Baur–Fischer–Lenz book). Lehmann gave Hitler a copy of the third edition (1923) of Günther's work on the German race, so Hitler might have had it while in Landsberg Prison. Günther was committed to a Darwinian explanation of human evolution and claimed that his study was a scientific study of race, the aim of which was to "awaken our attention to the racially determined character of the environment and history." He rejected the Marxist materialist conception of history, replacing it with the "racial conception of history." While specifically denying that racial purity exists anywhere in Europe, Günther portrayed racial mixture as harmful, and he hoped that the Nordic race would strive to purify itself of foreign racial elements. He claimed that mixed races possess a jumble of hereditary elements that produce confusion, including moral confusion, giving rise to ethical relativism. Günther was himself a moral relativist, however, in the sense that he thought each race had its own distinct morality, but he thought racial purity would strengthen the particular morality of each race. Thus Nordic morality could be fostered by purifying the race of its non-Nordic elements. Günther even formulated an overarching moral imperative, patterned on Kant's categorical imperative, but with different content: "Act so that at all times the tendency of your will can be the fundamental tendency of Nordic racial legislation."[54] As Michael Burleigh and Wolfgang Wippermann have indicated in *The Racial State*, this does seem to be the crux of Hitler's ethic.[55]

In a long appendix purporting to be a scientific study of the Jewish race, Günther thoroughly disparaged the Jews. While earlier he had depicted the Nordic race as the most creative race, he denied that Jews are creative, except perhaps in music. Anti-Semitism was not a product of cultural developments, according to Günther, but derived from an inherent repugnance in Nordic blood and was thus unavoidable. Günther decried intermarriage between Germans and Jews, calling such relationships "racial disgrace" (*Rassenschande*), a term the Nazis used later.[56]

Whether *Deutschlands Erneuerung* or the books Lehmann published merely confirmed views Hitler already held, or whether they actually exerted intellectual influence on Hitler, it is clear from examining Lehmann's ideology and his publishing program that eugenics, scientific racism, and anti-Semitism were circulating in Munich in circles close to Hitler. Whatever aspects of racist eugenics Hitler had not already incorporated into his worldview in Vienna were easily available in Munich.

Lehmann was not unusual in fusing racist eugenics, nationalism, and anti-Semitism. Many of the leading anti-Semitic thinkers in early twentieth-century Germany regarded racist eugenics a central part of their program for renewing the German nation. Theodor Fritsch certainly endorsed racist eugenics in his influential work, *Handbook on the Jewish Question*. By 1918 Fritsch was the most influential German publicist for anti-Semitism, not only through this book, but also through other writings issuing from his Hammer Publishing House. When the thirtieth edition of Fritsch's *Handbook* appeared in 1931, Hitler wrote a blurb to help advertise the book, stating, "Already in my youth in Vienna I thoroughly studied *The Handbook on the Jewish Question*. I am convinced that this work contributed in a special way to prepare the soil for the National Socialist anti-Semitic movement."[57] The Nazi Party Archive in Munich owned Fritsch's work, *A New Faith: Confession of the German Renewal Community*, wherein Fritsch promoted the "breeding of the strong and fit" Germans, while opposing race mixture as harmful.[58] Whether or not Hitler actually read Fritsch, as he later claimed, both of them shared common ideas about scientific racism, eugenics, and anti-Semitism.

Various scholars have suggested other possible influences on the development of Hitler's ideas about scientific racism and eugenics. Many are plausible, but unfortunately almost all lack certitude. In 1940 the famous geography professor in Munich, Karl Haushofer, claimed that Hitler thoroughly studied Ratzel's *Political Geography* while he was in Landsberg prison in 1923–24, during the same time he was composing *Mein Kampf*.[59] We know that Hitler began using Ratzel's concept of *Lebensraum* in the early 1920s to justify expansionism and racial struggle, so it would not be surprising if he read Ratzel's writings. However, we cannot be sure.

It seems unlikely that we will ever know the exact sources of Hitler's worldview. However, by examining Hitler's mature worldview and comparing it with the views of other leading scientists, physicians, professors, and social thinkers, it is apparent that social Darwinist racism, evolutionary ethics, and eugenics were not idiosyncratic views of radical "pseudoscientific" thinkers on the fringes. Even if Hitler imbibed the ideas from crass popularizers, the popularizers had derived these ideas from reputable scholars. Though not uncontested, they were mainstream ideas of respectable, leading thinkers in the German academic community. Many biologists and anthropologists at major universities, as we have seen, embraced ideas about racial inequality, racial struggle, eugenics, and euthanasia similar to Hitler's views. This helps make intelligible the willingness of "ordinary Germans," and even more so, leading physicians and scientists, to actively aid and abet Nazi atrocities.

Furthermore, if we look at the careers of leading scientists and physicians during the Nazi era, the complicity of scientists and physicians in the Nazi atrocities becomes even more obvious, as many recent studies have demonstrated.[60] Rüdin, who was named director of the Kaiser Wilhelm Institute for Psychiatry in Munich in 1932 (before the Nazis came to power), co-authored the official commentary on the Nazi sterilization law passed in July 1933. This compulsory sterilization law resulted in the sterilization of over 350,000 people on the basis of their alleged hereditary "defects," which included not only congenital mental and physical illnesses, but also alcoholism. Rüdin and many other psychiatrists and physicians participated in the Hereditary Health Courts set up to decide on sterilizations, and many more provided expert advice to the courts, usually based solely on previous medical records. Normally they did not even examine the individual in question. Rüdin was an enthusiastic supporter of Nazi policies, serving on official government committees related to eugenics. In a January 1943 article—a year-and-a-half after the beginning of the Holocaust—Rüdin not only praised the Nazi sterilization law and the Nuremberg racial laws, but he also extolled the Nazi's "combat against parasitic foreign-blooded races, like the Jews and Gypsies."[61]

Fischer and Lenz also enthusiastically supported the Nazi regime in its eugenics policies. Before the Nazis came to power Fischer had already served six years as director of the Kaiser Wilhelm Institute for Anthropology, Human Genetics, and Eugenics in Berlin. In 1933 he offered his expertise to help the regime frame population and eugenics policies, and, like Rüdin, he served on the Hereditary Health Courts. When Fischer retired in 1941, he was succeeded by a hand-picked successor, Otmar von Verschuer, who shared his views on racist eugenics, and three years later Fischer's institute was renamed the Eugen–Fischer–Institute in his honor. Lenz also advanced in his career because of his support for Nazi eugenics and racial policies. Shortly after the Nazis came to power he became director of the eugenics section of the Kaiser Wilhelm Institute in Berlin. Later Lenz even participated in the drafting of euthanasia legislation during the Nazi regime.

When the Nazis finally implemented their "euthanasia" program by Hitler's decree in 1939 after World War II began, they recruited many physicians with views on social Darwinism, eugenics, and racism similar to those of Rüdin, Lenz, Fischer, Ploetz, and other eugenicists we have already discussed. Nazi propaganda films, such as the documentary *Hereditary Illness* (1936) and the feature film *I Accuse* (1941), wooed Germans to the idea of euthanasia in the 1930s and 1940s (American eugenicists circulated some of these films in the United States, too, in the 1930s). If the title of one of these films, *All Life Is Struggle* (1937), is not Darwinian enough, then the commentary made it explicit. While showing a disfigured handicapped person, the narrator in *Victim of the Past* (1937) declared, "Everything in the natural world that is weak for life will ineluctably be destroyed. In the last few decades, mankind has sinned terribly against the law of natural selection. We haven't just maintained life unworthy of life, we have even allowed it to multiply. The descendants of these sick people look like this!"[62] Over 70,000 people perished in the "euthanasia" program at the hands of physicians, who were willing participants, because they were committed to a racist eugenics ideology that the Nazis favored.[63]

Many of the medical personnel participating in the "euthanasia" program later staffed the death camps, providing medical expertise to help the killing machinery run more smoothly. Physicians not only participated in the frequent "selections" of prisoners, deciding who would live or die, but they were also required to be on hand whenever prisoners were murdered in the gas chambers. The infamous Joseph Mengele, a protégé of the famous eugenicist Verschuer, even used his position as physician in Auschwitz to carry out experiments related to heredity. The tragic irony of Mengele and his like-minded colleagues is that while committing some of the worst atrocities in all of history, he hoped his experiments would help improve the human race.

Indeed Nazi barbarism was motivated by an ethic that prided itself on being scientific. The evolutionary process became the arbiter of all morality. Whatever promoted the evolutionary progress of humanity was deemed good, and whatever hindered biological improvement was considered morally bad. Multitudes must perish in this Malthusian struggle anyway, they reasoned, so why not improve humanity by speeding up the destruction of the disabled and the inferior races? According to this logic, the extermination of individuals and races deemed inferior and "unfit" was not only morally justified, but indeed, morally praiseworthy. Thus Hitler—and many other Germans—perpetrated one of the most evil programs the world has ever witnessed under the delusion that Darwinism could help us discover how to make the world better.

Conclusion ᕲ

S ince its advent in the mid-nineteenth century Darwinism has stirred up debate about many questions touching the very heart of human existence. Not least among these is: How should we live? While many philosophers and theologians ruled this question outside the purview of science, most prominent advocates of Darwinian theory—including biologists, physicians, social theorists, and popularizers—believed Darwinism had far-reaching ramifications for ethics and morality. Many argued that by providing a naturalistic account of the origin of ethics and morality, Darwinism delivered a death-blow to the prevailing Judeo-Christian ethics, as well as Kantian ethics and any other fixed moral code. If morality was built on social instincts that changed over evolutionary time, then morality must be relative to the conditions of life at any given time. Darwinism— together with other forms of historicism ascendant in the nineteenth century—thus contributed to the rise of moral relativism.

But, interestingly, many Darwinists were not willing to live with complete moral relativism. They still retained one fixed point of reference—the process of evolution itself. Since morality arose through evolution, they argued that the purpose of morality is to advance the evolutionary process. They thereby imported the nineteenth-century cult of progress into evolutionary theory. The problem with this is that it presupposed that some forms of morality are "better" than others. But, of course, Darwinism provided no basis to consider some form of morality "better" than any other, or for that matter, it gave no reason to think that morality was "better" in any real sense than immorality. Yet most used morally charged language quite freely, apparently oblivious to the contradiction this entailed.

Those Darwinists who made the evolutionary process the new criteria for morality radically altered the way that people thought about morality. Since they generally affirmed that good health and intelligence were key factors in the upward march of evolution, improving physical vitality and mental prowess—especially of future generations—became the highest moral virtue. The greatest sin was to contribute in some way to the decline of physical life or intellectual ability. This kind of evolutionary ethics flew in the face of Christian morality, in which one's health, vitality, and mental faculty play no role in determining moral or immoral behavior. While Christian morality demands a relationship of love toward God and one's neighbor, which involves self-sacrifice, evolutionary ethics focussed on breeding better humans, even if it meant sacrificing other people in the process. In some places the old and new morality might intersect, and indeed many proponents of evolutionary ethics carried a lot of baggage from traditional ethics into their "new morality." However, the foundations had shifted. This new stress on evolutionary progress and health as the norm for behavior spawned the eugenics movement around the turn of the twentieth century, which was overtly founded on Darwinian principles.

Darwinism also contributed to a rethinking of the value of human life in the late nineteenth century. In order to make human evolution plausible, prominent Darwinists argued that humans were not qualitatively different from animals. Also, the significance of the individual life did not seem all that great considering the mass death brought on by the Darwinian struggle for existence. Multitudes necessarily died before reproducing, and this was the key to evolutionary progress. Death was no longer a foe, as Christianity taught, but a beneficent force. Also, Darwinism stressed biological inequality, since evolution could not occur without significant variation. Humans were no exception, argued many Darwinists, so egalitarianism must be misguided.

These views on human inequality, the primacy of evolutionary progress, and the beneficence of death in furthering that process produced a worldview that devalued human life. Many used Darwinian arguments to assign some humans to the category of "inferior" or degenerate. Generally they considered two main categories of people "inferior": the handicapped and non-European races. Since they were "inferior," and since the death of the less fit in the struggle for existence will result in biological improvement, why not help evolution along by getting rid of the "inferior"?

To be sure, some Darwinists and eugenicists retained enough moral influence from their upbringing to resist the move to kill the "inferior." Instead, they often promoted a variety of measures of reproductive control to achieve their ends. Eugenicists could not agree on the best concrete reforms to improve the biological health of future generations, but they generally agreed that Christian sexual morality must be abandoned. Some proposed marriage reforms, others preferred free sex, while a few even supported polygamy. They agreed, however, that sexual morality must be subservient to the goal of increasing biological health and thus promoting evolution.

Following the lead of Haeckel, a number of Darwinists and eugenicists took these ideas in even more radical directions. They forthrightly promoted the killing of the handicapped and those of "lower" races. They rejected the dominant Christian attitude that placed value on the lives of the weak, the sick, and the handicapped. They denied that "lower" races could be elevated to the status of civilized people. Rather than allowing such people to drain the precious, limited resources of the earth, it would be better to kill them to make space for the healthy, vigorous, and intelligent to flourish.

Many Darwinists also believed that moral characteristics were hereditary. They thought normal Europeans were not only physically and mentally, but also morally, superior to the handicapped and non-Europeans. Thus ridding the world of these "inferior" people would actually result in the advancement of morality. Of course, they failed to notice that Darwinism offered no criteria by which morality could be judged, but they nonetheless affirmed the superiority of European morality (while ironically rejecting the very basis of that morality).

Some might object that Darwinism was not the sole factor producing this change of attitudes about morality and the value of human life. To this I heartily agree. Indeed, it is difficult to know what contributed most to the devaluing of human life—the naturalistic world view in general or biological evolution and Darwinism in particular. One could make a persuasive argument that it was philosophical materialism and monism that devalued human life rather than Darwinism. After all, the eighteenth-century French materialist Julien de La Mettrie called man a machine long before Darwin arrived on the scene.

However, why do we need to choose between Darwinism and philosophical naturalism to explain the devaluing of human life? Surely both were influential. The thinkers we have examined in this work saw Darwinism as an integral—indeed often as the foundational—aspect of their entire worldview. Certainly their view of the human condition relied heavily on their Darwinian understanding. Further, Darwinism played an integral role in the rise of materialism and positivism in the late nineteenth and twentieth centuries.

This study is important, not only because it shows the intersection of Darwinian biology and ethics in the past, especially the way that Darwinism influenced thinking about the value of human life, but also because these debates are still with us today. In the late twentieth and early twenty-first centuries sociobiology and evolutionary psychology are making similar claims about the implications of Darwinism for ethics and morality. Often these scientists and philosophers seem oblivious to the many earlier attempts to wed Darwinism to ethics. Also, many bioethicists today are articulating positions quite similar to the views of the figures in this study. Peter Singer and James Rachels, for example, are contemporary philosophers who argue that Darwinism has effectively undermined the Judeo-Christian doctrine of the sanctity of human life, thereby making involuntary euthanasia permissible in some circumstances, such as in the case of a severely handicapped infant.

Another reason this study is so important is because it gives further insight into the roots of Hitler's worldview and his genocidal mentality. It also helps explain why so many educated Germans would cooperate with the Nazis and participate in the Holocaust, including many medical personnel. When he embraced eugenics, involuntary euthanasia for the handicapped, and racial extermination, Hitler was drawing on ideas that were circulating widely among the educated elites. Klaus Fischer has rightly stated, "Adolf Hitler's racial image of the world was not simply the product of his own delusion but the result of the findings of 'respectable' science in Germany and in other parts of the world, including the United States."[1] These ideas were not dominant in German society, but they were reputable and mainstream in scholarly circles, especially among the medical and scientific elites.

It would be foolish to blame Darwinism for the Holocaust, as though Darwinism leads logically to the Holocaust. No, Darwinism by itself did not produce Hitler's worldview, and many Darwinists drew quite different conclusions from Darwinism for ethics and social thought than did Hitler. Eugenics and scientific racism were prominent in scholarly circles in many European countries and also in the United States, but obviously in none of them did Darwinism lead to the Holocaust. It did lead, however, to the compulsory sterilization of hundreds of thousands in the United States, Sweden, and other countries in the mid-twentieth century. As a result of the resurgence of eugenics in the late twentieth century, China passed the Maternal and Infant Health Law, which required premarital health exams and strongly encouraged sterilization for those deemed unfit to reproduce (while the sterilization measure was theoretically voluntary, Dikötter points out that in practice the Chinese government usually gets its way).[2] Darwinism also spawned debate on euthanasia and infanticide, and even though these are still illegal in most countries (the biggest exception is the Netherlands), they are practiced more widely than many suspect.[3]

To deny the influence of Darwinism on Hitler would also be foolish, however, especially since almost all scholars of Nazism acknowledge it. Richard J. Evans highlights the importance of social Darwinist discourse not only for Hitler, but also for those cooperating with

Hitler's genocidal program:

> Hitler took up this rhetoric and used his own version of the language of social Darwinism as a central element in the discursive practice of extermination. . . . The language of social Darwinism in its Nazi variant had come to be a means of legitimizing terror and extermination against deviants, opponents of the regime, and indeed anyone who did not appear to be wholeheartedly devoted to the war effort. The language of social Darwinism helped to remove all restraint from those who directed the terroristic and exterminatory policies of the regime, and it legitimized these policies in the minds of those who practiced them by persuading them that what they were doing was justified by history, science, and nature.[4]

Darwinism by itself did not produce the Holocaust, but without Darwinism, especially in its social Darwinist and eugenics permutations, neither Hitler nor his Nazi followers would have had the necessary scientific underpinnings to convince themselves and their collaborators that one of the world's greatest atrocities was really morally praiseworthy. Darwinism—or at least some naturalistic interpretations of Darwinism—succeeded in turning morality on its head.

Notes

INTRODUCTION

1. Adam Sedgwick to Charles Darwin, November 24, 1859 [in some works this letter is misdated as December 24, 1859], in *The Correspondence of Charles Darwin*, vol. 7: *1858–1859* (Cambridge, 1991), 397. I have added punctuation to the original.
2. Rudolf Schmid, *Die Darwinischen Theorien* (Stuttgart, 1876), 3.
3. Robby Kossmann, "Die Bedeutung des Einzellebens in der Darwinistischen Weltanschauung," *Nord und Süd* 12 (1880): 420–1. Emphasis is mine.
4. The Haeckel-to-Hitler (and Haeckel-to-Fascism) thesis is pursued relentlessly by Daniel Gasman in *The Scientific Origins of National Socialism* (London, 1971), and *Haeckel's Monism and the Birth of Fascist Ideology* (New York, 1998). Gasman's work is not highly regarded by most historians, and with good cause. However, other scholars have linked social Darwinism to Hitler in a more reasonable fashion: Mike Hawkins, *Social Darwinism in European and American Thought* (Cambridge, 1997), ch. 11; Richard J. Evans, "In Search of German Social Darwinism: The History and Historiography of a Concept," in *Medicine and Modernity* (Washington, 1997), 55–79; Peter Emil Becker, *Wege ins Dritte Reich*, (vol. 1: *Zur Geschichte der Rassenhygiene*; vol. 2: *Sozialdarwinismus, Rassismus, Antisemitismus und volkischer Gedanke*) (Stuttgart, 1988–90); Jürgen Sandmann, "Ansätze einer biologistischen Ethik bei Ernst Haeckel und ihre Auswirkungen auf die Ideologie des Nationalsozialismus," in *Heilen—Verwahren—Vernichten*, ed. G. Wahl and W. Schmitt (Reichenbach, 1997), 83–92; Jürgen Sandmann, *Der Bruch mit der humanitären Tradition* (Stuttgart, 1990); Günter Altner, *Weltanschauliche Hintergründe* (Zürich, 1968), chs. 1–2; Mario di Gregorio, "Entre Méphistophélès et Luther: Ernst Haeckel et la réforme de l'univers," in *Darwinisme et Société*, ed. Patrick Tort (Paris, 1992), 237–83; C. Vogel, "Rassenhygiene—Rassenideologie—Sozialdarwinismus: die Wurzeln des Holocaust," in *Dienstbare Medizin* (Göttingen, 1992), 11–31; Gerhard Baader, "Sozialdarwinismus—Vernichtungsstrategien im Vorfeld des Nationalsozialismus," in *"Bis endlich der langersehnte Umschwung kam . . ."* (Marburg, 1991), 21–35.

236 *Notes*

5. Hans-Günter Zmarzlik, "Social Darwinism in Germany, Seen as a Historical Problem," in *Republic to Reich: The Making of the Nazi Revolution*, ed. Hajo Holborn, trans. Ralph Manheim (New York, 1972), 466; Peter Weingart, Jürgen Kroll, and Kurt Bayertz make very similar statements about Darwinism and ethics in *Rasse, Blut, und Gene* (Frankfurt, 1988), 16–18.

6. Evans, "In Search of German Social Darwinism," 55–79. The revisionist receiving much of his criticism is Alfred Kelly, *The Descent of Darwin* (Chapel Hill, 1981).

7. Marx to Engels, December 19, 1860, in *Marx–Engels Werke* (Berlin, 1959 ff.), XXX: 131. For more on the socialist reception of Darwinism in Germany, see Richard Weikart, *Socialist Darwinism* (San Francisco, 1999); Kelly, *Descent of Darwin*, ch. 7; Ted Benton, "Social Darwinism and Socialist Darwinism in Germany: 1860 to 1900," *Rivista di filosofia* 73 (1982): 79–121; Kurt Bayertz, "Naturwissenschaft und Sozialismus. Tendenzen der Naturwissenschaft-Rezeption in der deutschen Arbeiterbewegung des 19. Jahrhundert," *Social Studies of Science* 13 (1983): 355–94; Hans-Josef Steinberg, *Sozialismus und deutsche Sozialdemokratie* (Hanover, 1967), ch. 3.

8. Atina Grossmann, *Reforming Sex* (New York, 1995), vii. On the German socialists' receptivity to eugenics, see Michael Schwartz, *Sozialistische Eugenik* (Bonn, 1995); and Doris Byer, *Rassenhygiene und Wohlfahrtspflege* (Frankfurt, 1988).

9. Steven E. Aschheim, *In Times of Crisis* (Madison, 2001), 111.

10. Kevin Repp, *Reformers, Critics, and the Paths of German Modernity, 1890–1914* (Cambridge, MA, 2000), 322; Paul Weindling also shows the ambiguities of eugenics, which promoted social reform, but also increased the authoritarian social control of health professionals, laying the groundwork for Nazi health measures; see Weindling, *Health, Race and German Politics* (Cambridge, 1989). Richard F. Wetzell shows similar ambiguities among criminologists in *Inventing the Criminal* (Chapel Hill, 2000).

11. Repp, *Reformers*, 130. Geoff Eley makes a similar point in the introduction to *Society, Culture, and the State in Germany, 1870–1930*, ed. Geoff Eley (Ann Arbor, 1996), 14.

12. Andrew Zimmerman, *Anthropology and Antihumanism in Imperial Germany* (Chicago, 2001), 10–11, 242–5; see also Pascal Grosse, *Kolonialismus, Eugenik und Bürgerliche Gesellschaft in Deutschland, 1850–1918* (Frankfurt, 2000), who shows how eugenicists' influence on colonial policy linked to the rise of the Nazi racial state (though he denies it is a straight line from one to the other).

13. Heinz-Georg Marten does, however, trace the some connections between Social Darwinism, racism, and anti-Semitism in "Racism, Social Darwinism, Anti-Semitism and Aryan Supremacy," *International Journal of the History of Sport* 16, 2 (1999): 23–41. `

14. For more on Jews' involvement in race science, see John Efron, *Defenders of the Race* (New Haven, 1994).

15. Sheila Faith Weiss, *Race Hygiene and National Efficiency* (Berkeley, 1987), 158; see also Weiss, "The Race Hygiene Movement in Germany" *Osiris*, 2nd series,

3 (1987): 236; Detlev Peukert, "The Genesis of the 'Final Solution' from the Spirit of Science," in *Reevaluating the Third Reich*, ed. Thomas Childers and Jane Caplan (New York, 1993); Christian Pross and Götz Aly, eds., *Der Wert des Menschen* (Berlin, 1989); Weingart et al., *Rasse, Blut, und Gene*.

16. Renate Bridenthal et al., eds., *When Biology Became Destiny: Women in Weimar and Nazi Germany* (New York: Monthly Review Press, 1984).

17. Many scholars discuss the complex issue of Nazism and modernity, but see especially Michael Burleigh and Wolfgang Wippermann, *The Racial State: Germany, 1933–1945* (Cambridge, 1991); Norbert Frei, "Wie modern war der Nationalsozialismus," *Geschichte und Gesellschaft* 19 (1993): 367–87; Jeffrey Herf, *Reactionary Modernism* (Cambridge, 1984); Zygmunt Bauman, *Modernity and the Holocaust* (Ithaca: Cornell University Press, 1989); Michael Prinz and Rainer Zitelmann, eds., *Nationalsozialismus und Modernisierung* (Darmstadt, 1991); Richard Rubenstein, "Modernization and the Politics of Extermination," in *A Mosaic of Victims*, ed. Michael Berenbaum (New York, 1990), 3–19; and Rainer Zitelmann, *Hitler: Selbstverständnis eines Revolutionärs* (Hamburg, 1987). Mitchell G. Ash, *Gestalt Psychology in German Culture, 1880–1967: Holism and the Qauest for Objectivity* (Cambridge, 1995), and Anne Harrington, *Reenchanted Science: Holism in German Culture from Wilhelm II to Hitler* (Princeton, 1996), both show that scientific thought in early-twentieth-century Germany contained ambiguities in relation to modernity and rationalism.

18. Michael Burleigh, "The Legacy of Nazi Medicine in Context," in *Medicine and Medical Ethics in Nazi Germany*, ed. Francis Nicosia and Jonathan Huener (New York, 2002), 119.

19. Fritz Lenz, *Die Rasse als Wertprinzip, Zur Erneuerung der Ethik* (Munich, 1933), 39, 7.

20. Adolf Hitler, *Mein Kampf*, 2 vols. in 1 (Munich, 1943), 420–1. Emphasis is mine.

21. Ian Kershaw in his magisterial biography of Hitler repeatedly mentions social Darwinism as a key component of Hitler's world view; see *Hitler*, 2 vols. (New York, 1998–2000), 1:78, 134–7, 290; 2:19, 208, 405, 780; Eberhard Jäckel also sees the Darwinian racial struggle as an important part of Hitler's ideology in *Hitler's Weltanschauung* (Middleton, CN, 1972), ch. 5; Brigitte Hamann discusses Darwinian racial thought in Hitler's Viennese milieu in *Hitler's Vienna* (New York, 1999), 82, 84, 102, 202–3. See also Burleigh and Wippermann, *The Racial State*, ch. 2.

22. Hamann, *Hitler's Vienna*, 233; Joachim Fest, *Hitler*, trans. Richard and Clara Winston (New York, 1974), 54, likewise calls Hitler's social Darwinism pseudoscientific. Gasman in *Scientific Origins* and *Haeckel's Monism* makes a similar move by distancing Haeckel from Darwin (and by wrongly portraying Haeckel's philosophy as mystical and vitalistic).

23. For example, Robert Proctor, *Racial Hygiene: Medicine under the Nazis* (Cambridge, 1988); Ute Deichmann, *Biologists under Hitler* (Cambridge, MA, 1996); Benno Müller-Hill, *Murderous Science* (Oxford, 1988). Michael

Burleigh shows an awareness of this problem when he recently stated that "we may be overstating Nazism as an aberrant branch of the scientific imagination, thereby overlooking the extent to which even the scientists were informed by what might be described as historical fantasizing" (Burleigh, "The Legacy of Nazi Medicine in Context," 119).

24. Richard Weikart, "Progress through Racial Extermination: Social Darwinism, Eugenics, and Pacifism, 1860–1918," *German Studies Review* 26 (2003): 273–94.
25. Quoted in Friedrich Hellwald, *Naturgeschichte des Menschen*, 2 vols. (Stuttgart, 1880), 1:66.
26. Ian Dowbiggin, *A Merciful End* (Oxford, 2003), 8.
27. N. D. A. Kemp, "*Merciful Release*" (Manchester, 2002), 19.
28. Greta Jones, *Social Darwinism and English Thought* (Sussex, 1980); Hawkins, *Social Darwinism in European and American Thought*; Daniel J. Kevles, *In the Name of Eugenics* (Berkeley, 1985); Diane B. Paul, *Controlling Human Heredity, 1865 to the Present* (Atlantic Highlands, NJ, 1995); Gunnar Broberg, and Nils Roll-Hansen, eds., *Eugenics and the Welfare State* (East Lansing, MI, 1996); Nancy Stepan, *"The Hour of Eugenics": Race, Gender and Nation in Latin America* (Ithaca, NY, 1991); William H. Schneider, *Quality and Quantity* (Cambridge, 1990); Frank Dikötter, *Imperfect Conceptions* (New York, 1998).
29. Quoted in Paul, *Controlling Human Heredity*, 17; Phillip Gassert and Daniel S. Mattern, eds. *The Hitler Library: A Bibliography* (Westport, CT, 2001).
30. Stefan Kühl, *The Nazi Connection* (Oxford, 1994).
31. *The Life and Letters of Charles Darwin*, ed. Francis Darwin (New York, 1919), 2:270.
32. David Friedrich Strauss, *Der alte und der neue Glaube* (Leipzig, 1872); Friedrich Albert Lange, *Die Arbeiterfrage in ihrer Bedeutung für Gegenwart und Zukunft* (Duisburg, 1865; rprt. Duisburg, 1975).
33. Ludwig Woltmann, *Die Darwinsche Theorie und der Sozialismus* (Düsseldorf, 1899). For a more precise definition of social Darwinism, see Hawkins, *Social Darwinism in European and American Thought*, ch. 1. For a more restricted definition, see Kurt Bayertz, "Darwinismus als Politik: Zur Genese des Sozialdarwinismus in Deutschland 1860–1900," in *Welträtsel und Lebenswunder* (Linz, 1998), 241.
34. Richard Goldschmidt, *Portraits from Memory* (Seattle, 1956), 34.
35. Quoted in Jürgen Sandmann, "Ernst Haeckels Entwicklungslehre als Teil seiner biologistischen Weltanschauung," in *Die Rezeption von Evolutionstheorien im 19. Jahrhundert*, ed. Eve-Marie Engels (Frankfurt, 1995), 330.
36. Ernst Haeckel, *Natürliche Schöpfungsgeschichte* (Berlin, 1868), 16, 125–8, 206; some scholars wrongly claim that because Haeckel was a Lamarckian, he rejected natural selection—see Peter Bowler, *The Non-Darwinian Revolution* (Baltimore, 1988), 83; and Britta Rupp-Eisenreich, "Le darwinisme social en Allemagne," in *Darwinisme et Société*, ed. Patrick Tort (Paris, 1992), 169–236. Probably the most balanced portrait of Haeckel is Erika Krausse, *Ernst Haeckel*, 2nd ed. (Leipzig, 1984).

37. Haeckel, *Natürliche Schöpfungsgeschichte*, 5, quote at 487.
38. Ernst Haeckel to Charles Darwin, February 9, 1879, in Georg Uschmann, *Ernst Haeckel: Biographie in Briefen* (Gütersloh, 1983), 156.
39. For one example, see Haeckel, *Die Welträthsel*, Volksausgabe (Bonn, 1903), 11.
40. Ludwig Büchner to Hermann Schaffhausen, July 11, 1863, in Autographen, S 2620a, Handschriftenabteilung, University of Bonn Library; see also Büchner, *Der Mensch und seine Stellung*, 2nd ed. (Leipzig, 1872), 5–6.
41. Alfred Grotjahn, *Erlebtes und Erstrebtes* (Berlin, 1932), 44.
42. Heinrich Schmidt, ed., *Was wir Ernst Haeckel verdanken*, 2 vols. (Leipzig, 1914); Kelly, *Descent of Darwin*.
43. Ernst Haeckel, *Der Monismus als Band zwischen Religion und Wissenschaft* (Bonn, 1892), 10; Haeckel, *Die Welträthsel*, 116–17. Gasman incorrectly interprets Haeckel's pantheism as mystical in *Scientific Origins of National Socialism*, xiii–xiv; and *Haeckel's Monism and the Birth of Fascist Ideology*, 15, 34. Niles Holt also stresses Haeckel's pantheistic side in "Ernst Haeckel's Monistic Religion," *Journal of the History of Ideas* 32 (1971): 265–80. Gasman's position is odd, since Haeckel devoted an entire speech to the annual Monist Congress in 1913 to voice his antipathy to mysticism—see Ernst Haeckel, "Monismus und Mystik," in *Der Düsseldorfer Monistentag,* ed. Wilhelm Blossfeldt (Leipzig, 1914), 93–8. Jürgen Sandmann not only interprets Haeckel's monism as essentially materialistic, but notes that almost all the secondary literature follows this line—see Sandmann, *Der Bruch mit der humanitären Tradition*, 52, 60–1. An excellent discussion of the tension in Haeckel's thought between materialism and pantheism is O. Breidbach, "Monismus um 1900—Wissenschaftspraxis oder Weltanschauung?" in *Welträtsel und Lebenswunder: Ernst Haeckel—Werk, Wirkung, und Folgen,* ed. Erna Aescht et al. (Linz, 1998), 289–316. For further discussion of this issue, see Richard Weikart, " 'Evolutionäre Aufklärung'? Zur Geschichte des Monistenbundes," in *Wissenschaft, Politik, und Öffentlichkeit,* ed. Mitchell G. Ash and Christian H. Stifter (Vienna, 2002), 131–48.
44. Frederick Gregory, *Nature Lost? Natural Science and the German Theological Traditions of the Nineteenth Century* (Cambridge: Harvard University Press, 1992).
45. On Dilthey, see Rudolf A. Makkreel, *Dilthey: Philosopher of the Human Studies* (Princeton: Princeton University Press, 1992); on neo-Kantianism, see Klaus Christian Köhnke, *The Rise of Neo-Kantianism* (Cambridge, 1991).
46. Annette Wittkau-Horgby, *Materialismus* (Göttingen, 1998), chs. 7–8, concurs with my view.
47. Weindling, *Health, Race and German Politics*; Weingart et al., *Rasse, Blut, und Gene*; Paul, *Controlling Human Heredity*, ch. 2; Kevles, *In the Name of Eugenics*; Becker, *Wege ins Dritte Reich*.
48. Alfred Ploetz to Carl Hauptmann, January 14, 1892, in Carl Hauptmann papers, K 121, Akademie der Künste Archives, Berlin; Alfred Ploetz to Ernst Haeckel, April 18, 1902, in Ernst Haeckel papers, Ernst-Haeckel-Haus, Jena; for more on Ploetz, see W. Doeleke, "Alfred Ploetz (1860–1940): Sozialdarwinist und Gesellschaftsbiologe," dissertation, University of Frankfurt, 1975.

49. Alfred Ploetz to Ernst Haeckel, October 19, 1903, in Alfred Ploetz papers, privately held by Wilfried Ploetz, Herrsching am Ammersee; "Aufforderung zum Abonnement. Archiv für Rassen- und Gesellschafts-Biologie," August 1903, in Felix von Luschan papers (among Alfred Ploetz's letters), in Staatsbibliothek Preussischer Kulturbesitz, Berlin.

50. Heinrich Ernst Ziegler, "Einleitung zu dem Sammelwerke Natur und Staat," in *Natur und Staat*, vol. 1 (bound with Heinrich Matzat, *Philosophie der Anpassung*) (Jena, 1903), 1–2; Klaus-Dieter Thomann and Werner Friedrich Kümmel, "Naturwissenschaft, Kapital und Weltanschauung: Das Kruppsche Preisausschreiben und der Sozialdarwinismus," *Medizinhistorisches Journal* 30 (1995): 99–143, 205–43. Sheila Faith Weiss provides a good discussion of the Krupp Prize competition in *Race Hygiene and National Efficiency*, 64–74.

51. Wilhelm Schallmayer, *Vererbung und Auslese* (Jena, 1903), ix–x.

52. Wilhelm Schallmayer to Alfred Grotjahn, June 3, 1910, in Alfred Grotjahn papers, Humboldt University Archives, Berlin; Schallmayer makes a similar claim in "Rassedienst," *Sexual-Probleme* 7 (1911): 547.

53. Schallmayer, *Vererbung und Auslese*, ix–x; see also Schallmayer, *Beiträge zu einer Nationalbiologie* (Jena, 1905), 124.

54. Hannsjoachim W. Koch, *Der Sozialdarwinismus Denken* (Munich, 1973), 64.

55. For a good philosophical discussion of the implications of body–soul dualism for biomedical ethics, see J. P. Moreland and Scott Rae, *Body and Soul: Human Nature and the Crisis in Ethics* (Downers Grove, IL, 2000).

56. Darwin, *The Origin of Species* (London: Penguin, 1968), 459.

57. Works stressing accommodationism include David N. Livingstone, *Darwin's Forgotten Defenders* (Grand Rapids, 1987); James R. Moore, *The Post-Darwinian Controversies* (Cambridge, 1979); while rejecting the warfare thesis, some scholars rightly caution against overreacting—see John Hedley Brooke, *Science and Religion: Some Historical Perspectives* (Cambridge, 1991); Holmes Rolston III, *Science and Religion: A Critical Survey* (Philadelphia, 1987); David Lindberg and Ronald Numbers, eds., *God and Nature* (Berkeley, 1986); David N. Livingstone, D. G. Hart, and Mark A. Noll, eds., *Evangelicals and Science in Historical Perspective* (New York, 1999).

58. Peukert, "The Genesis of the 'Final Solution' from the Spirit of Science," 240, 247.

1   THE ORIGIN OF ETHICS AND THE RISE OF MORAL RELATIVISM

1. Charles Darwin, *Autobiography* (New York: Norton, 1969), 94.

2. David Hull, *The Metaphysics of Evolution* (Albany: State University of New York Press, 1989), 75.

3. Darwin, *The Descent of Man* (London, 1871; rprt. Princeton, 1981), 1:106.

4. Adrian Desmond and James Moore, *Darwin* (London, 1991), xxi, 243, 263, 269.

5. Charles Darwin, *Metaphysics, Materialism, and the Evolution of Mind: Early Writings of Charles Darwin*, ed. Paul H. Barrett (Chicago: University of Chicago Press, 1974).

6. Darwin, *Descent*, 1:71–80.

7. Neal Gillespie, *Charles Darwin and the Problem of Creation* (Chicago, University of Chicago Press, 1979).

8. Robert J. Richards, *Darwin and the Emergence* (Chicago, 1987), 108–9; Robert J. Richards, "Darwin's Romantic Biology: The Foundations of His Evolutionary Ethics," in *Biology and the Foundation of Ethics*, ed. Jane Maienschein and Michael Ruse (Cambridge: Cambridge University Press, 1999), 113–53. See also Paul Lawrence Farber, *The Temptations of Evolutionary Ethics* (Berkeley, 1994), ch. 1.

9. Michael Bradie, *The Secret Chain: Evolution and Ethics* (Albany: Suny Press, 1994), ch. 2.

10. Richards, *Darwin and the Emergence* (Chicago, 1987), 116–17.

11. Paul Lawrence Farber, *The Temptations of Evolutionary Ethics* (Berkeley, 1994); C. M. Williams, *A Review of the Systems of Ethics founded on the Theory of Evolution* (London, 1893).

12. Kurt Bayertz, "Darwinismus als Politik: Zur Genese des Sozialdarwinismus in Deutschland 1860–1900," in *Welträtsel und Lebenswunder: Ernst Haeckel— Werk, Wirkung und Folgen*, ed. Erna Aescht et al. (Linz, 1998), 247–9.

13. Ernst Haeckel to Anna Sethe, July 14, 1860, in *Ernst Haeckel: Forscher, Künstler, Mensch*, ed. Georg Uschmann (Leipzig, 1961), 66. For an excellent account of Haeckel's ethics, see Jürgen Sandmann, *Der Bruch mit der humanitären Tradition* (Stuttgart, 1990).

14. Ernst Haeckel, "Ueber die heutige Entwickelungslehre im Verhältnisse zur Gesamtwissenschaft," in *Amtlicher Bericht der 50* (Munich, 1877), 19–20.

15. Ernst Haeckel, *Der Monismus* (Bonn, 1892), 30.

16. Ernst Haeckel, *Die Welträthsel*, Volksausgabe (Bonn, 1903), 134–40; quote at 135–6.

17. Ibid., 55, quote at 140.

18. Ernst Haeckel, *Gott–Natur (Theophysis)*, in *Gemeinverständliche Werke*, ed. Heinrich Schmidt (Leipzig: Alfred Kröner, 1924), 3: 468.

19. Haeckel, *Welträthsel*, 165; Ernst Haeckel to Arnold Dodel, June 26, 1906, in Arnold Dodel papers, Zürich Zentralbibliothek.

20. Bartholomäus von Carneri to Ernst Haeckel, September 4, 1883, in *Bartholomäus von Carneri's Breifwechsel mit Ernst Haeckel und Friedrich Jodl* (Leipzig, 1922), 30.

21. Bartholomäus von Carneri, *Sittlichkeit und Darwinismus* (Vienna, 1871), 123–4.

22. Bartholomäus von Carneri, *Grundlegung der Ethik*, Volksausgabe (Stuttgart, 1881), 1–9; Carneri, "Zur Glückseligkeitslehre," *Kosmos* 11 (1882): 242–3.

23. Carneri, *Sittlichkeit und Darwinismus*, 7, 275, quote at 308.

24. Ibid., 186.

25. Bartholomäus von Carneri. "Die Entwickelung der Sittlichkeitsidee," *Kosmos* 1 (1884): 406.

26. Bartholomäus von Carneri, *Der moderne Mensch*, Volksausgabe (Stuttgart, 1901), vii–viii.

27. Bartholomäus von Carneri to Prof. Spitzer, August 28, 1886 (Abschrift), in Bartholomäus von Carneri papers (Kryptonachlass), in Wilhelm Börner papers, Karton 1, in Handschriftensammlung, Wiener Stadt- und Landes-Bibliothek.

28. Carneri, *Der moderne Mensch*, 7–8; see also Carneri to Ludwig Fleischer, n.d. (Abschrift), in Bartholomäus Carneri papers in Wilhelm Börner papers, in Handschriftensammlung, Wiener Stadt- und Landes-Bibliothek.

29. Friedrich Jodl, "Bartholomaus von Carneri," in *Vom Lebenswege: Gesammelte Vorträge und Aufsätze*, ed. Wilhelm Börner, 2 vols. (Stuttgart, 1916–17), 1: 451–2.

30. Arnold Dodel to Bartholomäus von Carneri, May 16, 1886, in Aut. 178.299, Handschriftensammlung, Wiener Stadt- und Landes-Bibliothek.

31. Albert E. F. Schäffle, "Darwinismus und Sozialwissenschaft," in *Gesammelte Aufsätze* (Tübingen: H. Laupp'sche Buchhandlung, 1885), 1:2.

32. Albert E. F. Schäffle, *Bau und Leben des Socialen Korpers*, 4 vols. (Tübigen, 1881), 1: 585; Schäffle, "Darwinismus und Sozialwissenschaft," in *Gesammelte Aufsatze*, 1: 1–36; Schäffle, "Ueber die Entstehung der Gesellschaft nach den Anschauungen einer sociologischen Zuchtwahltheorie," *Vierteljahrsschrift für wissenschaftliche Philosophie* 1 (1877): 540–53.

33. Albert E. F. Schäffle, "Ueber Recht und Sitte vom Standpunkt der sociologischen Erweiterung der Zuchtwahltheorie," *Vierteljahrsschrift für wissenschaftliche Philosophie* 2 (1878): 66.

34. Schäffle, *Bau und Leben*, 2: 494–5.

35. Ibid., 2: 494.

36. Ibid., 1: 586–8.

37. Max Nordau, *Die konventionellen Lügen der Kulturmenschheit* (Leipzig, 1909), 26.

38. Max Nordau, *Morals and the Evolution of Man* [trans. of *Biologie und Ethik*] (London, 1922), 56–73; quote at 59.

39. Ibid., 48–55, 113, 142, 239.

40. Ibid., 70, quote at 73.

41. Georg von Gizycki, *Philosophische Consequenzen der Lamarck–Darwin'schen Entwicklungstheorie* (Leipzig, 1876), 39; see also Gizycki, "Darwinismus und Ethik," *Deutsche Rundschau* 43 (1885): 266.

42. Gizycki, *Philosophische Consequenzen*, quote at 49, see also 70; Gizycki, "Darwinismus und Ethik," 267.

43. Gizycki, "Darwinismus und Ethik," 279–80, 263–4, 269–71.

44. Ibid., 263–4.

45. Georg von Gizycki, *Grundzüge der Moral* (Leipzig, 1883).

46. Georg von Gizycki, *Moralphilosophie gemeinverständlich dargestellt*, 2nd ed. (Leipzig, 1895), 514–17.

47. Friedrich Jodl, "Morals in History," *International Journal of Ethics* 1 (1891): 208–9; emphasis mine.

48. Friedrich Jodl, *Allgemeine Ethik*, ed. Wilhelm Börner (Stuttgart, 1918), 134–5.

49. Ibid., 210.

50. Friedrich Jodl, "Ueber das Wesen des Naturrechts und seine Bedeutung in der Gegenwart," in *Vom Lebenswege: Gesammelte Vorträge und Aufsätze*, 2: 70–1.

51. Friedrich Jodl to Wilhelm Bolin, September 1, 1890, in *Unter uns gesagt: Friedrich Jodls Briefe an Wilhelm Bolin*, ed. Georg Gimpl (Vienna, 1990), 98.

52. Friedrich Jodl to Carl von Amira, April 13, 1884 (Abschrift), in Friedrich Jodl papers (Kryptonachlass), in Wilhelm Börner papers, Karton 2, in Wiener Stadt- und Landes-Bibliothek, Handschriftensammlung . For more on Jodl, see Wilhelm Börner, *Friedrich Jodl* (Stuttgart, 1911).

53. David Friedrich Strauss, *Der alte und der neue Glaube* (Leipzig, 1872), 202–7, 230–1, 252–4.

54. David Friedrich Strauss to Ernst Haeckel, August 24, 1873, in *Ausgewählte Briefe von David Friedrich Strauss*, ed. Eduard Zeller (Bonn, 1895), 554.

55. Friedrich von Hellwald, *Culturgeschichte* (Augsburg, 1875), quote at 27, see also 278, 569.

56. Ibid., 44–5.

57. Friedrich Jodl, *Die Culturgeschichtsschreibung* (Halle, 1878), 66–81.

58. Alexander Tille to Ernst Haeckel, June 12, 1896, in Ernst Haeckel papers, Ernst-Haeckel-Haus, Jena.

59. Alexander Tille, *Von Darwin bis Nietzsche* (Leipzig, 1895), 204; see also Tille, *Volksdienst* (Berlin, 1893), 82; and Tille, "Die Ueberwindung des Malthusianismus," *Die Zukunft* 9 (1894): 164.

60. Tille, *Volksdienst* (Berlin, 1893), 36.

61. Tille, *Von Darwin bis Nietzsche*, 21. For more on Tille, see Wilfried Schungel, *Alexander Tille (1866–1912): Leben und Ideen eines Sozialdarwinisten* (Husum, 1980); and Peter Emil Becker, *Wege ins Dritte Reich*, vol. 2: *Sozialdarwinismus, Rassismus, Antisemitismus und volkischer Gedanke* (Stuttgart, 1990), ch.10.

62. Wilhelm Schallmayer, *Vererbung und Auslese* (Jena, 1903), 213–14.

63. August Forel, "August Forel" [autobiography], in *Führende Psychiater in Selbstdarstellungen* (Leipzig: Felix Meiner, 1930), 55; see also Forel, *Out of My Life and Work*, trans. Bernard Miall (New York, 1937), 53.

64. August Forel, *Die sexuelle Frage* (Munich, 1905), 440; see also Forel, "Ueber Ethik," *Zukunft* 28 (1899): 578.

65. Forel, *Out of My Life and Work*, 53, 256; Forel, "August Forel," 55.

66. August Forel, *Kulturbestrebungen der Gegenwart* (Munich, 1910), 8–9, quote at 15.

67. Ludwig Büchner, *Die Macht der Vererbung und ihr Einfluss auf den moralischen und geistigen Fortschritt der Menschheit* (Leipzig, 1882), 39.

68. Niles Holt and Jürgen Sandmann both call Büchner an ethical relativist—see Holt, "The Social and Political Ideas of the German Monist Movement, 1871–1914" (Ph.D. dissertation, Yale University, 1967), 70; Jürgen Sandmann, Jürgen, *Der Bruch mit der humanitären Tradition*, 36. For more on

Büchner, see Frederick Gregory, *Scientific Materialism in Nineteenth Century Germany* (Dordrecht, 1977); and Richard Weikart, *Socialist Darwinism* (San Francisco, 1999), ch. 3.

69. In addition to Büchner's many published statements on this matter, see Ludwig Büchner to Ernst Haeckel, December 14, 1895, in Ernst Haeckel papers, Ernst-Haeckel-Haus, Jena; for Haeckel, see Ernst Haeckel to Richard Semon, September 27, 1905 and October 31, 1906, in Richard Semon papers, Bayerische Staatsbibliothek, Munich.

70. August Forel to Richard Semon, December 5, 1905, in Richard Semon's papers, Bayerische Staatsbibliothek, Munich.

71. Darwin, *Descent*, 1: 173.

72. Mariacarla Gadebusch Bondio, *Die Rezeption der kriminalanthropologischen Theorien von Cesare Lombroso in Deutschland von 1880–1914* (Husum, 1995); Hans Kurella, *Cesare Lombroso* (New York, n.d. [1911]); Richard F. Wetzell, *Inventing the Criminal* (Chapel Hill, 2000), ch. 2.

73. Kurella, *Cesare Lombroso*, 116–18; Kurella, *Die Grenzen der Zurechnungsfähigkeit und die Kriminal-Anthropologie* (Halle, 1903), 38.

74. Hans Kurella, *Naturgeschichte des Verbrechers: Grundzüge der criminellen Anthropologie und Criminalpsychologie* (Stuttgart, 1893), 255.

75. Ibid., 13–25.

76. Kurella, *Die Grenzen der Zurechnungsfähigkeit*, 98.

77. Robert Sommer, *Kriminalpsychologie und strafrechtliche Psychopathologie auf naturwissenschaftliche Grundlage* (Leipzig, 1904), 309.

78. Emil Kraepelin, *Lebenserinnerungen* (Berlin, 1983), 27.

79. Emil Kraepelin, *Psychiatrie: Ein kurzes Lehrbuch*, 4th ed. (Leipzig, 1893), 670–2, 60, 189.

80. Wetzell, *Inventing the Criminal*, 59.

81. Otto Binswanger, "Geistesstörung und Verbrechen," *Deutsche Rundschau* 57 (1888): 423.

82. Eugen Bleuler, *Lehrbuch der Psychiatrie* (Berlin, 1916), 139–43, 422–5; see also Bleuler, *Der geborene Verbrecher: Eine kritische Studie* (Munich, 1896), 1–2, 20.

83. Felix von Luschan, "Culture and Degeneration," n.d. [after 1911], in Felix von Luschan papers, Kasten 14, Staatsbibliothek Preussischer Kulturbesitz, Berlin.

84. Wetzell, *Inventing the Criminal*, chs. 2–3.

85. Maurice Mandelbaum, *History, Man, and Reason: A Study in Nineteenth-Century Thought* (Baltimore, 1971), chs. 2–7. Many works discuss how historicism undermined the natural law tradition of morality, including Georg Iggers, *The German Conception of History: The National Tradition of Historical Thought from Herder to the Present* (Middleton: CT, 1983); Iggers, "Historicism: The History and Meaning of the Term," *Journal of the History of Idea* 56 (1995): 129–52; Leonard Krieger, *The German Idea of Freedom: History of a Political Tradition* (Boston: Beacon Press, 1957); Friedrich Meinecke, *Historism: The Rise of a New Historical Outlook* (New York: Herder and Herder, 1972).

86. James T. Kloppenberg, *Uncertain Victory: Social Democracy and Progressivism in European and American Thought, 1870–1920* (Oxford, 1986), 4, 133–41.

2 EVOLUTIONARY PROGRESS AS THE HIGHEST GOOD

1. Arnold Dodel, *Aus Leben und Wissenschaft* (Stuttgart, 1896–1905), 2: 48.
2. Willibald Hentschel to Christian von Ehrenfels, n.d., in Christian von Ehrenfels papers, Forschungsstelle und Dokumentationszetrum für Oesterreichische Philosophie, Graz.
3. Richard Weikart, "The Origins of Social Darwinism in Germany, 1859–1895," *Journal of the History of Ideas* 54 (1993): 469–88.
4. Gustav Ratzenhofer, *Wesen und Zweck der Politik* (Leipzig, 1893), 1: 3 and passim.
5. Gustav Ratzenhofer, *Positive Ethik* (Leipzig, 1901), 39, quote at 66.
6. Alexander Tille, "Zwei ethische Welten," *Die Zukunft* 4 (1893): 255; Tille, *Von Darwin bis Nietzsche* (Leipzig, 1895), 172–3.
7. Tille, *Von Darwin bis Nietzsche*, 23–4.
8. Alexander Tille, "Deutsche Darwinisten als Sozialethiker," *Die Zukunft* 8 (1894): 519.
9. Tille, *Von Darwin bis Nietzsche*, 214.
10. Ibid., 232–4.
11. See, e.g., Walter Kaufmann, *Nietzsche: Philosopher, Psychologist, Antichrist*, 4th ed. (Princeton, 1974); Alexander Nehamas, *Nietzsche: Life as Literature* (Cambridge, MA, 1985); Lewis Call, "Anti-Darwin, Anti-Spencer: Nietzsche's Critique of Darwin and 'Darwinism,' " *History of Science* 36 (1998): 1–22.
12. Jean Gayon, "Nietzsche and Darwin," in *Biology and the Foundation of Ethics*, eds. Jane Maienschein and Michael Ruse (Cambridge, 1999), 154–97.
13. Peter Weingart, Jürgen Kroll, and Kurt Bayertz, *Rasse, Blut, und Gene.* (Frankfurt, 1988), 70–2.
14. The best work on the selective reception of Nietzsche is Steven E. Aschheim, *The Nietzsche Legacy in Germany, 1890–1990* (Berkeley, 1992).
15. Rüdiger Safranski, *Nietzsche: A Philosophical Biography*, trans. Shelley Frisch (New York, 2002); Gregory Moore, *Nietzsche: Biology and Metaphor* (Cambridge, 2002); Jean Gayon, "Nietzsche and Darwin"; Steven E. Aschheim, *In Times of Crisis* (Madison, WI, 2001), chs. 1–2.
16. Friedrich Nietzsche, *David Strauss: in Unzeitgemässe Betrachtungen*, § 7. I will cite the sections in Nietzsche rather than page numbers because of the many different editions and translations of his work. All translations of Nietzsche are mine, except for citations from *The Will to Power*.
17. Thomas Brobjer, *Nietzsche's Ethics of Character* (Uppsala, 1995), 135.
18. Brendan Donnellan, "Friedrich Nietzsche and Paul Rée: Cooperation and Conflict," *Journal of the History of Ideas* 43 (1982): 595–612.
19. Paul Rée, *Der Ursprung der moralischen Empfindungen* (Chemnitz, 1877), 7, 12–13, 23, 29, 34, 42–4, 136–8; quote at vii–viii.
20. Quoted in Moore, *Nietzsche: Biology and Metaphor*, 53.
21. Friedrich Nietzsche, *Die fröhliche Wissenschaft*, § 349 (a passage added to the 1887 edition).

22. Thomas H. Brobjer, "Nietzsche's Reading and Private Library, 1885–1889," *Journal of the History of Ideas* 58 (1997): 663–93; Moore, *Nietzsche: Biology and Metaphor*, chs. 1–2.

23. Nietzsche, *The Will to Power*, trans. Walter Kaufmann (New York, 1967), § 734.

24. Safranski, *Nietzsche*, 262.

25. Nietzsche, *Zarathustra*, Part 1, "On Voluntary Death."

26. Nietzsche, *Die fröhliche Wissenschaft*, § 73.

27. Nietzsche, *Götzen-Dämmerung*, § IX, 36.

28. Nietzsche, *Ecce Homo*, "Die Geburt der Tragödie," § 4.

29. Wilhelm Schallmayer to Ludwig Schemann, September 6, 1912, in Ludwig Schemann papers, IV B 1/2, University of Freiburg Library, Handschriftensammlung.

30. Wilhelm Schallmayer, "Zum Einbruch der Naturwissenschaft in das Gebiet der Geisteswissenschaften," *Archiv für Rassen- und Gesellschaftsbiologie* 1 (1904): 586–97; Schallmayer, "Zur sozialwissenschaftlichen und sozialpolitischen Bedeutung der Naturwissenschaften, besonders der Biologie," *Vierteljahrsschrift für wissenschaftliche Philosophie* 29 (1905): 495–512.

31. Wilhelm Schallmayer, *Vererbung und Auslese im Lebenslauf der Völker* (Jena, 1903), 322.

32. Wilhelm Schallmayer, "Generative Ethik," *Archiv für Rassen- und Gesellschaftsbiologie* 6 (1909): 199–231; Schallmayer, "Ueber die Grundbedeutung der Ethik und ihr Verhältnis zu den Forderungen des Rassedienstes," *Die neue Generation* 6 (1910): 493.

33. Schallmayer, *Vererbung und Auslese*, ix–x, 381, quote at 243.

34. Ibid., 235, 245, 250; Schallmayer, "Generative Ethik," 201–10.

35. Schallmayer, "Ueber die Grundbedeutung der Ethik," 495–6.

36. Wilhelm Schallmayer, *Beiträge zu einer Nationalbiologie* (Jena, 1905), 123; Schallmayer, "Generative Ethik," 230.

37. Schallmayer, *Vererbung und Auslese*, 307, quote at 227; Schallmayer, "Generative Ethik," 200–1.

38. Schallmayer, "Generative Ethik," 209.

39. Alfred Ploetz, "Vorwort," *Archiv fur Rassen- u. Gesellschaftsbiologie* 1 (1904): vi; see also "Aufforderung zum Abonnement. Archiv für Rassen- und Gesellschafts-Biologie," August 1903, in Felix von Luschan papers, Staatsbibliothek Preussischer Kulturbesitz, Berlin.

40. Alfred Ploetz, "Die Begriffe Rasse und Gesellschaft und die davon abgeleiteten Disziplinen," *Archiv für Rassen- und Gesellschaftsbiologie* 1 (1904): 2.

41. Alfred Ploetz, *Die Tüchtigkeit* (Berlin, 1895), 10–11.

42. Ibid., 11, 146, 197–8, 224–5; Alfred Ploetz, "Die Begriffe Rasse und Gesellschaft und einige zusammenhängende Probleme," *Verhandlungen des Ersten Deutschen Soziologentages* 1 (1911): 135–6.

43. Alfred Ploetz, "Denkschrift über die Gründung der Internationalen Gesellschaft für Rassen-Hygiene," n.d., in Alfred Ploetz papers, privately held by Wilfried Ploetz, Herrsching am Ammersee.

44. A. Nordenholz, "Über die Gefährdung unserer nationalen Tüchtigkeit im modernen Staat," *Annalen der Naturphilosohie* 11 (1912): 80.

45. Ibid., 71.

46. Ibid., 80–1.

47. Christian von Ehrenfels, "Werdende Moralität," *Freie Bühne* 3 (1892): 1049, 1053–6.

48. Christian von Ehrenfels, "Gedanken über die Regeneration der Kulturmenschheit" (privately published, 1901), 3, 7, 25–9; available at the Forschungsstelle und Dokumentationszentrum für Oesterreichische Philosophie, Graz.

49. Ibid., 52–3, 60, 72.

50. Christian von Ehrenfels, "Entwicklungsmoral," *Politisch–anthropologische Revue* 2 (1903–4): 218.

51. Christian von Ehrenfels, *Grundbegriffe der Ethik* (Wiesbaden, 1907), 9, 13, 17, 20, 22.

52. Felix von Luschan, "Die gegenwärtigen Aufgaben der Anthropologie," *Verhandlungen der Gesellschaft* (Leipzig, 1910) 2:201–8, quote at 206.

53. Felix von Luschan, "Culture and Degeneration," n.d. [after 1911], p. 15, in Felix von Luschan papers, Kasten 14, Staatsbibliothek Preussischer Kulturbesitz, Berlin.

54. Hugo Ribbert, *Heredity, Disease and Human Evolution* (New York, 1918), 214, 237.

55. Ibid., 40, 241–2.

56. Theodor Fritsch, "Die rechte Ehe. Ein Wort zum Züchtungs-Gedanken und Mittgart-Problem" (n.d.), p. 6, in Reichshammerbund Flugblätter und Flugschriften, ZSg. 1—263/3, Bundesarchiv Koblenz.

57. Theodor Fritsch, *Vom neuen Glauben* (Leipzig, 1914), 240; available in Deutsche Erneuerungsgemeinde papers, ZSg. 1—263/6, Bundesarchiv Koblenz; interestingly the copy at the Bundesarchiv has a stamp inside indicating that it belonged to the Haupt-Archiv der NSDAP München.

58. Advertisement for Kathreiners Malzkaffee, *Die neue Generation* 4 (1908): 398.

59. Gustav von Bunge to Alfred Ploetz, April 29, 1903, in Alfred Ploetz papers, privately held by Wilfried Ploetz, Herrsching am Ammersee; "1. und 2. Jahresbericht der Gesellschaft für Rassenhygiene" (1906), in Deutsche Kongress Zentrale, Box 255, in Hoover Institution Archives, Stanford University.

60. August Forel to Carl Hauptmann, June 30, 1888, in Carl Hauptmann papers, K58, in Archiv der Akademie der Künste, Berlin.

61. August Forel to Ernst Haeckel, May 14, 1906, in Ernst Haeckel papers, Bestand A-Abteilung 1, Nr. 1339, in Ernst-Haeckel-Haus, Jena.

62. August Forel to Richard Semon, May 1, 1909, in Richard Semon papers, Bayerische Staatsbibliothek, Munich.

63. Personal conversation with Wilfried Ploetz (Alfred Ploetz's son), June 16, 1999, Herrsching am Ammersee.

3   ORGANIZING EVOLUTIONARY ETHICS

1. James T. Kloppenberg, *Uncertain Victory* (New York, 1986), intro. and ch. 4.
2. Ernst Haeckel to Wilhelm Bölsche, October 26, 1892, in Wilhelm Bölsche papers, Böl.Hae. 21, in University of Wroclaw Library; Wilhelm Foerster, *Lebenserinnerungen und Lebenshoffnungen* (Berlin, 1911), 225–9; Ernst Haeckel, "Ethik und Weltanschauung" *Die Zukunft* 1 (1892): 309–15.
3. "Ethische Kultur" [Georg von Gizycki] to Bertha von Suttner, May 18, 1893, in Suttner-Fried Collection, League of Nations Archives, United Nations Library, Geneva.
4. Wilhelm Foerster, *Lebenserinnerungen und Lebenshoffnungen* (Berlin, 1911), 229.
5. Friedrich Wilhelm Foerster, *Erlebte Weltgeschichte, 1869–1953. Memoiren* (Nurnberg, 1953), 65–9, 78–9, 112, 155.
6. Rudolph Penzig, "Darwinismus und Ethik," in *Darwin. Seine Bedeutung im Ringen um Weltanschauung und Lebenswert*, ed. Max Apel (Berlin, Hilfe, 1909), 89–91.
7. Friedrich Jodl to Wilhelm Bolin, June 24–25, 1892, in *Unter uns gesagt*, 124.
8. "Autobiographie von Prof. Dr. Friedrich Jodl," in Sammlung Darmstaedter, 2a 1878 (10), in Staatsbibliothek Preussischer Kulturbesitz, Berlin.
9. Friedrich Jodl to Wilhelm Bolin, June 11, 1895 and December 11, 1895, in *Unter uns gesagt*, 159, 162.
10. Moritz von Egidy, "Weltanschauung; Pfingsten" (1895), in *M[oritz]. von Egidy. Sein Leben und Wirken*, ed. Heinrich Driesmans, 2 vols. (Dresden, E. Pierson's, 1900), 1: 110.
11. Georg von Gizycki to Bartholomäus von Carneri, September 14, 1892; Bartholomäus von Carneri to Ludwig Fleischer, February 12, 1892 (Abschrift); and Bartholomäus von Carneri to [Wiener] Ethische Gesellschaft, January26, 1895, in Bartholomäus von Carneri papers (in Wilhelm Börner papers), Karton 1, in Stadt- und Landes-Bibliothek, Vienna.
12. Ernst Haeckel to Wilhelm Bölsche, March 24, 1899, in Wilhelm Bölsche papers, Böl.Hae. 53, in Wroclaw University Library; Ernst Haeckel to Frida von Uslar-Gleichen, December 20, 1899, in *Franziska von Altenhausen. Ein Roman aus dem Leben eines berühmten Mannes in Briefen aus den Jahren 1898/1903*, ed. Johannes Werner (Biberach, 1927), 171.
13. Heinrich Ernst Ziegler to Ernst Haeckel, March 14, 1900, March 25, 1900, November 11, 1900, in Ernst Haeckel papers, Ernst-Haeckel-Haus, Jena.
14. "Statuten der Ethophysis Gesellschaft" in Ernst Haeckel to Friedrich Ratzel, n.d. [November 1901], in Friedrich Ratzel papers, Ratzeliana Supplement, Bayerische Staatsbibliothek, Munich.
15. Samson's Testament, July 20, 1904, in Albert-Samson-Stiftung papers, II–XI, 123, p. 26, at Archiv der Berlin-Brandenburgischen Akademie der Wissenschaften, Berlin; Wilhelm Waldeyer to Königliche preussischer Akademie der Wissenschaften, September 8, 1908, in Albert-Samson-Stiftung papers, II–XI, 115, p. 49.

16. "Auszug aus dem Protokoll der Sitzung der Gesamt-Akademie vom 16. Februar 1905," in Albert-Samson-Stiftung papers, II–XI, 115, pp. 1–2, 24–5 in Archiv der Berlin Brandenburgische Akademie der Wissenschaften, Berlin.

17. Wilhelm Waldeyer to Königliche preussischer Akademie der Wissenschaften, September 8, 1908, in Albert-Samson-Stiftung papers, II–XI, 115, p. 49 in Archiv der Berlin Brandenburgische Akademie der Wissenschaften, Berlin.

18. Wilhelm Waldeyer to Königliche preussische Akademie der Wissenschaft, July 23, 1913, in Albert-Samson-Stiftung papers, II–XI, 115, p. 82, in Archiv der Berlin Brandenburgischen Akademie der Wissenschaften, Berlin.

19. Wilhelm Waldeyer to Albert Samson, May 5, 1905, June 2, 1905, February 24, 1908, in Albert-Samson-Stiftung papers, II–XI, 117 in Archiv der Berlin Brandenburgischen Akademie der Wissenschaften, Berlin.

20. Heinrich Ernst Ziegler. "Einleitung zu dem Sammelwerke Natur und Staat," in *Natur und Staat*, vol. 1 (Jena, 1903) (bound with Heinrich Matzat, *Philosophie der Anpassung*).

21. Albert Hesse, *Natur und Gessellschaft* (Jena, 1904), 72–5, 112–14, 130–4; Arthur Ruppin, *Darwinismus und Sozialwissenschaft* (Jena, 1903), 12, 60–5, 89–91.

22. Ernst Haeckel, "Thesen zur Organisation des Monismus," *Das freie Wort 4*, 13 (October 1904): 481–9.

23. "Was will der Deutsche Monistenbund?" *Blätter. des Deutschen Monistenbundes*, Nr. 1 (July 1906): 7–9.

24. Johannes Unold, *Aufgaben und Ziele des Menschenlebens* (Leipzig, 1899), quotes at 38, 44; see also Unold, *Organische und soziale Lebensgesetze* (Leipzig, 1906), v–vii, 3, 75–7.

25. Johannes Unold, *Der Monismus und seine Ideale* (Leipzig, 1908), quotes at 90–1.

26. Friedrich Jodl, "Der Monismus und die Kulturprobleme der Gegenwart," in *Der erste internationale Monisten-Kongress in Hamburg vom 8.–11. September 1911*, ed. Wilhelm Blossfeldt (Leipzig, 1912), 117–30, quote at 120.

27. Wilhelm Ostwald to Friedrich Jodl, September 13, 1913, Aut. 177.869, in Handschriftensammlung, Stadt- und Landes-Bibliothek, Vienna; Wilhelm Blossfeldt, ed., *Der Magdeburger Monistentag vom 6.–9. Sept. 1912* (Munich, 1913), 7.

28. Wilhelm Ostwald to Friedrich Jodl, September 21, 1911, Aut. 177.868, and September 29, 1911, Aut. 177.848 in Handschriftensammlung, Stadt- und Landes-Bibliothek, Vienna.

29. Heinrich Ernst Ziegler to Friedrich Jodl, February 1, 1912, Aut. 177.786, in Handschriftensammlung, Stadt- und Landes-Bibliothek, Vienna.

30. Wilhelm Ostwald to Margarete Jodl, February 1, 1914, Aut. 178.167, in Handschriftensammlung, Stadt- und Landes-Bibliothek, Vienna.

31. Andreas W. Daum, *Wissenschaftspopularisierung im 19. Jahrhundert* (Munich, 1998), 216–19. Gasman's work on the Monist League is rather one-sided, but see also Richard Weikart, " 'Evolutionäre Aufklärung'? Zur Geschichte des Monistenbundes," in *Wissenschaft, Politik, und Öffentlichkeit* (Vienna, 2002),

131–48; Frank Simon-Ritz, *Die Organisation einer Weltanschauung* (Gütersloh, 1997); Horst Groschopp, *Dissidenten: Freidenkerei und Kultur in Deutschland* (Berlin, 1997), 245–92; Gangolf Hübinger, "Die monistische Bewegung: Sozialingenieure und Kulturprediger," in *Kultur und Kulturwissenschaften um 1900* (Stuttgart, 1997), 246–59.

32. August Forel to Paul Gerson Unna, November 9, 1911, in *August Forel: Briefe, Correspondance, 1864–1927* (Bern, 1968), 416–17; see also August Forel, "Aufruf" [für den Internationalen Orden für Ethik und Kultur], in Adele Schreiber papers, Lfd. Nr. 18, Bundesarchiv Koblenz; August Forel to Wilhelm Ostwald, November 20, 1911, in Wilhelm Ostwald papers, 790, Berlin-Brandenburgische Akademie der Wissenschaft, Berlin; August Forel, "Internationaler Orden für Ethik und Kultur," *Nachrichten des Internationalen Ordens für Ethik und Kultur*, Nr. 2–3 (March 1911): 9–10.

33. "Protokoll der Aussschusssitzung des Bundes für Mutterschutz am 6. Febr. 08," in Adele Schreiber papers, Lfd. 25, Bundesarchiv Koblenz; "Zur Aufklärung" March 20, 1908, in Adele Schreiber papers, Lfd. 24, Bundesarchiv Koblenz; Helene Stöcker, "Zur Reform der sexuellen Ethik," *Mutterschutz* 1 (1905): 3–12; Stöcker, "Von neuer Ethik," *Mutterschutz* 2 (1906): 3–11; Stöcker, "Unsere Sache," *Die neue Generation* 1 (1908): 1–6.

34. Kevin Repp, *Reformers, Critics, and the Paths of German Modernity, 1890–1914* (Cambridge, MA, 2000).

35. Emil Kraepelin, *Die psychiatrische Aufgaben des Staates* (Jena, 1900), 2, is only one example among many.

36. Theodor Fritsch's Rundbrief to Ludwig Schemann, October 14, 1908; "Vorläufige Grundsätze [of Deutsche Erneuerungsgemeinde]," n.d.; "Mitteilungen der Erneuerungs-Gemeinde," July 31, 1909, in Deutsche Erneuerungsgemeinde papers, II D, in Ludwig Schemann papers, University of Freiburg Library.

37. *Vom neuen Glauben: Bekenntnis der Deutschen Erneuerungs-Gemeinde* (Leipzig: Hammer-Verlag, 1914), 240; available at Bundesarchiv Koblenz, ZSg. 1— 263/6, Deutsche Erneuerungsgemeinde; emphasis in original.

38. "Aufruf der Reichsarbeitsgemeinschaft freigeistiger Verbände," *Die Stimme der Vernunft* 17 (1932): 54–6.

39. Private conversation with Wilfried Ploetz (Alfred Ploetz's son), June 16, 1999.

40. Gasman ignores these issues in *The Scientific Origins of National Socialism* (London, 1971).

4   THE VALUE OF LIFE AND THE VALUE OF DEATH

1. Darwin, *Origin of Species* (London: Penguin, 1968), 129, 263.

2. Ibid., 459.

3. I Corinthians 15: 54.

4. Max von Gruber, "Vererbung, Auslese und Hygiene," *Deutsche Medizinische Wochenschrift* 35 (1909): 1993.

5. Adrian Desmond, *Huxley: From Devil's Disciple to Evolution's High Priest* (Reading, MA, 1997), 271.

6. Darwin, *Origin of Species*, 230.

7. Edward J. Larson and Darrel W. Amundsen, *A Different Death* (Downers Grove, IL, 1998); Alexander Murray, *Suicide in the Middle Ages*, vol. 2: *The Curse on Self-Murder* (New York: Oxford University Press, 2000).

8. Udo Benzenhöfer, *Der gute Tod?* (Munich, 1999), 66, 71.

9. The best works discussing this issue are Jürgen Sandmann, *Der Bruch mit der humanitären Tradition* (Stuttgart, 1990), esp. chs. 4–6; Benzenhöfer, *Der gute Tod?*, ch. 4; and Kurt Nowak, *"Euthanasie" und Sterilisierung im "Dritten Reich"* (Göttingen, 1977), 11–26; brief discussions are also found in Hans Walter Schmuhl, *Rassenhygiene, Nationalsozialismus, Euthanasie* (Göttingen, 1987), 18–19; and Michael Schwartz, " 'Euthanasie'—Debatten in Deutschland (1895–1945)" *Vierteljahrshefte für Zeitgeschichte* 46 (1998): 618–19, 622, 653.

10. Haeckel called this the "question of all questions," in "Ueber die heutige Entwickelungslehre im Verhältnisse zur Gesamtwissenschaft," in *Amtlicher Bericht* (Munich, 1877), 14.

11. Ernst Haeckel, *Die Lebenswunder* (Stuttgart, 1904), 445.

12. Ernst Haeckel to his father, March 21, 1864, quoted in Heinrich Schmidt, *Ernst Haeckel. Leben und Werke* (Berlin, 1926), 203–4.

13. Ernst Haeckel, *Natürliche Schöpfungsgeschichte* (Berlin, 1868), 546; David Friedrich Strauss, *Der alte und der neue Glaube* (Leipzig, 1872), 200–2.

14. Haeckel, *Natürliche Schöpfungsgeschichte*, 189.

15. Haeckel, *Die Welträthsel* (Bonn, 1903), 39–40.

16. August Forel, *Out of My Life and Work* (New York, 1937), 53, quote at 256; see also August Forel to Paul Gerson Unna, November 9, 1911, in *August Forel: Briefe, Correspondance* (Bern, 1968), 414–15.

17. August Forel, "August Forel" (autobiography), in *Führende Psychiater in Selbstdarstellungen* (Leipzig: Felix Meiner, 1930), 60.

18. August Forel to Ernst Haeckel, August 23, 1905, in Ernst Haeckel papers, Ernst-Haeckel-Haus, Jena; see also August Forel to Paul Gerson Unna, November 9, 1911, in *August Forel: Briefe, Correspondance*, 414–15.

19. Emil Kraepelin, *Lebenserinnerungen* (Berlin, 1983), 2–3; Kraepelin, *Psychiatrie: Ein kurzes Lehrbuch*, 4th ed. (Leipzig, 1893), 1–2.

20. Interestingly, many Christian philosophers today who accept Darwinian evolution reject the notion of body–soul dualism.

21. Robby Kossmann, "Die Bedeutung des Einzellebens in der Darwinistischen Weltanschauung," *Nord und Süd* 12 (1880): 414.

22. Ibid., 420–1. Emphasis is mine.

23. Robby Kossmann, *Züchtungspolitik* (Schmargendorf, 1905).

24. Arnold Dodel, *Die Neuere Schöpfungsgeschichte* (Leipzig, 1875), 417. For more on Dodel and his influence, see Werner Beyl, *Arnold Dodel (1843–1908)* (Frankfurt, 1984).

25. Hans von Hentig, *Strafrecht und Auslese* (Berlin, 1914), 1, quote at 218.
26. Otto Ammon, *Die Gesellschaftsordnung und ihre natürlichen Grundlagen*, 3rd ed. (Jena, 1900), 288.
27. Ernst Haeckel, *Freie Wissenschaft und freie Lehre* (Stuttgart, 1878), 73–4.
28. Ibid., 74.
29. Wilhelm Preyer, *Die Concurrenz in der Natur* (Breslau, 1882), 13, 27; Preyer, *Der Kampf um das Dasein* (Bonn, 1869).
30. Ludwig Büchner, *Die Macht der Vererbung* (Leipzig, 1882), 100.
31. Friedrich Hellwald, *Culturgeschichte in ihrer natürlichen Entwicklung bis zur Gegenwart* (Augsburg, 1875), 58, 27; "Der Kampf ums Dasein im Menschen- und Völkerleben," *Das Ausland* 45 (1872): 105.
32. Bartholomäus von Carneri, *Sittlichkeit und Darwinismus* (Vienna, 1871), 19.
33. Georg von Gizycki, *Moralphilosophie gemeinverständlich dargestellt* (Leipzig, 1895), 363, 357–61.
34. Max von Gruber, *Ursachen und Bekämpfung* (Munich, 1914); Alfred Grotjahn, *Geburtenrückgang und Geburtenregelung* (Berlin, 1921).
35. Alexander Tille, *Volksdienst* (Berlin, 1893), 28.
36. August Forel, "Ueber Ethik," *Zukunft* 28 (1899): 584; see also Forel, *Die sexuelle Frage* (Munich, 1905), 439.
37. Johannes Unold, *Der Monismus und seine Ideale* (Leipzig, 1908), 83.
38. August Weismann, "Ueber die Dauer des Lebens," in *Aufsätze über Vererbung* (Jena, 1892), 11, 27–8. Emphasis in original.
39. August Weismann, "Ueber Leben und Tod," in ibid., 183–4.
40. Wilhelm Schallmayer, *Vererbung und Auslese* (Jena, 1903), 241–2; Schallmayer, *Beiträge zu einer Nationalbiologie* (Jena, 1905), 129–31.
41. Arnold Dodel, *Moses oder Darwin?* (Zurich, 1889), 90–1.
42. Arnold Dodel, *Die Neuere Schöpfungsgeschichte* (Leipzig, 1875), 150.
43. Eduard David, "Darwinismus und soziale Entwicklung," in Max Apel, ed., *Darwin* (Berlin, 1909), 48.
44. Haeckel, *Natürliche Schöpfungsgeschichte* (1870), 154–5.
45. See Daniel Pick, *Faces of Degeneration: A European Disorder, c.1848–c.1918* (Cambridge, 1989).
46. August Weismann, "Über den Rückschritt in der Natur," in *Aufsätze über Vererbung* (Jena, 1892), 549, 566, quote at 574–5.
47. Pick, *Faces of Degeneration*, 2.
48. Hans Kurella, *Naturgeschichte des Verbrechers* (Stuttgart, 1893), 11–13, 259–60.
49. Mariacarla Gadebusch Bondio, *Die Rezeption der kriminalanthropologischen Theorien von Cesare Lombroso in Deutschland von 1880–1914* (Husum, 1995), 197–8, 182.
50. Dirk Blasius, *Umgang mit Unheilbarem* (Bonn, 1986), 60.
51. On degeneration and eugenics, see Peter Weingart, Jürgen Kroll, and Kurt Bayertz, *Rasse, Blut, und Gene* (Frankfurt, 1988), ch. 2.
52. Wilhelm Schallmayer, *Die drohende physische Entartung der Culturvölker*, 2nd ed. [1st ed. had title, *Ueber die drohende körperliche Entartung der Culturmenschheit*], (Berlin, 1891), 1, 4, 7, 17–19.

53. August Forel, *Kulturbestrebungen der Gegenwart* (Munich, 1910), 19; Forel, *Hygiene of Nerves and Mind* (New York, 1907), 229–30.
54. On the influence of Darwinism and eugenics on Gerhart Hauptmann, see Günter Schmidt, *Die Literarische Rezeption des Darwinismus* (Berlin, 1974).
55. Wilhelm Bölsche, *Stirb und Werde!* (Jena, 1921), 147.
56. Heinrich Driesmans, *Dämon Auslese* (Berlin, 1907), xiv, 264.

5 THE SPECTER OF INFERIORITY: DEVALUING THE DISABLED AND "UNPRODUCTIVE"

1. Jennifer Michael Hecht, "The Solvency of Metaphysics: The Debate over Racial Science and Moral Philosophy in France, 1890–1919," *Isis* 90 (1999): 5–6.
2. Quoted in Mike Hawkins, *Social Darwinism in European and American Thought, 1860–1945* (Cambridge, 1997), 129.
3. Ernst Haeckel, *Natürliche Schöpfungsgeschichte* (Berlin, 1868), 546.
4. Rudolf Virchow, *Die Freiheit der Wissenschaft im modernen Staat* (Berlin, 1877), 6–8, 12–15, 31–2.
5. Ernst Haeckel, *Freie Wissenschaft und freie Lehre* (Stuttgart, 1878), 72.
6. Oscar Schmidt, "Darwinismus und Socialdemokratie," *Deutsche Rundschau* 17 (1878): 289–90.
7. Wilhelm Preyer, "Briefe von Darwin," *Deutsche Rundschau* 67 (1891): 382; see also Preyer, *Der Kampf um das Dasein* (Bonn, 1869), and *Die Concurrenz in der Natur* (Breslau, 1882).
8. Friedrich Hellwald, *Culturgeschichte in ihrer natürlichen Entwicklung bis zur Gegenwart* (Augsburg, 1875), 720.
9. Heinrich Ernst Ziegler, *Die Naturwissenschaft und die Socialdemokratische Theorie* (Stuttgart, 1893), 158; "Zu den Kritikern über das Jenenser Preisausschreiben," *Politisch-anthropologische Revue* 3 (1904–05): 438.
10. Ludwig Büchner to Ernst Haeckel, October 21, 1878, in Ernst Haeckel papers, Ernst-Haeckel-Haus, Jena.
11. Ludwig Büchner to J. Ph. Becker, August 28, 1866, in J. Ph. Becker papers, D 324, International Institute for Social History, Amsterdam.
12. Ludwig Büchner, *Darwinismus und Sozialismus* (Leipzig, 1894), passim; Büchner, *Fremdes und Eigenes* (Leipzig, 1890), 226–30; quote at 251.
13. Max Nordau, *Die konventionellen Lügen* (Leipzig, 1909), 112–16.
14. Alfred Ploetz, *Die Tüchtigkeit* (Berlin, 1895), 194–6.
15. Alfred Ploetz, "Denkschrift über die Gründung der Internationalen Gesellschaft für Rassen-Hygiene"; and "Satzungen der Deutschen Gesellschaft für Rassen-Hygiene," March 7, 1911, in Alfred Ploetz papers, privately held by Wilfried Ploetz, Herrsching am Ammersee.
16. Wilhelm Schallmayer, *Vererbung und Auslese* (Jena, 1903), 371.
17. Christian von Ehrenfels, *Sexualethik* (Wiesbaden, 1907), 95.
18. Arnold Dodel, *Moses oder Darwin?* (Zurich, 1889), 102–3.

19. Alfred Blaschko, "Natürliche Auslese und Klassentheilung," *Die neue Zeit* 13, 1 (1894–95): 615.

20. Emil Reich, "Evolutionistische Ausblicke," *Dokumente des Fortschritts* 1 (1908): 711.

21. Karl Kautsky, *Vermehrung und Entwicklung in Natur und Gesellschaft* (Stuttgart, 1910), 206, 260–4.

22. Lily Braun, *Memoiren einer Sozialistin*, vol. 2: *Kampfjahre* (Munich, 1911), 474, 589, 635.

23. Karl Kautsky, "Medizinisches," *Die neue Zeit* 10, 1 (1891–92): 644–5; Kautsky, *Vermehrung und Entwicklung in Natur und Gesellschaft* (Stuttgart, 1910), 206, 260–4.

24. Alexander Tille, *Von Darwin bis Nietzsche* (Leipzig, 1895), 21–2. Emphasis in original.

25. Alexander Tille, *Volksdienst* (Berlin, 1893), 57.

26. Felix von Luschan, "Die gegenwärtigen Aufgaben der Anthropologie," *Verhandlungen der Gesellschaft* (Leipzig, 1910), 2:205.

27. Carl Vogt, *Vorlesungen über den Menschen* (Giessen, 1863), 1:214, 256.

28. Andrew Zimmerman, *Anthropology and Antihumanism in Imperial Germany* (Chicago, 2001), 75.

29. Vogt, *Vorlesungen über den Menschen*, 1:295.

30. Zimmerman, *Anthropology and Antihumanism*, 77–83; see also Nigel Rothfels, *Savages and Beasts* (Baltimore, 2002), ch. 3.

31. Erwin Baur, *Einführung in die experimentelle Vererbungslehre*, 2nd ed. (Berlin, 1914), 342–5.

32. Kurt Goldstein, *Über Rassenhygiene* (Berlin, 1913), 74–5.

33. Georg von Gizycki, *Grundzüge der Moral* (Leipzig, 1883), 90–1.

34. Hugo Ribbert, *Heredity, Disease and Human Evolution* (New York: Critic and Guide, 1918), 57.

35. Alfred Grotjahn, "Das Problem der Entartung," *Archiv für soziale Hygiene* 6 (1910): 74.

36. Edward Ross Dickinson, "Reflections on Feminism and Monism in the Kaiserreich, 1900–1913," *Central European History* 34 (2001): 191–230; Ann Taylor Allen, "German Radical Feminism and Eugenics, 1900–1918," *German Studies Review* 11 (1988): 31–56; Kevin Repp, " 'More Corporeal, More Concrete': Liberal Humanism, Eugenics, and German Progressives at the Last Fin de Siècle," *Journal of Modern History* 72 (2000): 683–730; Anette Herlitzius, *Frauenbefreiung und Rassenideologie* (Wiesbaden, 1995).

37. Adele Schreiber, "Missbrauchte und unwillkommene Mutterschaft," in *Mutterschaft*, ed. Adele Schreiber (Munich, 1912), 202.

38. Quotes in Ruth Bré, "Der erste deutsche Bund für Mutterschutz," *Die neue Heilkunst* (May 8, 1905), in Adele Schreiber papers, N 1173, 29; see also Bré, "Ruth Bré und der 'Bund für Mutterschutz,' " March 1905; and "Aufruf" [des Bund für Mutterschutz], January 1905, in Adele Schreiber papers, N 1173, 17, Bundesarchiv Koblenz.

39. "Bund für Mutterschutz. Protokoll der Ausschuss–Sitzung vom 15. Mai 1905," in Adele Schreiber papers, N 1173, 25.

40. Helene Stöcker, "Geburtenrückgang und Monismus," in *Der Düsseldorfer Monistentag*, ed. Wilhelm Blossfeldt (Leipzig, 1914), 40.

41. "Referat über den Vortrag von Dr. phil. Helene Stöcker, Mutterschutz und Abtreibungsstrafe," April 1909; and "Leitsätze zum Referat von Dr. phil. Helene Stöcker auf der Generalversammlung des Deutschen Bundes für Mutterschutz in Hamburg 1909," in Adele Schreiber papers, N 1173, 28.

42. Quote in Helene Stöcker, "Geburtenrückgang und Monismus," in *Der Düsseldorfer Monistentag*, 47; see also Helene Stöcker, "Von neuer Ethik," *Mutterschutz* 2 (1906): 3–11; Stöcker, "Mutterschutz und Sexualreform," *Das monistische Jahrhundert* 2 (1913): 14–18; Stöcker, "Staatlicher Gebärzwang oder Rassenhygiene?" *Die neue Generation* 10 (1914): 134–49; Stöcker, "Rassenhygiene und Eheatteste," *Die neue Generation* 13 (1917): 49–51.

43. Richard F. Wetzell, *Inventing the Criminal* (Chapel Hill, 2000), 48–9.

44. Dirk Blasius, *Umgang mit Unheilbarem* (Bonn, 1986), 60–6; Paul Weindling *Health, Race and German Politics* (Cambridge, 1989).

45. "Preisausschreiben," *Die Umschau* 15 (1911): 127–8.

46. Max von Gruber, "Vererbung, Auslese und Hygiene," *Deutsche Medizinische Wochenschrift* 35 (1909): 1993, 2051–2.

47. Ludwig Jens, "Was kosten die schlechten Rassenelemente den Staat und die Gesellschaft," *Archiv für soziale Hygiene* 8 (1913): 322.

48. Quoted in Karl Sablik, *Julius Tandler* (Vienna, 1983), 122.

49. Ignaz Kaup, "Was kosten die minderwertigen Elemente dem Staat und der Gesellschaft?" *Archiv für Rassen- und Gesellschaftsbiologie* 10 (1913): 747. For more on Kaup, see Doris Byer, *Rassenhygiene und Wohlfahrtspflege* (Frankfurt, 1988).

50. Hans Kurella, "Soziale Anthropologie," *Zukunft* 12 (1895): 312–14.

51. Hans Kurella, *Naturgeschichte des Verbrechers* (Stuttgart, 1893), 210.

52. Alfred Ploetz, "Die Begriffe Rasse und Gesellschaft und einige zusammenhängende Probleme," *Verhandlungen des Ersten Deutschen Soziologentages* 1 (1911): 146–8.

53. Heinz Potthoff, "Schutz der Schwachen?" *Archiv für Rassen- und Gesellschaftsbiologie* 8 (1911): 89–90.

54. August Forel, *Die sexuelle Frage*, 457.

55. August Forel, "Vom Neomalthusianismus," *Mutterschutz* 1 (1905): 479.

56. Wilhelm Schallmayer, "Rassedienst," *Menschheitsziele* 3 (1909): 296.

57. Rudolf Goldscheid, *Höherentwicklung und Menschenökonomie* (Leipzig, 1911), 565, quotes at 594, 578; see also Rudolf Goldscheid, *Entwicklungstheorie, Entwicklungsökonomie, Menschenökonomie* (Leipzig, 1908), 17–22, 205.

58. Rudolf Goldscheid, *Darwin als Lebenselement unserer modernen Kultur* (Vienna, 1909), 16, 49, 52, quote at 66; Goldscheid, *Höherentwicklung und Menschenökonomie*, xv–xvi.

59. "3. Bericht der Internationale Gesellschaft für Rassen-Hygiene" (1.1.1908–2.3.1909), in Deutsche Kongress Zentrale, Box 255, Hoover Institution Archives.

60. Letter by Paul Kammerer and Julius Tandler, March 1914 (contained in file with letters from Rudolf Goldscheid to Max Hirsch), Max Hirsch papers, Staatsbibliothek Preussischer Kulturbesitz, Berlin.

6   THE SCIENCE OF RACIAL INEQUALITY

1. Some scholars claim that European imperialism produced social Darwinist racism—see Woodruff D. Smith, *The Ideological Origins of Nazi Imperialism* (New York, 1986), ch. 7; Hans-Ulrich Wehler, *Das deutsche Kaiserreich 1871–1918*, 2nd ed. (Göttingen, 1975), 179–81; Hannah Arendt, *The Origins of Totalitarianism* (New York, 1973), 183–4; others argue that social Darwinist racism encouraged imperialism—see H. W. Koch, "Social Darwinism as a Factor in the 'New Imperialism'," in *The Origins of the First World War*, ed. H. W. Koch (New York, 1972); Koch shows that these ideas were not exclusively German in *Der Sozialdarwinismus* (Munich, 1973).

2. Quoted in Leon Poliakov, *The Aryan Myth* (London, 1974), 158. I recognize that Herder was not exactly an exemplar of dominant Enlightenment ideas, but in this case, he did reflect views widespread among the philosophes.

3. Darwin, *The Descent of Man*, (London, 1871; rprt., Princeton, 1981), 1: 35, 97; quote at 1: 3. On how preexisting racism may have impacted Darwin, see Barry W. Butcher, "Darwinism, Social Darwinism and the Australian Aborigines: A Reevaluation," in *Darwin's Laboratory* (Honolulu, 1994), 371–394. The extent of Darwin's own racism is hotly debated. Many scholars admit that Darwin was racist and included racial conflict in the human struggle for existence—see Adrian Desmond and James Moore, *Darwin* (London, 1991); John Greene, "Darwin as a Social Evolutionist," in *Science, Ideology and World View* (Berkeley, 1981), 95–127; Robert Young, "Darwinism *Is* Social," in *The Darwinian Heritage*, ed. David Kohn (Princeton, 1985), 609–38; Heinz-Georg Marten, "Racism, Social Darwinism, Anti-Semitism and Aryan Supremacy" *International Journal of the History of Sport* 16, 2 (1999): 32; Other scholars wrongly claim that Darwin was not really racist and later manifestations of social Darwinist racism were departures from Darwin's own views—see Robert C. Bannister, *Social Darwinism* (Philadelphia, 1979), 15; Gloria McConnaughey, "Darwin and Social Darwinism," *Osiris* 9 (1950): 412; and many others.

4. Ernst Haeckel, *Generelle Morphologie*, 2 vols. (Berlin, 1866), II: 435. Emphasis in original.

5. Ernst Haeckel, *Natürliche Schöpfungsgeschichte* (Berlin, 1868), 546, 548–9; Haeckel, *Der Monismus als Band zwischen Religion und Wissenschaft* (Bonn, 1892), 11; Haeckel, *Welträthsel* (Bonn, 1903), 13; Haeckel, *Die Lebenswunder* (Stuttgart, 1904), 327.

6. Haeckel, *Natürliche Schöpfungsgeschichte* (1868), 549.

7. Ibid., frontispiece, 555.
8. Ibid., 514, 546–9; quote at 547.
9. Ibid., 511.
10. Haeckel, *Natürliche Schöpfungsgeschichte* 11th ed. (Berlin, 1911), lxviii–lxix.
11. Haeckel, *Welträthsel*, 53; he made a similar comment in *Lebenswunder*, 364.
12. Haeckel, *Natürliche Schöpfungsgeschichte* (1911), 754–6.
13. Haeckel, *Lebenswunder*, 450.
14. Ibid., quotes at 449, 451–2. Rolf Winau's presentation of Haeckel's racism is roughly equivalent to mine in "Ernst Haeckels Vorstellungen von Wert und Werden menschlicher Rassen und Kulturen," *Medizinhistorisches Journal* 16 (1981): 270–9 (though it is not always clear where he is writing his own prose and where he is quoting from Haeckel's works).
15. Piet de Rooy, "In Search of Perfection: The Creation of a Missing Link," in *Ape, Man, Apeman*, ed. Raymond Corbey and Bert Theunissen (Leiden, 1995), pp. 195–207.
16. Ernst Haeckel to Karl Vogt, August 8, 1896, in Karl Vogt papers, Ms. fr. 2189, Geneva University and Public Library Archive.
17. Carl Vogt, *Vorlesungen über den Menschen*, 2 vols. (Giessen, 1863), 1: 242–4, 2: 282; quote at 2: 277.
18. Bartholomäus von Carneri, *Sittlichkeit und Darwinismus* (Vienna, 1871), 29.
19. Ibid., 313.
20. Ludwig Büchner, *Der Mensch und seine Stellung* 2nd ed. (Leipzig, 1872), 138–9.
21. Ludwig Büchner, in Friedrich Hellwald, *Kulturgeschichte*, 4th ed. (Leipzig, 1896), 1: 97.
22. Ludwig Büchner, *Die Macht der Vererbung* (Leipzig, 1882), 49–50, quote at 59.
23. Oscar Peschel, "Eine neue Lehre über die Schöpfungsgeschichte der organischen Welt," *Das Ausland* 33, 5 (January 29, 1860): 97–101 and 33, 6 (February 5, 1860): 135–40.
24. Oscar Peschel, "Mensch und Affe," *Das Ausland* 36 (1863): 521. The earlier article was "Mensch und Affe," *Das Ausland* 34 (1861): 833–5.
25. Oscar Peschel, *Völkerkunde*, 2nd ed. (Leipzig, 1875), 149, 341, 516.
26. Friedrich Hellwald, *Culturgeschichte in ihrer natürlichen Entwicklung* (Augsburg, 1875), 55, 61, 63–5, 134–5, 221–2.
27. Friedrich Ratzel to Ernst Haeckel, September 25, 1869, February 15, 1891, February 15, 1894, quote in March 24, 1872, in Ernst Haeckel papers, Ernst-Haeckel-Haus, Jena.
28. Friedrich Ratzel, Vorlesungen über "Völkerkunde, insbesondere Völkerbewegungen," ca. 1895–1900, in Friedrich Ratzel papers, folder 80, Institut für Länderkunde, University of Leipzig.
29. Friedrich Ratzel, *Sein und Werden der organischen Welt* (Leipzig, 1869), 507.
30. Friedrich Ratzel, *The History of Mankind*, trans. A. J. Butler, 3 vols. (London, 1896), 1: 3–20, 342.
31. Friedrich Ratzel, review of S. R. Steinmetz's article, "Der erbliche Rassen- und Volkscharakter," in Friedrich Ratzel papers, Ratzeliana III, Bayerische Staatsbibliothek, Munich.

32. Friedrich Ratzel, Vorlesungen über "Biogeographie: Biogeographie und Anthropogeographie," 1896/97–1902, in Friedrich Ratzel papers, folder 153, Institut für Länderkunde, University of Leipzig.

33. Friedrich Ratzel, review of Heinrich Driesmans, *Rasse und Milieu* in *Historische Vierteljahrsschrift* (1904): 394; in Friedrich Ratzel papers, Ratzeliana III, Bayerische Staatsbibliothek; see also Heinz Gollwitzer, *Die gelbe Gefahr* (Göttingen, 1962), 203.

34. Mark Bassin, "Imperialism and the Nation State in Friedrich Ratzel's Political Geography," *Progress in Human Geography* 11 (1987): 489, n. 12, acknowledges the heavy influence of Darwinism on Ratzel, but incorrectly denies that Ratzel was racist in his later writings.

35. Adolf Heilborn, in Hermann Klaatsch, *Grundzüge der Lehre Darwins*, 4th ed. (Mannheim, 1919), 7–8.

36. Hermann Klaatsch, "Die Morphologie und Psychologie der niederen Menschenrassen in ihrer Bedeutung für die Probleme der Kriminalistik," in *Bericht über den VII. Internationalen Kongress für Kriminalanthropologie, Köln a. Rhein, 9.–13. Oktober 1911*, ed. Gustav Aschaffenburg and Dr. Partheimer (Heidelberg, 1912), 58–71, 73.

37. Ibid.

38. Quoted in Benoit Massin, "From Virchow to Fischer: Physical Anthropology and 'Modern Race Theories' in Wilhelmine Germany," in *Volksgeist as Method and Ethic*, ed. George W. Stocking (Madison, 1996), 120.

39. Klaatsch, "Die Morphologie und Psychologie," 71–3.

40. Massin, "From Virchow to Fischer," 99. Massin discusses the relationship between Darwinism and racism in German anthropology in considerable detail. Andrew Zimmerman also discusses the shift toward Darwinism in German anthropology in *Anthropology and Antihumanism in Imperial Germany* (Chicago, 2001), 7, 202, 214.

41. Quoted in Gunter Mann, "Rassenhygiene—Sozialdarwinismus," in *Biologismus im 19. Jahrhundert*, ed. Gunter Mann (Stuttgart, 1973), 83.

42. Alfred Ploetz to Carl Hauptmann, June 23, 1890, in Carl Hauptmann papers, K 121, Archiv der Akademie der Künste, Berlin.

43. Alfred Ploetz, *Die Tüchtigkeit* (Berlin, 1895), 5.

44. Ibid., 10

45. Alfred Ploetz, "Denkschrift über die Gründung der Internationalen Gesellschaft für Rassen-Hygiene," n.d., in Alfred Ploetz papers, privately held by Wilfried Ploetz, Herrsching am Ammersee.

46. Alfred Ploetz, "Unser Weg," 1911, in Alfred Ploetz papers, privately held by Wilfried Ploetz, Herrsching am Ammersee.

47. Ploetz, Tüchtigkeit, 78, 91–5, 130–2.

48. Alfred Ploetz, "Sozialpolitik und Rassenhygiene in ihrem prinzipiellen Verhältnis," *Archiv für soziale Gesetzgebung und Statistik* 17 (1902): 399.

49. Alfred Ploetz, "Zum deutsch-polnischen Kampf," *Archiv für Rassen- und Gesellschaftsbiologie* 3 (1906): 922.

50. Alfred Ploetz, "Die Begriffe Rasse und Gesellschaft und einige zusammenhängende Probleme," *Verhandlungen des Ersten Deutschen Soziologentages* 1 (1911): 163.

51. Wilhelm Schallmayer, "Rassedienst," *Sexual-Probleme* 7 (1911): 435.

52. Wilhelm Schallmayer, "Der Krieg als Züchter," *Archiv für Rassen- und Gesellschaftsbiologie* 5 (1908): 367, 387; Schallmayer, *Vererbung und Auslese* (Jena, 1903), 178.

53. Wilhelm Schallmayer to David Starr Jordan, November 2, 1913, in David Starr Jordan papers, Hoover Institution Archives, Stanford University.

54. Wilhelm Schallmayer to Ludwig Schemann, April 21, 1912 and May 3, 1912, in Ludwig Schemann papers, IV B 1/2, University of Freiburg Library Archive.

55. Heinrich Driesmans, *Kulturgeschichte der Rasseninstinkte*, 2 vols. (Leipzig, 1900–01), 2:v–vii; Driesmans, *Rasse und Milieu* (Berlin, 1902), 4–5, 21, 40; Driesmans, *Dämon Auslese* (Berlin, 1907), 240–1, 253, 260.

56. Moritz Wagner, *Die Entstehung der Arten* (Basel, 1889), 169–72.

57. See Alfred Grotjahn's description of Woltmann in *Erlebtes und Erstrebtes: Erinnerungen eines sozialistischen Arztes* (Berlin, 1932), 65.

58. Richard Weikart, *Socialist Darwinism* (San Francisco, 1999) , 210–13; Ludwig Woltmann, "Marxismus und Rassetheorie," *Politisch-Anthropologische Revue* 4 (1905–06): 268–280.

59. Ludwig Woltmann to Alfred Ploetz, June 6, 1901, March 14, 1902, March 30, 1902 [no date given] December 1902, in Alfred Ploetz papers, privately held by Wilfried Ploetz, Herrsching am Ammersee.

60. Ludwig Woltmann, *Politische Anthropologie* (Jena, 1903), 289.

61. Ludwig Woltmann, *Die Germanen und die Renaissance in Italien* (Leipzig, 1905); Woltmann, *Die Germanen in Frankreich* (Jena, 1907).

62. Otto Ammon, *Die Natürliche Auslese beim Menschen* (Jena, 1893), 177, 185.

63. Ludwig Wilser, *Die Ueberlegenheit der germanischen Rasse* (Stuttgart, 1915), 7–8.

64. Ernst Rüdin, review of Woltmann, *Die Germanen und die Renaissance in Italien*, in *Archiv für Rassen- und Gesellschaftsbiologie* 1 (1904): 309; review of Woltmann, *Politische Anthropologie*, in *Archiv für Rassen- und Gesellschaftsbiologie* 2 (1905): 609–19; review of Woltmann, *Die Germanen in Frankreich*, in *Archiv für Rassen- und Gesellschaftsbiologie* 4 (1907): 234–8.

65. Eugen Fischer to Ludwig Schemann, January 16, 1910, in Ludwig Schemann papers, IV B 1/2, University of Freiburg Library Archives.

66. Eugen Fischer, *Sozialanthropologie und ihre Bedeutung für den Staat* (Freiburg, 1910), 18–19; Eugen Fischer, "Sozialanthropologie," in *Handwörterbuch der Naturwissenschaften* (Jena, Gustav Fischer, 1912–13), 9: 177, 184–5.

67. Eugen Fischer, *Die Rehobother Bastards* (Jena, 1913), 296–306, quote at 302. For more on Fischer, see Niels Lösch, *Rasse als Konstrukt* (Frankfurt, 1997).

68. Fritz Lenz, *Die Rasse als Wertprinzip* (Munich, 1933), 5–7, quote at 39.

69. Fritz Lenz to Ludwig Schemann, December 1, 1919, in Ludwig Schemann papers, IV B 1/2.

70. Fritz Lenz, "Zur Rasse- und Rassenwertung," *Archiv für Rassen- und Gesellschaftsbiologie* 11 (1914–15): 500–1. For more on Lenz, see Renate Rissom, *Fritz Lenz und die Rassenhygiene* (Husum, 1983).

71. Ludwig Schemann to Wilhelm Schallmayer, May 8, 1912; Ludwig Schemann to Otto Ammon, April 18, 1909, in Ludwig Schemann papers, University of Freiburg Library Archives, IV B 1/2; "Erster Bericht über die Gobineau–Vereinigung," (1895); Gobineau–Vereinigung Mitgleider Verzeichnis, February 1902–September 1903; Gobineau–Vereinigung Mitglieder Verzeichnis, February 1905–December 1906; "Elfter Bericht über die Gobineau–Vereinigung" (1911);Gobineau–Vereinigung Mitglieder Verzeichnis, January 1911–September 1912; in Gobineau–Vereinigung papers, D II, University of Freiburg Library Archives; Ludwig Schemann, *Lebensfahrten eines Deutschen* (Leipzig, 1925), 295–7, 346–7; Ludwig Schemann, *Die Rasse in den Geisteswissenschaften: Studien zur Geschichte des Rassengedankens*, vol. 3: *Die Rassenfragen im Schrifttum der Neuzeit*, 2nd ed. (Munich, 1943), xvi, 229–35, 249, 434. For more on Schemann, see Peter Emil Becker, *Wege ins Dritte Reich*, vol. 2: *Sozialdarwinismus, Rassismus, Antisemitismus und volkischer Gedanke* (Stuttgart, 1990), ch. 3.

72. Gary Stark, *Entrepreneurs of Ideology* (Chapel Hill, 1981).

73. Willibald Hentschel to Carl Hauptmann, August 23, 1903 and April 16, 1904, in Carl Hauptmann papers, K 91, Archiv der Akademie der Künste, Berlin.

74. Willibald Hentschel, *Mittgart*, 2nd ed. (Leipzig, 1906), 5.

75. Willibald Hentschel, *Varuna* (Leipzig, 1907), 1–2, quote at 600. For more on Hentschel, see Dieter Löwenberg, "Willibald Hentschel (1858–1947): Seine Pläne zur Menschenzüchtung, sein Biologismus und Antisemitismus" (dissertation, University of Mainz, 1978).

76. Theodor Fritsch, *Handbuch der Judenfrage*, 27th ed. (Hamburg, 1910), 238, quote at 6; Theodor Fritsch to Ludwig Schemann, February 11, 1908, in Ludwig Schemann papers, IV B 1/2, Freiburg University Library Archives. For more on Fritsch, see Michael Bönisch, "Die 'Hammer Bewegung,'" in *Handbuch zur "Völkischen Bewegung" 1871–1918*, ed. Uwe Puschner, Walter Schmitz, and Justus H. Ulbricht (Munich, 1996); Richard S. Levy, *The Downfall of the Anti-Semitic Political Parties in Imperial Germany* Yale (New Haven: University Press, 1975).

77. George Mosse, *The Crisis of German Ideology* (New York, 1964); George Mosse, *Towards the Final Solution* (New York, 1978); Fritz Stern, *The Politics of Cultural Despair: A Study in the Rise of the Germanic Ideology* (Berkeley, 1963). Mosse stresses the irrationalist roots of Nazi racism and underemphasizes the important role of scientific racism in shaping Nazi ideology.

78. Houston Stewart Chamberlain, *Die Grundlagen des neunzehnten Jahrhunderts*, 2 vols. (Munich, 1899), 2: 717, quote at 1: 531.

79. Ibid., 1: 265–6, 278–9, 284.

7 CONTROLLING REPRODUCTION: OVERTURNING TRADITIONAL SEXUAL MORALITY

1. On Freud's reception of Darwinism, see Lucille B. Ritvo, *Darwin's Influence on Freud* (New Haven, 1990); and Frank Sulloway, *Freud: Biologist of the Mind* (New York, 1979).
2. Edward O. Wilson, *Sociobiology: The New Synthesis* (Cambridge, MA: Harvard University Press, 1975), 3.
3. Helene Stöcker, "Zur Emanzipation des Mannes" (1904), in *Die Liebe und die Frauen*, 2nd ed. (Minden,1906), 122.
4. Max von Gruber, *Kreig, Frieden und Biologie* (Berlin, 1915), 26.
5. August Forel, *Out of My Life and Work*, trans. Bernard Miall (New York, 1937), 191–2.
6. August Forel, *Hygiene of Nerves and Mind*, trans. Herbert Austin Aikins (New York, 1907), 277.
7. August Forel, *Die sexuelle Frage* (Munich, 1905), 4, 444, quote at 441.
8. Ibid., 389–90, 395, 445–6, 526.
9. Wilhelm Schallmayer, "Ueber die Grundbedeutung der Ethik und ihr Verhältnis zu den Forderungen des Rassedienstes," *Die neue Generation* 6 (1910): 495–6.
10. Gustav Bunge, *Die zunehmende Unfähigkeit der Frauen*, 7th ed. (Munich, 1914), 36–7.
11. Helene Stöcker, "Von neuer Ethik," *Mutterschutz* 2 (1906): 3.
12. *Der erste internationale Monisten-Kongress*, ed. Wilhelm Blossfeldt (Leipzig, 1912), 45.
13. Helene Stöcker, "Geburtenrückgang und Monismus," in *Der Düsseldorfer Monistentag. 7. Hauptversammlung des Deutschen Monistenbunedes vom 5.–8. September 1913*, ed. Wilhelm Blossfeldt (Leipzig, 1914), 40–1.
14. Helene Stöcker, "Staatlicher Gebärzwang oder Rassenhygiene?" *Die neue Generation* 10 (1914): 144
15. Alfred Ploetz, "Bund für Mutterschutz," *Archiv für Rassen- und Gesellschaftsbiologie* 2 (1905): 164–6; Ploetz, "Bund für Mutterschutz," *Archiv für Rassen- und Gesellschaftsbiologie* 2 (1905): 316–17.
16. Helene Stöcker, "Staatlicher Gebärzwang oder Rassenhygiene?" *Die neue Generation* 10 (1914): 139. For more on Stöcker, see Christl Wickert, *Helene Stöcker, 1869–1943: Frauenrechtlerin, Sexualreformerin und Pazifistin* (Bonn, 1991), and Edward Ross Dickinson, "Reflections on Feminism and Monism in the Kaiserreich, 1900–1913," *Central European History* 34 (2001): 191–230.
17. Alfred Ploetz, *Die Tüchtigkeit* (Berlin, 1895), 64.
18. Alfred Ploetz, "Neo-Malthusianism and Race Hygiene," in *Problems in Eugenics: Report of Proceedings of the First International Eugenics Congress*, vol. 2 (London: Eugenics Education Society, 1913).

19. Alfred Ploetz to Max von Gruber, November 19, 1911, in Alfred Ploetz papers, privately held by Wilfried Ploetz, Herrsching am Ammersee.

20. Max von Gruber and Ernst Rüdin, *Fortpflanzung, Vererbung, Rassenhygiene*, 2nd ed. (Munich, 1911), 171.

21. Max von Gruber, *Ursachen und Bekämpfung des Geburtenrückgangs im Deutschen Reich*, 3rd ed. (Munich, 1914).

22. Max von Gruber to Alfred Grotjahn, June 9, 1910, in Alfred Grotjahn papers, Humboldt University Archives, Berlin.

23. Felix von Luschan, "Culture and Degeneration," n.d., p. 10, in Felix von Luschan papers, Kasten 14, Staatsbibliothek Preussischer Kulturbesitz, Berlin.

24. Hugo Ribbert, *Heredity, Disease and Human Evolution* (New York, 1918), 224.

25. Adele Schreiber, "Missbrauchte und unwillkommene Mutterschaft," *Mutterschaft: Ein Sammelwerk* ed. Adele Schreiber (Munich, 1912), 202.

26. Henriette Fürth, "Der Neomalthusianismus und die Soziologie," *Sozialistische Monatshefte* 17 (1911): 1665–72.

27. Henriette Fürth, "Die Lage der Mutter und die Entwicklung des Mutterschutzes in Deutschland," in *Mutterschaft: Ein Sammelwerk*, 284.

28. Diane B. Paul, *Controlling Human Heredity, 1865 to the Present* (Atlantic Highlands, NJ, 1995), 20.

29. Ann Taylor Allen, "Feminism and Eugenics in Germany and Britain, 1900–1948: A Comparative Perspective," *German Studies Review* 23 (2000): 477–505; Ann Taylor Allen, "German Radical Feminism and Eugenics, 1900–1918," *German Studies Review* 11 (1988): 31–56; Dickinson, "Reflections on Feminism and Monism," *Central European History* 34 (2001).

30. Max von Gruber to Alfred Grotjahn, June 9, 1910, in Alfred Grotjahn papers; Alfred Grotjahn, "Das Problem der Entartung," *Archiv für soziale Hygiene* 6 (1910): 80.

31. Alfred Grotjahn, *Geburtenrückgang und Geburtenregelung*, 2nd ed. (Berlin, 1921), 186.

32. Forel, *Sexuelle Frage*, 455–6.

33. Hugo Ribbert, *Heredity, Disease and Human Evolution*, trans. Eden and Cedar Paul (New York, 1918), 224.

34. Arthur Ruppin, *Darwinismus und Sozialwissenschaft* (Jena, 1903), 89–91.

35. On Kautsky's views on eugenics, see Richard Weikart, *Socialist Darwinism: Evolution in German Socialist Thought from Marx to Bernstein* (San Francisco, 1999), ch. 6.

36. Emil Kraepelin, *Die psychiatrische Aufgaben des Staates* (Jena, 1900), 16–17.

37. Ignaz Kaup, "Was kosten die minderwertigen Elemente dem Staat und der Gesellschaft?" *Archiv für Rassen- und Gesellschaftsbiologie* 10 (1913): 747.

38. Hugo Ribbert, *Rassenhygiene: Eine gemeinverständliche Darstellung* (Bonn, 1910), 64.

39. Felix von Luschan, "Culture and Degeneration" (n.d.), p. 4, in Felix von Luschan papers.

40. Paul Näcke, "Die Kastration bei gewissen Klassen von Degenerirten als ein wirksamer socialer Schutz," *Archiv für Kriminal—Anthroplogie und Kriminalistik* 3 (1900): 58–84.

41. Ernst Rüdin, "Der Alkohol im Lebensprozess der Rasse," *Politisch-Anthropologische Revue* 2 (1903–04): 553–66.
42. Geza von Hoffmann, *Die Rassenhygiene in den Vereinigten Staaten von Nordamerika* (Munich, 1913), 108–09 and passim.
43. Felix von Luschan, "Culture and Degeneration" (n.d.), in Felix von Luschan papers, Kasten 14, p. 5; Max Hirsch, *Fruchtabtreibung und Präventivverkehr* (Würzburg, 1914), 210–13; Hugo Ribbert, *Heredity, Disease and Human Evolution* (New York, 1918), 222; see also Joachim Müller, *Sterilisation und Gesetzgebung bis 1933* (Husum, 1985).
44. Helene Stöcker, "Von neuer Ethik," *Mutterschutz* 2 (1906): 9.
45. One example among many is Wilhelm Schallmayer, *Die drohende physische Entartung der Culturvölker*, 2nd ed. (Berlin, 1891), 18–19.
46. F. Müller-Lyer, "Die Ehe," in *Mutterschaft: Ein Sammelwerk*, 151–55.
47. "Satzungen der Deutschen Gesellschaft für Rassen-Hygiene," March 7, 1911, in Alfred Ploetz papers.
48. "Die Jahresversammlung des Deutschen Monistenbund," *Monismus* 3 (1908): 378–80; *Der Magdeburger Monistentag vom 6.–9. Sept. 1912,* ed. Wilhelm Blossfeldt (Munich, 1913), 110.
49. "Bund für Mutterschutz," *Sexualreform* 2 (1907): 2–3.
50. *Das Monistische Jahrhundert* 2 (1913): 1026–8.
51. On Tandler's eugenics, see Doris Byer, *Rassenhygiene und Wohlfahrtspflege* (Frankfurt, 1988); and Karl Sablik, *Julius Tandler* (Vienna, 1983).
52. Eduard David, "Darwinismus und soziale Entwicklung," in *Darwin*, ed. Max Apel (Berlin, 1909), 58.
53. Alfred Hegar, *Der Geschlechtstrieb: Eine Social-Medicinische Studie* (Stuttgart, 1894), 145–7.
54. For one example, see Hirsch, *Fruchtabtreibung und Präventivverkehr*, 208–09.
55. Christian von Ehrenfels, "Gedanken über die Regeneration der Kulturmenschheit," (privately published, 1901, available at the Forschungsstelle und Dokumentationszentrum für Oesterreichischer Philosophie, Graz), 25, 28, quote at 52–3. For more on Ehrenfels, see Reinhard Fabian, "Leben und Wirken von Christian v. Ehrenfels. Ein Beitrag zur Intellektuellen Biographie," in *Christian von Ehrenfels: Leben und Werk*, ed. Reinhard Fabian (Amsterdam, 1986), 1–64.
56. Max Nordau, *Die konventionellen Lügen der Kulturmenschheit* (Leipzig, 1909), 291–303.
57. Christian von Ehrenfels, *Sexualethik* (Wiesbaden, 1907), 10–11; Ehrenfels, "A Program for Breeding Reform" (December 23, 1908) in *Minutes of the Vienna Psychoanalytic Society* ed. Herman Nunberg and Ernst Federn (New York, 1967), 2: 96–7.
58. Willibald Hentschel to Christian von Ehrenfels, July 23, 1907 and August 15, 1907, in Christian von Ehrenfels papers, Forschungsstelle und Dokumentationszentrum für Oesterreichischer Philosopie, Graz.
59. Willibald Hentschel, *Mittgart: Ein Weg zur Erneuerung der germanischen Rasse*, 2nd ed. (Leipzig, 1906).

60. Hentschel, *Mittgart*, 169–71.
61. Willibald Hentschel to Christian von Ehrenfels, July 23, 1907 and October 27, 1907; Alfred Ploetz to Christian von Ehrenfels, February 17, 1908, in Christian von Ehrenfels papers; Theodor Fritsch to Ludwig Schemann, February 17, 1908, in Hammer–Verlag papers, , IV A 1/2, Freiburg University Library Archives. For more on Hentschel, see Dieter Löwenberg, "Willibald Hentschel (1858–1947): Seine Pläne zur Menschenzüchtung, sein Biologismus und Antisemitismus" (dissertation, University of Mainz, 1978).
62. Theodor Fritsch, "Die rechte Ehe. Ein Wort zum Züchtungs–Gedanken und Mittgart–Problem," n.d., in Reichshammerbund Flugblätter und Flugschriften, ZSg. 1—263/3, Bundesarchiv Koblenz.
63. Wilhelm Schallmayer, *Vererbung und Auslese* (Jena, 1903), 341.
64. Wilhelm Schallmayer, "Rassedienst," *Sexual-Probleme* 7 (1911): 440–1.
65. Fritz Lenz, *Über die krankhaften Erbanlagen des Mannes und die Bestimmung des Geschlechtes beim Menschen* (Jena, 1912), 138.
66. F. Müller-Lyer, "Die Ehe," in *Mutterschaft: Ein Sammelwerk*, 153–4.
67. Max von Gruber, *Ursachen und Bekämpfung*, 3rd ed. (Munich, 1914), 51, 54, 59, 63–4; Max von Gruber and Ernst Rüdin, *Fortpflanzung, Vererbung, Rassenhygiene*, 2nd ed. (Munich, 1911), 159, 162.
68. Alfred Ploetz to Max von Gruber, November 19, 1911, in Alfred Ploetz papers; Alfred Ploetz to Christian von Ehrenfels, July 14, 1907, in Christian von Ehrenfels papers.
69. Alfred Ploetz to Gerhart Hauptmann, October 4, 1904, in Gerhart Hauptmann papers, Staatsbibliothek Preussischer Kulturbesitz, Berlin.
70. Luschan, "Culture and Degeneration," in Felix Luschan papers, Kasten 14.
71. Hirsch, *Fruchtabtreibung und Präventivverkehr*, 247.

8 KILLING THE "UNFIT"

1. Edward J. Larson and Darrel W. Amundsen, *A Different Death: Euthanasia and the Christian Tradition* (Downers Grove, IL, 1998).
2. Richard Weikart, "Darwinism and Death: Devaluing Human Life in Germany, 1860–1920," *Journal of the History of Ideas* 63 (2002): 323–44.
3. Hans Walter Schmuhl, *Rassenhygiene, Nationalsozialismus, Euthanasie* (Göttingen, 1987), 18–19, quote at 106.
4. Michael Schwartz, " 'Euthanasie'—Debatten in Deutschland (1895–1945)," *Vierteljahrshefte für Zeitgeschichte* 46 (1998), 623.
5. Udo Benzenhöfer, *Der gute Tod?* (Munich, 1999), ch. 4; on the connections between Darwinism and euthanasia, see also Ernst Klee, *"Euthanasie" im NS-Staat* (Frankfurt, 1985), ch. 1; Kurt Nowak, *"Euthanasie" und Sterilisierung* (Göttingen, 1977), 25–7; Bruce Fye, "Active Euthanasia: An Historical Survey of Its conceptual Origins and Introduction into Medical Thought," *Bulletin of the History of Medicine* 52 (1978): 497–500.

6. Ian Dowbiggin, *A Merciful End* (Oxford, 2003); N. D. A. Kemp, *"Merciful Release"* (Manchester, 2002).

7. Ernst Haeckel, *Natürliche Schöpfungsgeschichte*, 2nd ed. (Berlin, 1870), 152–5; quote at 155.

8. Ernst Haeckel, *Die Lebenswunder* (Stuttgart, 1904), 135–6; Haeckel, *Ewigkeit*, (Berlin, 1917), 33–4.

9. Haeckel, *Lebenswunder* (Stuttgart, 1904), 22.

10. Haeckel, *Ewigkeit* (Berlin, 1917), 33–4.

11. Haeckel, *Lebenswunder* (Stuttgart, 1904), 136.

12. Ernst Haeckel, *Der Monismus als Band zwischen Religion und Wissenschaft* (Bonn, 1892), 26; Ernst Haeckel to Richard Semon, May 20, 1917, in Richard Semon papers, Bayerische Staatsbibliothek, Munich.

13. Haeckel, *Lebenswunder* (Stuttgart, 1904) 375.

14. Ernst Haeckel to Frieda von Uslar-Gleichen, January 22, 1900, in *Franziska van Altenhausen*. ed. Johannes Werner, 6th ed. (Leipzig, 1927), 145.

15. Haeckel, *Lebenswunder*, 128, quote at 132.

16. Ibid., 21–2, 134–5.

17. Hugo Ribbert, *Heredity, Disease and Human Evolution* (New York, 1918), 179–80.

18. Alfred Grotjahn, "Wahre und falsche Euthanasie," n.d. (ca. 1919), in Alfred Grotjahn papers, 303, Humboldt University Archives, Berlin.

19. Ludwig Gumplowicz, *Sozialphilosophie im Umriss* (1910), in *Ludwig Gumplowicz oder die Gesellschaft als Natur*, ed. Emil Brix (Vienna, 1986), 272–3.

20. Brix, ed., *Ludwig Gumplowicz oder die Gesellschaft als Natur* (Vienna, 1986), 42–3.

21. Bartholomäus Carneri, *Der moderne Mensch*, Volksausgabe (Stuttgart, 1901), 142.

22. Wilhelm Schallmayer, *Beiträge zu einer Nationalbiologie* (Jena, 1905), 46.

23. Christian von Ehrenfels, "Gedanken über die Regeneration der Kulturmenschheit" (privately published, 1901; available at the Forschungsstelle und Dokumentationszentrum für Österreichische Philosophie, Graz, Austria), 41.

24. Hans von Hentig, *Strafrecht und Auslese* (Berlin, 1914), 24–46.

25. Susanne Hahn, " 'Minderwertige, widerstandlose Individuen . . .': Der erste Weltkrieg und das Selbstmordproblem in Deutschland," in *Die Medizin und der Ersten Weltkrieg*, ed. Wolfgang Eckert et al. (Pfaffenweiler, 1996), 273–97.

26. Roland Gerkan, "Euthanasie," *Das monistische Jahrhundert* 2 (1913): 169–74.

27. Ibid., Wilhelm Ostwald, "Euthanasie," *Das monistische Jahrhundert* 2 (1913): 337–41.

28. Eugen Wolfsdorf, "Euthanasie und Monismus," *Das monistische Jahrhundert* 2 (1913): 305–09.

29. A. Braune, "Euthanasie und Arzt," *Das monistische Jahrhundert* 2 (1913–14): 871–3; Friedrich Siebert, in *Was wir Ernst Haeckel verdanken*, ed. Heinrich Schmidt (Leipzig, 1914), I: 339–51.

30. Wilhelm Börner, "Euthanasie (Eine Erwiderung)," *Das monistische Jahrhundert* 2 (1913): 249–54.

31. Friedrich Hellwald, *Culturgeschichte in ihrer natürlichen Entwicklung* (Augsburg, 1875), 276.

32. Arnold Dodel, *Moses oder Darwin?* (Zurich, 1889), 72–3.

33. Alfred Ploetz, *Die Tüchtigkeit* (Berlin, 1895), 144–5.

34. Alfred Ploetz, "Ableitung einer Rassenhygiene und ihrer Beziehung zur Ethik," *Vierteljahresschrift für wissenschaftliche Philosophie* 19 (1895): 370.

35. Adolf Jost, *Das Recht auf den Tod* (Göttingen, 1895), 4, 19, quote at 18.

36. Oda Olberg, "Das Recht auf den Tod," *Zukunft* 18 (1897): 493–502; quote at 494.

37. August Forel, *Die sexuelle Frage* (Munich, 1905), 399–400.

38. August Forel, *Leben und Tod* (Munich, 1908), 3.

39. August Forel, *Kulturbestrebungen der Gegenwart* (Munich, 1910), 26–7.

40. Wilhelm Schallmayer, *Vererbung und Auslese* (Jena, 1903), 210.

41. Wilhelm Schallmayer, "Rassedienst," *Sexual-Probleme* 7 (1911): 437.

42. Agnes Bluhm, "Eugenics and Obstetrics," in *Problems in Eugenics: Papers Communicated to the First International Eugenics Congress* (London: Eugenics Education Society, 1912; rprt. New York: Garland, 1984), 395.

43. Alfred Hegar, "Die Wiederkehr des Gleichen und die Vervollkommnung des Menschengeschlechtes," *Archiv für Rassen- und Gesellschaftsbiologie* 8 (1911): 72–85; quote at 80–1.

44. Alfred Hoche, *Jahresringe: Innenansicht eines Menschenlebens* (Munich, 1935), 22.

45. Ibid., 291.

46. Ibid., 289–90.

47. Karl Binding and Alfred Hoche, *Die Freigabe der Vernichtung lebensunwerten Lebens:* (Leipzig, 1920), 49–62.

48. Siegfried Weinberg, "Die Vernichtung des keimenden Lebens," *Mutterschutz* 1 (1905): 312–19.

49. Otto Ehinger, "Der Grund der Abtreibungsbestrafung," *Die neue Generation* 8 (1912): 237–8.

50. Bund für Mutterschutz, untitled declaration on abortion, n.d.; "Referat über den Vortrag von Dr. phil. Helene Stöcker am 27.11.08, Zwangsmutterschaft und Kultur"; "Leitsätze zum Referat von Dr. phil. Helene Stöcker auf der Generalversammlung des Deutschen Bundes für Mutterschutz in Hamburg 1909," April 1909; "Referat über den Vortrag von Dr. phil. Helene Stöcker, Mutterschutz und Abtreibungsstrafe," April 1909; in Adele Schreiber papers, N 1173, 28, Bundesarchiv Koblenz; Helene Stöcker, "Strafrechtreform und Abtreibung," *Die neue Generation* 4 (1908): 399–410.

51. Eduard David, "Darwinismus und soziale Entwicklung," in *Darwin*, ed. Max Apel (Berlin, 1909), 59.

52. Quoted in Alfred Meyer, *The Feminism and Socialism of Lily Braun* (Bloomington, 1985), 147.

53. Alfred Ploetz to Helene Stöcker, n.d., in Alfred Ploetz papers, privately held by Wilfried Ploetz, Herrsching am Ammersee.

54. Agnes Bluhm, "Frauenbewegung, Strafrecht, und Rassenhygiene," *Archiv für Rassen- und Gesellschaftsbiologie* 6 (1909): 134–9; Anna Pappritz, "Die Vernichtung des keimenden Lebens," *Sexual-Probleme* 5 (1909): 492; see also Barbara Greven-Aschoff, *Die bürgerliche Frauenbewegung in Deutschland 1894–1933* (Göttingen, 1981), 115–16; Richard J. Evans, *The Feminist Movement in Germany, 1894–1933* (London, 1976), 158–66.

55. Anna Pappritz, "Die Vernichtung des keimenden Lebens," *Sexual-Probleme* 5 (1909): 495–6, 498.

56. Doris Byer, *Rassenhygiene und Wohlfahrtspflege* (Frankfurt, 1988), 121–3.

57. Ernst Rüdin, "Rassenhygiene und kommunaler Schutz der Minderwertigen," *Archiv für Rassen- und Gesellschaftsbiologie* 5 (1908): 153–4; Rüdin, "Der Alkohol im Lebensprozess der Rasse," *Politisch-Anthropologische Revue* 2 (1903–4): 564.

58. Max Hirsch, *Fruchtabtreibung und Präventivverkehr* (Würzburg, 1914), 210–13; quote at 214.

59. Ernst Haeckel, *Natürliche Schöpfungsgeschichte*, 4th ed. (Berlin, 1873), 155.

60. Willibald Hentschel, *Mittgart*, 2nd ed. (Leipzig, 1906), 14.

61. On dueling, see Keven McAleer, *Dueling: The Cult of Honor in Fin-de-Siecle Germany* (Princeton: Princeton University Press, 1997).

9   WAR AND PEACE

1. William Jennings Bryan, *In His Image* (Freeport, NY, 1922, rprt. 1971), 124–5.

2. Quoted in Antonello La Vergata, "Evolution and War, 1871–1918," *Nuncius* 9 (1994): 148.

3. Vernon Kellogg, *Headquarters Nights* (Boston, 1917), 22–32.

4. H. W. Koch, "Social Darwinism as a Factor in the 'New Imperialism'," *The Origins of the First World War* (New York, 1972), 329–54; Daniel Pick, *War Machine* (New Haven, 1993), ch. 8.

5. La Vergata, "Evolution and War," 143–163; Klaus Schwabe, *Wissenschaft und Kriegsmoral* (Göttingen, 1969), 34; Roland N. Stromberg, *Redemption by War: The Intellectuals and 1914* (Lawrence, 1982), 78–9.

6. Roger Chickering, *Imperial Germany and a World without War* (Princeton, 1975). In making his case for Anglo-American "peace biology," Paul Crook underestimates the influence of social Darwinism in general and of social Darwinist militarism in particular—see *Darwinism, War and History* (Cambridge, 1994). On the British scene, see also Nancy Leys Stepan, " 'Nature's Pruning Hook': War, Race and Evolution, 1914–1918," in *The Political Culture of Modern Britain*, ed. J. M. W. Bean (London, 1987).

7. Hannsjoachim W. Koch, *Der Sozialdarwinismus* (Munich, 1973), ch. 8.

8. Mike Hawkins, *Social Darwinism in European and American Thought, 1860–1945* (Cambridge, 1997), 110.

9. Friedrich Rolle, *Der Mensch* (Frankfurt, 1866), 109–19, 142–4; quote at 112.

10. Oscar Peschel, "Ein Rückblick auf die jüngste Vergangenheit," *Das Ausland* 39, 36 (September 1866): 874.

11. Gustav Jaeger, "Ein biologisches Moment der neueren Völkergeschichte," *Das Ausland* 39 (1866): 1024–6; quote at 1024.

12. Gustav Jaeger, "Naturwissenschaftliche Betrachtungen über den Krieg," *Das Ausland* 43 (1870): 1161.

13. Friedrich Hellwald, "Der Kampf ums Dasein im Menschen- und Völkerleben" *Das Ausland* 45 (1872): 103–6, 140–4; Ingo Wiwjorra, "Die deutsche Vorgeschichtsforschung," in *Handbuch zur "Völkischen Bewegung" 1871–1918*, eds. Uwe Puschner, Walter Schmitz, and Justus H. Ulbricht (Munich, 1996), 195.

14. David Friedrich Strauss to Christian Käserle, January 16, 1869, in *Ausgewählte Briefe von David Friedrich Strauss* (Bonn: Emil Straus, 1895), 506.

15. David Friedrich Strauss, *Der alte und der neue Glaube* (Leipzig, 1872), 185–7, 252–6.

16. Ernst Haeckel to Friedrich Hellwald, February 24, 1874, in *Briefwechsel zwischen Ernst Haeckel und Friedrich von Hellwald* (Ulm, 1901), 9; David Friedrich Strauss to Ernst Haeckel, August 24, 1873, in *Ausgewählte Briefe von David Friedrich Strauss* (Bonn: Emil Straus, 1895), 554.

17. Friedrich Hellwald, *Culturgeschichte* (Augsburg, 1875), 44–5.

18. Ibid., 58.

19. Friedrich Hellwald, "Der Kampf ums Dasein im Menschen- und Völkerleben," *Das Ausland* 45 (1872): 105.

20. Hellwald, *Culturgeschichte* 325, 656.

21. Albert Schäffle, "Darwinismus und Socialwissenschaft," in *Gesammelte Aufsätze* (Tübingen, H. Laupp'sche Buchhandlung, 1885), 1: 3–6; quote in *Bau und Leben* (Tübigen, 1881), 2: 353.

22. Schäffle, *Bau und Leben*, 2: 228–9, 248–51, 361, 369, quote at 373.

23. Heinrich Ernst Ziegler, *Die Naturwissenschaft und die Socialdemokratische Theorie, ihr Verhältnis dargelegt auf Grund der Werke von Darwin und Bebel* (Stuttgart, 1893), 167–8.

24. Felix von Luschan, "Die gegenwärtigen Aufgaben der Anthropologie," *Verhandlungen* (Leipzig, 1910), Part 2, 205–06.

25. Alfred Kirchhoff, *Darwinismus angewandt auf Völker und Staaten* (Frankfurt, 1910), 87; see also Alfred Kirchhoff, "Darwinismus in der Völkerentwickelung," *Nord und Süd* 31 (1884): 367–77.

26. Sebald Rudolf Steinmetz, *Die Philosophie des Krieges* (Leipzig, 1907), 20–3, 43–5.

27. Ibid., ch. 7; quotes at 222, 252–3,

28. Klaus Wagner, *Krieg* (Jena, 1906), 53.

29. Ibid., 109.

30. Otto Schmidt-Gibichenfels, "Der Krieg als Kulturfaktor," *Politisch-Anthropologische Revue* 11 (1912): 393–407, 449–61.

31. Istvan Deak, *Beyond Nationalism* (New York, 1990), 73.

32. Conrad von Hötzendorf, *Aus meiner Dienstzeit,* vol. 1 (Vienna, 1921), 7–14.

33. Lawrence Sondhaus, *Franz Conrad von Hötzendorf* (Boston, 2000), 5, 15–16, 82, passim.

34. Conrad von Hötzendorf, *Private Aufzeichnungen*, ed. Kurt Peball (Vienna, 1977), 317, quotes at 148, 305, 307.

35. Friedrich von Bernhardi, *Deutschland und der nächste Krieg*, 6th ed. (Stuttgart, 1913), 11–21; quotes at 11, 57.

36. Max von Gruber, *Kreig, Frieden und Biologie* (Berlin, 1915), 7–11.

37. Ernst Haeckel, *Natürliche Schöpfungsgeschichte* (Berlin, 1868), 218–19.

38. Haeckel, *Natürliche Schöpfungsgeschichte*, 2nd ed. (Berlin, 1870), 153–4.

39. Ernst Haeckel to Bertha von Suttner, October 31, 1891, in Bertha von Suttner, *Memoirs of Bertha von Suttner: The Records of an Eventful Life* (Boston: Ginn, 1910), 347. Strangely, this letter is not in the Suttner–Fried Collection at the League of Nations Archive in Geneva.

40. "Das Monistische Jahrhundert" (signed by W. Blossfeldt, Redaktion) to Alfred Fried, June 18, 1912 and October 3, 1912, in Suttner–Fried Collection, League of Nations Archives, United Nations Library, Geneva. See also "Aufruf zur Begründung eines Verbandes für internationale Verständigung," *Der Monismus* 5 (1910): 411–12, for Monist League support for the peace movement.

41. Wilhelm Blossfeldt, ed., *Der Magdeburger Monistentag* (Munich, 1913), 9–13.

42. Ernst Haeckel, "Englands Blutschuld am Weltkriege," *Das monistische Jahrhundert* 3 (1914–15): 538–48; Haeckel, "Weltkrieg und Naturgeschichte," *Nord und Süd* 151 (November 1914): 140–7.

43. Ernst Haeckel to Richard Hertwig, September 19, 1917 and January 12, 1918, in *Ernst Haeckel. Sein Leben, Denken, und Wirken* (Jena, 1943–44), 1: 64, 66.

44. Heinrich Fick, "Ueber den Einfluss der Naturwissenschaft auf das Recht," in Helene Fick, *Heinrich Fick. Ein Lebensbild*, 2 vols. (Zurich, 1897–1908), 2: 292, 304–5. Fick's speech was also published in *Jahrbücher für Nationalökonomie und Statistik* 18 (1872): 248–77.

45. Alexander Tille, *Volksdienst* (Berlin, 1893), 5–7, 11–13, 21, 25–7, 36, 38–9, 43.

46. Wilhelm Schallmayer, *Vererbung und Auslese* (Jena, 1903), 111–15, 177–8, 245, 250, 296; Schallmayer, "Die Erbentwicklung bei Völkern," *Menschheitsziele* 1 (1907): 95; Wilhelm Schallmayer to Ludwig Schemann, November 12, 1914, in Ludwig Schemann papers, IV B 1/2, University of Freiburg Library Archives.

47. Wilhelm Schallmayer, "Die Auslesewirkungen des Krieges," *Menschheitsziele* 2 (1908): 381–5.

48. Wilhelm Schallmayer, "Krieg als Züchter," 395–6.

49. In her excellent work on Schallmayer, Sheila Faith Weiss, *Race Hygiene and National Efficiency* (Berkeley, 1987), 38–9, 47–8, 141–2, emphasizes his anti-militarism, but for a more balanced view, see Pascal Grosse, *Kolonialismus, Eugenik und Bürgerliche* (Frankfurt, 2000), 211–14.

50. Alfred Ploetz, *Die Tüchtigkeit* (Berlin, 1895), 61–3, quote at 147; Robert Proctor in *Racial Hygiene* (Cambridge, 1988), 28, stresses Ploetz's antimilitarism.

51. Arnold Dodel, "Charles Robert Darwin, sein Leben, seine Werke und sein Erfolg," *Die neue Zeit* 1 (1883): 117–18.

52. Arnold Dodel to Bertha von Suttner, October 8, 1889, April 16, 1890, etc., in Suttner–Fried Collection, League of Nations Archives, United Nations Library, Geneva.

53. Bertha Suttner, *Lebenserinnerungen*, ed. Fritz Böttger, 4th ed. (Berlin, 1972), 187, 351.

54. Beatrix Kempf, *Woman for Peace* (Park Ridge, NJ, 1973), 173.

55. Bertha von Suttner, *Die Waffen nieder!*, Volksausgabe (Dresden, 1900), 31, 204; Bertha von Suttner, *Daniela Dormes* (Dresden, 1906), 98–101.

56. Bertha von Suttner, *Maschinenzeitalter* (1898; rprt. Düsseldorf, 1983), 270.

57. Bertha von Suttner, *Inventarium einer Seele* (Leipzig, 1883), 108.

58. Ibid., 108.

59. Brigitte Hamann, *Bertha von Suttner* (Munich, 1986), 254–6; see also Suttner, *Inventarium einer Seele*, 127.

60. Alfred Fried, "Unser Jahrhundert," *Die Friedenswarte* 2 (1900): 2.

61. Alfred Fried to David Starr Jordan, December 16, 1910, in David Starr Jordan papers, Box 20, folder 47, Hoover Institution Archives, Stanford University.

62. Alfred Fried to Alfred Ploetz, September 14, 1904, in Fried–Suttner Collection, League of Nations Archive, United Nations Library, Geneva; Fried (published anonymously), *Experimental-Ehen: Ein "Document humain" als Beitrag zur Eherechtsreform* (Munich: Ernst Reinhardt, 1906).

63. Alfred Grotjahn to Alfred Fried, December 3, 1909, in Fried–Suttner Collection, League of Nations Archive, United Nations Library, Geneva.

64. Alfred Fried, *Die Grundlagen des revolutionaren Pacifismus* (Tübingen, 1908), 35; Fried, *Handbuch der Friedensbewegung* (Vienna, 1905), 34–5; Fried, *Die moderne Friedensbewegung* (Leipzig, 1907), 2–4.

65. Hamann, *Bertha von Suttner*, 254–6.

66. Alfred Fried, "Und wieder ein Krieg!" *Die Friedenswarte* 2 (1900): 97–9.

67. Alfred Fried, *Der kranke Krieg* (Leipzig, 1909), 141.

68. Fried, *Handbuch der Friedensbewegung*, 23–4.

69. Ibid., 22.

10   RACIAL STRUGGLE AND EXTERMINATION

1. Friedrich Hellwald, *Kulturgeschichte* (Leipzig, 1896), 4: 615–16.

2. Charles Darwin, *Descent of Man*, 2 vols. (London, 1871; rprt. Princeton, 1981), 1: 166.

3. Alfred Kirchhoff, *Darwinismus* (Frankfurt, 1910), 73, 86–7.

4. Heinz Gollwitzer, *Die gelbe Gefahr* (Göttingen, 1962), 169; Benoit Massin's survey of the German press seems to confirm this point; see "From Virchow to

Fischer: Physical Anthropology and 'Modern Race Theories' in Wilhelmine Germany," in *Volksgeist as Method and Ethic* (Madison, 1996), 126; Brigitte Hamann in *Hitler's Vienna* (New York, 1999), ch. 7, confirms this point in relation to the Viennese press.

5. Quoted in Diane B. Paul, *Controlling Human Heredity* (Atlantic Highlands, NJ, 1995), 75; see also Hannsjoachim W. Koch, *Der Sozialdarwinismus* (Munich, 1973), ch. 9.

6. Jacob Katz discusses this in relation to anti-Semitic racism in *From Prejudice to Destruction* (Cambridge, MA, 1980), 20, 323.

7. Woodruff D. Smith, *Ideological Origins of Nazi Imperialism* (New York, 1986).

8. Darwin, *Descent*, 1: 201.

9. Francis Darwin, *Charles Darwin: His Life Told in an Autobiographical Chapter, and in a Selected Series of His Published Letters* (London, John Murray, 1902), 64.

10. Quoted in Adrian Desmond and James Moore, *Darwin* (London, 1991), 521.

11. Ernst Haeckel, "Ueber die Entwickelungstheorie Darwins," in *Amtlicher Bericht über die 38. Versammlung Deutscher Naturforscher und Ärzte* (Stettin, 1864), 28.

12. Ernst Haeckel, *Natürliche Schöpfungsgeschichte* (Berlin, 1868), 218–19.

13. Ibid., 206; see also Haeckel, *Natürliche Schöpfungsgeschichte*, 2nd ed. (Berlin, 1870), 153.

14. Ibid., 520.

15. Ernst Haeckel, *Welträthsel* (Bonn, 1903), 141; Haeckel, *Indische Reisebriefe* (Berlin, 1883), 354–5.

16. Haeckel, *Ewigkeit* (Berlin, 1917), 35–6, 110–11, 120–3; quotes at 85–6, 35.

17. Oscar Peschel, "Ursprung und Verschiedenheit der Menschenrassen," *Das Ausland* 33 (1860): 393.

18. Oscar Peschel, *Völkerkunde*, 2nd ed. (Leipzig, 1875), 153–5.

19. Friedrich Hellwald, *Naturgeschichte des Menschen*, 2 vols. (Stuttgart, 1880), I:54–66; quote at 66; Oscar Peschel, "Nekrolog der Tasmanier," *Das Ausland* 43 (1870): 189.

20. Friedrich Hellwald, *Culturgeschichte in ihrer natürlichen Entwicklung bis zur Gegenwart* (Augsburg, 1875), 44–5.

21. Quoted in ibid., 797–8.

22. Ibid., 656, 755.

23. Friedrich Hellwald, *Die Erde und Ihre Völker*, 2 vols. 2nd ed. (Stuttgart, 1877–78), 1: 118.

24. Friedrich Rolle, *Der Mensch, seine Abstammung und Gesittung im Lichte der Darwin'schen Lehre* (Frankfurt, 1866), 109–19, 142–4; quote at 112.

25. Oscar Schmidt, "Darwinismus und Socialdemokratie," *Deutsche Rundschau* 17 (1878): 284.

26. Oscar Schmidt, *The Doctrine of Descent and Darwinism*, 4th ed. (London, C. Kegan Paul, 1881), 296–7.

27. Ludwig Büchner, *Der Mensch und seine Stellung in der Natur*, 2nd ed. (Leipzig, 1872), 53, 137–9, 154, quote at 147.

28. Carus Sterne (pseudonym of Ernst Krause), *Werden und Vergehen*, 6th ed. (Berlin, 1905), 2: 381.

29. Ludwig Gumplowicz, *Der Rassenkampf* (Innsbruck, 1883), 65–6, 160–94, 234–8; quote at 218; see also Emil Brix, ed., *Ludwig Gumplowicz oder die Gesellschaft als Natur* (Vienna, 1986).

30. Ludwig Gumplowicz to Lester F. Ward, June 10, 1907, in *Letters of Ludwig Gumplowicz* (Leipzig, 1933), 24–5.

31. Friedrich Ratzel, Vorlesungen über "Politische Geographie: Grundzüge der politischen Ethnographie: Völker und Rasse: Rassenfragen" 1895–99, in Friedrich Ratzel papers, Folder 324, Insitut für Länderkunde Archives, University of Leipzig.

32. Friedrich Ratzel, *Sein und Werden der organischen Welt* (Leipzig, 1869), 469.

33. Friedrich Ratzel to Eisig, May 20, 1885, quoted in Gerhard H. Müller, *Friedrich Ratzel* (Stuttgart, 1996), 74; see also Woodruff D. Smith, *The Ideological Origins of Nazi Imperialism*, ch. 5; Mark Bassin, "Imperialism and the Nation State in Friedrich Ratzel's Political Geography," *Progress in Human Geography* 11 (1987): 473–95.

34. Friedrich Ratzel, "Moritz Wagner" (1896), in *Kleine Schriften* (Munich, 1906), 1: 468.

35. Moritz Wagner, "Die Darwin'sche Theorie und das Migrationsgesetz der Organismen" (1868), in *Die Entstehung der Arten durch räumliche Sonderung* (Basel, 1889), 78–9. On the relationship between Wagner and Ratzel, see Wolfgang J. Smolka, *Völkerkunde in München* (Berlin, 1994).

36. Friedrich Ratzel, *Lebensraum* (Tübingen, 1901), 1, 51–60.

37. Friedrich Ratzel, *Politische Geographie*, 2nd ed. (Munich, 1903), ch. 6.

38. Quoted in Gerhard Müller, *Friedrich Ratzel* (Stuttgart, 1996), 76.

39. Ratzel, *Politische Geographie*, 44, 129–53, 371–4; quotes at 153, 143.

40. Ludwig Gumplowicz to Lester F. Ward, August 7, 1902, in *Letters of Ludwig Gumplowicz* (Leipzig, 1933), 10–11.

41. Gustav Ratzenhofer, *Wesen und Zweck der Politik* (Leipzig, 1893), 1: 3; 2: 251, 362; 3: 6; quote at 1: 127.

42. Gustav Ratzenhofer, *Positive Ethik* (Leipzig, 1901), 319–20.

43. Gustav Ratzenhofer, *Soziologie* (Leipzig, 1907), 47, 68–70, 228; quote at 16.

44. Ratzenhofer, *Positive Ethik* (Leipzig, 1901) 319–20.

45. Karl Peters, *Deutsch-national: Kolonialpolitische Aufsätze*, 2nd ed. (Berlin, 1887), 46, 52–3, 99; quotes at 8–9 and 103. In the second quote, Peters was quoting Karl von der Heydt, with whom he fully agreed.

46. Otto Seeck, *Geschichte des Untergangs der antiken Welt*, 4th ed. (Stuttgart, 1921; rprt. 1966), vol. 1, ch. 3.

47. Ludwig Woltmann, *Politische Anthropologie* (Jena, 1903), passim, esp. 261–7; quotes at 267, 298.

48. Quoted in Jennifer Michael Hecht, "Vacher de Lapouge and the Rise of Nazi Science," *Journal of the History of Ideas* 61 (2000): 287.

49. Otto Ammon, *Die Gesellschaftsordnung*, 3rd ed. (Jena, 1900), 164. Emphasis in original.

50. Otto Ammon to Ludwig Schemann, November 27, 1899, in Ludwig Schemann papers, IV B 1/2, University of Freiburg Library Archives.

51. Quoted in Gollwitzer, *Die gelbe Gefahr* (Göttingen, 1962) 169–70.

52. Otto Ammon, *Die Natürliche Auslese beim Menschen* (Jena, 1893), 315.

53. Otto Ammon to Matthäus Much, January 22, 1897, Aut. 124.941, in Wiener Stadt- und Landesbibliothek. For more on Ammon, see Peter Emil Becker, *Wege ins Dritte Reich*, vol. 2 (Stuttgart, 1990), ch. 7.

54. Eugen Fischer, *Die Rehobother Bastards* (Jena, 1913), 296–306; quote at 302.

55. "First Universal Races Congress" announcement and program, in Deutsche Kongress Zentrale, Box 256, Hoover Institution Archives, Stanford University.

56. Felix von Luschan, "Anthropological View of Race," in *Papers on Inter-Racial Problems* (London, 1911), 23; Luschan took a similar position in "Die gegenwärtigen Aufgaben der Anthropologie," *Verhandlungen der Gesellschaft deutscher Naturforscher und Ärzte* (Leipzig, 1910), Part 2, 201–8.

57. Sebald Rudolf Steinmetz, *Die Philosophie des Krieges* (Leipzig, 1907), 246–66; quote at 254.

58. Klaus Wagner, *Krieg* (Jena, 1906), 150–1.

59. Max von Gruber, *Kreig, Frieden und Biologie* (Berlin, 1915), 14–17, quote on 15; Gruber, *Ursachen und Bekämpfung*, 3rd ed. (Munich, 1914); Gruber, "Rassenhygiene, die wichtigste Aufgabe völkischer Innenpolitik," *Deutschlands Erneuerung* 2 (1918): 17–32.

60. Heinrich Driesmans, *Dämon Auslese* (Berlin, 1907), xiv, 260–4; Willibald Hentschel, *Varuna* (Leipzig, 1907), 14–15, 33.

61. Jörg Lanz-Liebenfels, "Rasse und Wohlfahrtspflege, ein Anruf zum Streik der wahllosen Wohltätigkeit," *Ostara* Heft 18 (December 1907), 3–5, 16; Nicholas Goodrick-Clark, *The Occult Roots of Nazism* (New York, 1992), 85 and passim.

62. Roger Chickering, *Imperial Germany and a World without War* (Princeton, 1975), 306–7, 320–5.

63. In scanning the leading pacifist journal in Germany, *Friedenswarte*, edited by Alfred Fried, I could not find a single article discussing the major German colonial wars—the Herero Revolt and the Maji Maji Revolt—in the first decade of the twentieth century.

64. Alfred Ploetz, "Unser Weg," in Alfred Ploetz papers, privately held by Wilfried Ploetz, Herrsching am Ammersee.

65. "Münchener Bogenklub Satzungen," in Alfred Ploetz papers.

66. Fritz Lenz to Ludwig Schemann, December 15, 1916, in Ludwig Schemann papers, IV B 1/2.

67. Alfred Ploetz to Gerhart Hauptmann, December 24, 1913, in Gerhart Hauptmann papers, Staatsbibliothek Preussischer Kulturbesitz, Berlin.

68. Alfred Ploetz to Felix von Luschan, November 18, 1911, in Felix von Luschan papers, Staatsbibliothek Preussischer Kulturbesitz, Berlin.

69. Alfred Ploetz, "Neo-Malthusianism and Race Hygiene," in *Problems in Eugenics: Report of Proceedings of the First International Eugenics Congress* (London: Eugenics Education Society, 1913), 2: 189.

70. Wilhelm Schallmayer, "Die Erbentwicklung bei Völkern," *Menschheitsziele* 1 (1907): 95.
71. Wilhelm Schallmayer, *Vererbung und Auslese* (Jena, 1903), 178.
72. August Forel, *Die sexuelle Frage* (Munich, 1905), 584; see also 519.
73. August Forel, *Out of My Life and Work* (NY, 1937), 193.
74. August Forel, "Ueber Ethik," *Zukunft* 28 (1899): 580–1.
75. Forel, *Die Sexuelle Frage* (Munich, 1905) 440.
76. Helene Stöcker, "Staatlicher Gebärzwang oder Rassenhygiene?" *Die neue Generation* 10 (1914): 145.
77. Helene Stöcker, "Rassenhygiene und Mutterschutz," *Die neue Generation* 13 (1917): 138–42.
78. Ibid., 141; the neurologist Kurt Goldstein took a similar position in *Über Rassenhygiene* (Berlin, 1913), 92–5.
79. Regina Braker, "Helene Stöcker's Pacifism in the Weimar Republic: Between Ideal and Reality," *Journal of Women's History* 13 (2001): 70–97.
80. Christian von Ehrenfels to his friends (Rundbrief), November 1920, in Christian von Ehrenfels papers, Forschungsstelle und Dokumentationszentrum für Österreichische Philosophie, Graz; Christian von Ehrenfels, "Gedanken über die Regeneration der Kulturmenschheit" (privately published, 1901; available at Forschungsstelle und Dokumentationszentrum für Österreichische Philosophie, Graz), 21.
81. Christian von Ehrenfels, "A Program for Breeding Reform" (December 23, 1908) in *Minutes of the Vienna Psychoanalytic Society*, ed. Herman Nunberg and Ernst Federn (New York, 1967), 2: 94–7, quote at 94; Ehrenfels, "Die gelbe Gefahr," *Sexual-Probleme* 4 (1908): 185–205; Ehrenfels, "Weltpolitik und Sexualpolitik," *Sexual-Probleme* 4 (1908): 472–89.
82. Christian von Ehrenfels, "Ein Züchtungsfanatiker," *Sexual-Probleme* 5 (1909): 921.
83. G. F. Nicolai, *Biologie des Krieges* (Zurich, 1917), 83.
84. Ibid., 82–5.
85. Massin, "From Virchow to Fischer," *Volksgeist* (Madison, 1996).
86. Pascal Grosse, *Kolonialismus* (Frankfurt, 2000), 135–6; Wolfgang Eckart, *Medizin und Kolonialimperialismus* (Paderborn, 1997), 57–66.
87. Daniel Goldhagen, *Hitler's Willing Executioners* (New York, 1996).
88. Theodor Fritsch, *Handbuch der Judenfrage*, 27th ed. (Hamburg, 1910), 238–40.
89. Theodor Fritsch (for the Deutsche Antisemitische Vereinigung), "Erklärung der deutschen Antisemiten und Deutsch-Sozialen," Flugblatt Nr. 27 (January 1890), ZSg. 1—617/5, Bundesarchiv Koblenz.
90. Adolf Harpf, *Zur Lösung der brennendsten Rassenfrage* (Vienna, 1898), 24–5, 39, 58, quote at 27.
91. Helmuth Stoecker, "The German Empire in Africa before 1914," in *German Imperialism in Africa* (London, 1986), 209–12.
92. William II to Theodore Roosevelt, September 4, 1905, quoted in Mehnert, *Deutschland, Amerika und die "gelbe Gefahr,"* 9; on the connection between the "Yellow Peril" and Darwinism, see p. 20.

93. Jon Bridgman and Leslie J. Worley, "Genocide of the Hereros," in *Genocide in the Twentieth Century* (New York, 1995), 10, 29; Jon Bridgman, *Revolt of the Hereros* (Berkeley, 1981), 60–3.

94. Quoted in Horst Drechsler, *Let Us Die Fighting* (London, 1980), 167–8, n. 6.

95. Quoted in Peter Schmitt-Egner, *Kolonialismus und Faschismus* (Giessen, 1975), 125.

96. Helmut Walser Smith, "The Talk of Genocide, the Rhetoric of Miscegenation: Notes on Debates in the German Reichstag concerning Southwest Africa, 1904–14," in *The Imperialist Imagination* (Ann Arbor, 1998), 107–23, esp. 121.

97. Helmut Bley, *South-West Africa under German Rule, 1894–1914* (Evanston, 1971), 163–8, quotes at 163, 165.

98. Tilman Dedering, "The German-Herero War of 1904: Revisionism of Genocide or Imaginary Historiography?" *Journal of Southern African Studies* 19 (1993): 80–8. Jon Swan explicitly connects the Herero extermination and the Holocaust in "The Final Solution in Southwest Africa," *MHQ: The Quarterly Journal of Military History* 3 (1991): 36–55.

11 HITLER'S ETHIC

1. Gisela Bock, *Zwangssterilisation im Nationalsozialismus* (Opladen, 1986); Bock, "Sterilization and 'Medical' Massacres in National Socialist Germany: Ethics, Politics, and the Law," in *Medicine and Modernity: Public Health and Medical Care in Nineteenth- and Twentieth-Century Germany*, ed. Manfred Berg and Geoffrey Cocks (Washington, 1997), 149–72.

2. Eberhard Jäckel, *Hitler's Weltanschauung* (Middleton, CN, 1972).

3. One scholar agreeing that Hitler had a consistent ethic is Peter J. Haas, *Morality after Auschwitz* (Philadelphia, 1988); however, Haas does not analyze Hitler's thought adequately.

4. Robert Gellately, *Backing Hitler* (Oxford, 2001).

5. Richard Steigmann-Gall's claim in *The Holy Reich: Nazi Conceptions of Christianity, 1919–1945* (Cambridge, 2003) that Hitler's social ethic was based on Christian principles is mistaken; see my review of Steigmann-Gall in *German Studies Review* (forthcoming).

6. Haas, *Morality after Auschwitz*, wrongly claims that Hitler's ethic of racial war was based on the just war theory and thus showed continuity with previous Christian morality. Haas fails to understand that Hitler's evolutionary ethic and racial war ideology had little or nothing in common with just war theory.

7. Mike Hawkins, *Social Darwinism in European and American Thought* (Cambridge, 1997), ch. 11; Richard J. Evans, "In Search of German Social Darwinism: The History and Historiography of a Concept," in *Medicine and Modernity* (Washington, 1997), 55–79.

8. Adolf Hitler, "Weltjude und Weltbörse, die Urschuldigen am Weltkriege" April 13, 1923, in *Hitler. Sämtliche Aufzeichnungen, 1905–1924*, ed. Eberhard Jäckel (Stuttgart, 1980), 887; see also *Hitlers Zweites Buch* (Stuttgart, 1961), 46.

9. Adolf Hitler, *Mein Kampf* (Munich, 1943), 315.

10. Ibid., 144–5.

11. Richard Weikart, "Progress through Racial Extermination: Social Darwinism, Eugenics and Pacifism in Germany, 1860–1918," *German Studies Review* 26 (2003): 273–94.

12. Hitler, *Mein Kampf,* 316–17.

13. Ibid., 325–7.

14. *The Testament of Adolf Hitler: The Hitler–Bormann Documents, February–April 1945,* ed. Francois Genoud (London, 1961), 51; Hitler expressed similar sentiments much earlier; see Hitler. *Sämtliche Aufzeichnungen, 1905–1924,* 156, 176–7.

15. Adolf Hitler, "Politik und Rasse. Warum sind wir Antisemiten?" in *Hitler. Sämtliche Aufzeichnungen, 1905–1924,* 909.

16. Hitler, *Mein Kampf,* 420–1. Emphasis is mine.

17. Ibid., 444.

18. *Hitlers Zweites Buch,* 56–7.

19. August Kubizek, *The Young Hitler I Knew* (Boston, 1955), 182–3.

20. Brigitte Hamann, *Hitler's Vienna* (New York, 1999), ch. 7.

21. *The Speeches of Adolf Hitler, April 1922–August 1939* (Oxford, 1942), 1: 464.

22. Mario Di Gregorio provides a more nuanced view of Haeckel's racism in "Reflections of a Nonpolitical Naturalist: Ernst Haeckel, Wilhelm Bleek, Friedrich Müller and the Meaning of Language," *Journal of the History of Biology* 35 (2002): 79–109.

23. J. Lanz-Liebenfels, "Revolution oder Evolution? Ein freikonservative Osterpredigt für das Herrentum europäischer Rasse," *Ostara* 3 (April 1906); Lanz-Liebenfels, "Moses als Darwinist, eine Einführung in die anthropologische Religion," *Ostara* 2nd ed., no. 46 (1917); Lanz–Liebenfels, "Die Kunst, schön zu lieben und glücklich zu heiraten, ein rassenhygienisches Brevier für Liebesleute," *Ostara* 2nd ed., no. 47 (1916).

24. J. Lanz von Liebenfels, "Adolf Harpf als Prediger der Rassenweisheit," *Ostara* 1 (March 1907): 38–9.

25. J. Lanz von Liebenfels, *Ostara* 66 (1913): 1.

26. Lanz-Liebenfels, "Revolution oder Evolution?", 8.

27. J. Lanz-Liebenfels, "Rasse und Wohlfahrtspflege, ein Anruf zum Streik der wahllosen Wohltätigkeit," *Ostara* Heft 18 (December 1907).

28. J. Lanz-Liebenfels, "Moses als Darwinist, eine Einführung in die anthropologische Religion," *Ostara* 2nd ed., no. 46 (1917): 3.

29. Hamann, *Hitler's Vienna,* 206–14.

30. Nicholas Goodrick-Clark, *The Occult Roots of Nazism* (New York, 1992), 202.

31. Peter Emil Becker, *Wege in Dritte Reich,* vol. 2 (Stuttgart, 1990), ch. 8 (Anhang).

32. Josef Ludwig Reimer, *Ein Pangermanisches Deutschland* (Berlin, 1905), 32.

33. Ibid., 2.

34. Hitler, *Mein Kampf* (Boston, 1943), 99.

35. Georg Schönerer, *Zehn Reden Georg Schönerers aus den Jahren 1882 bis 1888* (Vienna, 1914), 123.
36. Ibid., 162.
37. For Schönerer's politics, see Carl E. Schorske, *Fin-de-Siecle Vienna: Politics and Culture* (New York, 1981), 120–33.
38. William L. Shirer, *The Rise and Fall of the Third Reich: A History of Nazi Germany* (NY: Simon and Schuster, 1960), 104–09; Roderick Stackelberg, *Idealism Debased* (Kent: OH, 1981).
39. Ian Kershaw, *Hitler* (New York, 1998–2000), 1: 240.
40. Ibid., 1: 60–7; Hamann, *Hitler's Vienna*, 348–52.
41. Jäckel, *Hitler's Weltanschauung* (Middleton: CN, 1972).
42. Auerbach, "Hitlers politische Lehrjahre und die Münchener Gesellschaft 1919–1923," *Vierteljahrshefte für Zeitgeschichte* 25 (1977): 14.
43. Gary Stark, *Entrepreneurs of Ideology* (Chapel Hill, 1981), 113–31, quote on 120.
44. Melanie Lehmann, *Verleger J. F. Lehmann* (Munich, 1935), 36.
45. Bogenklub Mitgliederliste (n.d.), and Mitgliederliste von December 31, 1913, in Alfred Ploetz papers, privately held by Wilfried Ploetz, Herrsching am Ammersee.
46. Ian Kershaw, *Hitler*, 2: 138–9.
47. Auerbach, *Hitler's politische Lehrjahre*, *Vierteljahrshefte für Zeitgeschichte* 25 (1977): 8–9.
48. Listing of Books in Hitler's Personal Library, Hitler Collection, Box 4, Hoover Institution, Stanford University.
49. Hans Günther, "Hass," *Deutschlands Erneureung* 5 (1921): 398–400.
50. Adolf Hitler, "Warum musste ein 8. November kommen?" *Deutschlands Erneuerung* 8 (April 1924): 199–207; quotes at 200–01, 207.
51. Quoted in Ernst Klee, *Deutsche Medizin im Dritten Reich* (Frankfurt, 2001), 256; see also Hans F. K. Günther, *Mein Eindruck von Adolf Hitler* (Pähl, 1969), 93–4.
52. Erwin Baur, Eugen Fischer, and Fritz Lenz, *Grundriss der menschlichen Erblichkeitslehre* (Munich, 1923), 1: 407–27, 2: 129, 192; quote at 1: 131–2.
53. Heiner Fangerau, *Etablierung eines rassenhygienischen Standardwerkes 1921–1941* (Frankfurt, 2001).
54. Hans F. K. Günther, *Rassenkunde des deutschen Volkes*, 3rd ed. (Munich, 1923), 17, 23, 242–3, ch. 22; quotes at 15, 430.
55. Michael Burleigh and Wolfgang Wippermann, *The Racial State: Germany, 1933–1945* (Cambridge, 1991).
56. Hans F. K. Günther, *Rassenkunde*, 432, 487–91, 502.
57. Quoted in Dieter Löwenberg, "Willibald Hentschel (1858–1947): Seine Pläne zur Menschenzüchtung, sein Biologismus und Antisemitismus" (dissertation, University of Mainz, 1978), 45.
58. Deutsche Erneuerungsgemeinde, ZSg. 1—263/6, Bundesarchiv Koblenz.
59. Karl Haushofer, "Einleitung," in Friedrich Ratzel, *Erdenmacht und Völkerschicksal* (Stuttgart, 1940), xxvi.

60. Klee, *Deutsche Medizin im Dritten Reich*; Paul Weindling, *Health, Race and German Politics* (Cambridge, 1989); Robert Lifton, *The Nazi Doctors: Medical Killing and the Psychology of Genocide* (New York, 1986); Benno Müller-Hill, *Murderous Science* (Oxford, 1988); Robert Proctor, *Racial Hygiene* (Cambridge, MA, 1988).
61. Quoted in Klee, *Deutsche Medizin im Dritten Reich*, 331.
62. Michael Burleigh, *Death and Deliverance* (Cambridge, 1994), ch. 6; quote at 189.
63. Henry Friedlander, "Physicians as Killer in Nazi Germany: Hadamar, Treblinka, and Auschwitz," and Michael H. Kater, "Criminal Physicians in the Third Reich: Toward a Group Portrait," in *Medicine and Medical Ethics in Nazi Germany: Origins, Practices, Legacies*, ed. Nicosia, Francis R. and Jonathan Huener (New York, 2002), 59–92; Henry Friedlander, *The Origins of Nazi Genocide: From Euthanasia to the Final Solution* (Chapel Hill, NC, 1995).

CONCLUSION

1. Klaus P. Fischer, *The History of an Obsession* (New York, 1998), 118.
2. Frank Dikötter, *Imperfect Conceptions* (New York, 1998), 1–3, 174–5.
3. Wesley J. Smith, *Culture of Death: The Assault on Medical Ethics in America* (San Francisco, Encounter, 2000).
4. Richard J. Evans, "In Search of German Social Darwinism: The History and Historiography of a Concept," in *Medicine and Modernity* (Washington, 1997), 79. See also Michael Kater, "Criminal Physicians in the Third Reich: Toward a Group Portrait," in *Medicine and Medical Ethics in Nazi Germany*, ed. Francis Nicosia and Jonathan Huener (New York, 2002), 82, 87–8; Jeremy Noakes, "Nazism and Eugenics: The Background to the Nazi Sterilization Law of 14 July 1933," in *Ideas into Politics*, ed. Roger J. Bullen et al. (London, Croom Helm, 1984).

Bibliography

ARCHIVAL SOURCES

Akademie der Künste Archiv, Berlin
Carl Hauptmann papers
Alfred Ploetz papers
privately held by Wilfried Ploetz, Herrsching am Ammersee
Bayerische Staatsbibliothek Archiv
Friedrich Ratzel papers
Richard Semon papers
Berlin-Brandenburg Akademie der Wissenschaften Archiv
Albert-Samson-Stiftung papers
Wilhelm Ostwald papers
Bonn Universitätsbibliothek, Handschriftensammlung
Hermann Schaafhausen papers
Bundesarchiv Koblenz
Adele Schreiber papers (N 1173)
Reichshammerbund documents (ZSg. 1–263/3, 263/6)
Ernst-Haeckel-Haus, Jena
Ernst Haeckel papers
Forschungsstelle und Dokumentationszentrum für österreichische Philosophie, Graz
Christian von Ehrenfels papers
Freiburg Universitätsbibliothek, Handschriftensammlung
Ludwig Schemann papers
August Weismann papers
Hoover Institution Archives, Stanford University
Hitler Collection
Deutsche Kongress Zentrale
David Starr Jordan papers
Humboldt Universitätsbibliothek Archiv, Berlin
Alfred Grotjahn papers
Institut für Länderkunde Archiv, University of Leipzig
Friedrich Ratzel papers

International Institute for Social History, Amsterdam
  J. Ph. Becker papers
League of Nations Archives, United Nations Library, Geneva
  Suttner-Fried Collection
Staatsbibliothek Preussischer Kulturbesitz, Handschriftensammlung, Berlin
  Gerhart Hauptmann papers
  Max Hirsch papers
  Sammlung Darmstaedter
  Felix von Luschan papers
Stanford University Archives
  David Starr Jordan papers
University of Geneva Library Archive
  Karl Vogt papers
University of Wroclaw Library Archive
  Wilhelm Bölsche papers
Wiener (Vienna) Stadt- u. Landesbibliothek, Archiv
  Bartholomäus von Carneri papers (in Wilhelm Börner papers)
  Friedrich Jodl papers (in Wilhelm Börner papers)
  Friedrich Jodl letters
  Friedrich Hellwald letters
  Adolf Josef Lanz (von Liebenfels) letters
Zürich Zentralbibliothek
  Arnold Dodel papers

PERIODICALS (BEFORE 1945)

Annalen der Naturphilosohie
Archiv für Kriminal-Anthropologie und Kriminalistik
Archiv für Rassen- und Gesellschaftsbiologie
Archiv für soziale Gesetzgebung und Statistik
Archiv für soziale Hygiene
Das Ausland
Blätter des Deutschen Monistenbundes
Deutsche Medizinische Wochenschrift
Deutsche Rundschau
Deutschlands Erneuerung
Dokumente des Fortschritts
Freie Bühne
Das freie Wort
Die Friedenswarte
International Journal of Ethics
Kosmos
Menschheitsziele

Monismus
Das monistische Jahrhundert
Mutterschutz
Nachrichten des Internationalen Ordens für Ethik und Kultur
Die neue Generation
Die neue Zeit
Nord und Süd
Ostara
Politisch-anthropologische Revue
Sexual-Probleme
Sexualreform
Die Stimme der Vernunft
Die Umschau
Vierteljahrsschrift für wissenschaftliche Philosophie
Die Waffen Nieder
Die Zukunft

PUBLISHED PRIMARY SOURCES

Ammon, Otto. *Die Natürliche Auslese beim Menschen.* Jena: Gustav Fischer, 1893.
———. *Die Gesellschaftsordnung und ihre natürlichen Grundlagen.* 3rd ed. Jena: Gustav Fischer, 1900.
Apel, Max, ed. *Darwin. Seine Bedeutung im Ringen um Weltanschauung und Lebenswert.* Berlin: Buchverlag der "Hilfe," 1909.
Bahr, Hermann. *Der Antisemitismus. Ein internationales Interview.* Berlin: S. Fischer, 1894.
Bartholomäus von Carneri's Briefwechsel mit Ernst Haeckel und Friedrich Jodl. Leipzig: K. F. Koehler, 1922.
Baur, Erwin. *Einführung in die experimentelle Vererbungslehre.* 2nd ed. Berlin: Gebrüder Borntraeger, 1914.
Baur, Erwin, Eugen Fischer, and Fritz Lenz, *Grundriss der menschlichen Erblichkeitslehre und Rassenhygiene.* 2 vols. 2nd ed. Munich: J. F. Lehmanns Verlag, 1923.
Baynes, Norman H., ed. *The Speeches of Adolf Hitler, April 1922–August 1939.* 2 vols. Oxford: Oxford University Press, 1942.
Bernhardi, Friedrich V. *Unsere Zukunft: Ein Mahnwort an das deutsche Volk.* Stuttgart: J. G. Cotta, 1912.
———. *Deutschland und der nächste Krieg.* 6th ed. Stuttgart: J. G. Cotta, 1913.
Binding, Karl and Alfred Hoche. *Die Freigabe der Vernichtung lebensunwerten Lebens. Ihr Mass und Ihre Form.* Leipzig: Felix Meiner, 1920.
Blaschko, A. *Geburtenrückgang und Geschlechtskrankheiten.* Leipzig: Johann Ambrosius Barth, 1914.
Bleuler, Eugen. *Der geborene Verbrecher: Eine kritische Studie.* Munich: J. F. Lehmann, 1896.

Bleuler, Eugen. *Lehrbuch der Psychiatrie*. Berlin: Julius Springer, 1916.

Blossfeldt, Wilhelm, ed. *Der erste internationale Monisten-Kongress in Hamburg vom 8.–11. Sept., 1911*. Leipzig: Alfred Kröner, 1912.

———. ed. *Der Magdeburger Monistentag* [1912]. Munich: Ernst Reinhardt, 1913.

———. ed. *Der Düsseldorfer Monistentag [1913]*. Leipzig: Unesma, 1914.

Bölsche, Wilhelm. *Die Eroberung des Menschen*. 3rd. ed. Berlin: Franz Wunder, 1903.

———. *Das Liebesleben in der Natur: Eine Entwicklungsgeschichte der Liebe*. 2 vols. in 3. Revised ed. Jena: Eugen Diederichs, 1909.

———. *Der Mensch der Zukunft*. Stuttgart: Kosmos, 1915.

———. *Stirb und Werde!: Naturwissenschaftliche Plaudereien*. Jena: Eugen Diederichs, 1921.

Börner, Wilhelm. *Friedrich Jodl*. Stuttgart: J. G. Cotta, 1911.

Braun, Lily. *Memoiren einer Sozialistin*. 2 vols. Munich: Albert Langen, 1911.

Breitenbach, Wilhelm. *Die Gründung und erste Entwicklung des Deutschen Monistenbundes*. Brackwede: W. Breitenbach, 1913.

Briefwechsel zwischen Evnst Haeckel und Friedrich von Hellwald. Ulm: H. Kerler, 1901.

Brix, Emil, ed. *Ludwig Gumplowicz oder die Gesellschaft als Natur*. Vienna: Harmann Böhlaus Nachf., 1986.

Bryan, William Jennings. *In His Image*. Freeport, NY: Books for Libraries Press, 1922. Rprt. 1971.

Büchner, Ludwig. *Kraft und Stoff*. 3rd ed. Frankfurt: Meidinger Sohn, 1856.

———. *Der Mensch und seine Stellung in der Natur in Vergangenheit, Gegenwart, und Zukunft*. 2nd ed. Leipzig: Theodor Thomas, 1872.

———. *Die Macht der Vererbung und ihr Einfluss auf den moralischen und geistigen Fortschritt der Menschheit*. Leipzig: Ernst Günthers Verlag, 1882.

———. *Fremdes und Eignes aus dem geistigen Leben der Gegenwart*. 2nd ed. Leipzig: M. Spohr, 1890.

———. *Darwinismus und Sozialismus oder Der Kampf um das Dasein und die moderne Gesellschaft*. Leipzig: Ernst Günthers Verlag, 1894.

Bunge, Gustav. *Die zunehmende Unfähigkeit der Frauen, ihre Kinder zu stillen*. 7th ed. Munich: Ernst Reinhardt, 1914.

Cahnman, Werner J. "Scholar and Visionary: The Correspondence betweeen Herzl and Ludwig Gumplowicz," *Herzl Yearbook* 1 (1958): 165–80.

Carneri, Bartholomäus. *Sittlichkeit und Darwinismus. Drei Bücher Ethik*. Vienna: Wilhelm-Braunmüller, 1871.

———. *Grundlegung der Ethik*. Volksausgabe. Stuttgart: Alfred Kröner, 1881.

———. *Der moderne Mensch: Versuch über Lebensführung*. Volksausgabe. Stuttgart: Emil Strauss, 1901.

Chamberlain, Houston Stewart. *Die Grundlagen des neunzehnten Jahrhunderts*. 2 vols. Munich: F. Bruckmann, 1899.

———. *Lebenswege meines Denkens*. Munich: F. Bruckmann, 1919.

Darwin, Charles. *The Origin of Species*. London: Penguin, 1968.

———. *The Descent of Man, and Selection in Relation to Sex*. 2 vols in 1. Princeton: Princeton University Press, 1981.

————. *The Correspondence of Charles Darwin.* Vol. 7: *1858–1859.* Cambridge: Cambridge University Press, 1991.

Francis, Darwin, ed. *The Life and Letters of Charles Darwin.* 2 vols. New York: D. Appleton, 1919.

Dodel, Arnold. *Die Neuere Schöpfungsgeschichte nach dem gegenwärtigen Stande der Naturwissenschaften.* Leipzig: F. A. Brockhaus, 1875.

————. ed. *Konrad Deubler: Tagebücher, Biographie und Briefwechsel des oberöster-reichischen Bauernphilosoph.* 2 vols. Leipzig: B. Elischer, 1886.

————. *Moses oder Darwin? Eine Schulfrage.* Zurich: Cäsar Schmidt, 1889.

————. *Aus Leben und Wissenschaft. Gesammelte Vorträge und Aufsätze.* 2 vols. Stuttgart: J. H. W. Dietz, 1896–1905.

Driesmans, Heinrich. *Kulturgeschichte der Rasseninstinkte.* 2 vols. Leipzig: Eugen Diederichs, 1900–01.

————. *Rasse und Milieu.* Berlin: Johannes Räde, 1902.

————. *Menschenreform und Bodenreform.* Leipzig: Felix Dietrich, 1904.

————. *Dämon Auslese: Vom theoretischen zum praktischen Darwinismus.* Berlin: Vita, 1907.

————. ed. *M[oritz]. von Egidy. Sein Leben und Wirken.* 2 vols. Dresden: E. Pierson's, 1900.

Ehrenfels, Christian v. *Grundbegriffe der Ethik.* Wiesbaden: J. F. Bergmann, 1907.

————. *Sexualethik.* Wiesbaden: J. F. Bergmann, 1907.

————. "Presentation on *Die sexuelle Not* by Fritz Wittels," in *Minutes of the Vienna Psychoanalytic Society,* ed. Herman Nunberg and Ernst Federn. New York: International Universities Press, 1967. Vol. 2, pp. 82–92.

————. "A Program for Breeding Reform" (23.12.1908) in *Minutes of the Vienna Psychoanalytic Society,* ed. Herman Nunberg and Ernst Federn. New York: International Universities Press, 1967. Vol. 2, pp. 93–100.

Fick, Helene. *Heinrich Fick. Ein Lebensbild.* 2 vols. Zurich: J. Schaebelitz (Vol. 2 by J. Heemann), 1897–1908.

Fischer, Eugen. *Sozialanthropologie und ihre Bedeutung für den Staat.* Freiburg: Speyer and Kaerner, 1910.

————. *Die Rehobother Bastards und das Bastardierungsproblem beim Menschen.* Jena: Gustav Fischer, 1913.

Foerster, Friedrich Wilhelm. *Erlebte Weltgeschichte, 1869–1953. Memoiren.* Nuremberg: Glock und Lutz, 1953.

Foerster, Wilhelm. *Lebenserinnerungen und Lebenshoffnungen.* Berlin: Georg Reimer, 1911.

Forel, August. *Die sexuelle Frage. Eine naturwissenschaftliche, psychologische, hygienis-che und soziologische Studie für Gebildete.* Munich: Ernst Reinhardt, 1905.

————. *Hygiene of Nerves and Mind.* Trans. Herbert Aikins. New York: G. P. Putnam's Sons, 1907.

————. *Leben und Tod. Ein Vortrag.* Munich: Ernst Reinhardt, 1908.

————. *Kulturbestrebungen der Gegenwart.* Munich: Ernst Reinhardt, 1910.

————. *Out of My Life and Work.* Trans. Bernard Miall. New York: Norton, 1937.

Franz, Victor, ed. *Ernst Haeckel. Sein Leben, Denken, und Wirken.* Jena: Wilhelm Gronau, 1943–4.

Fried, Alfred (written anonymously). *Experimental-Ehen: Ein "Document humain" als Beitrag zur Eherechtsreform. Von einem Versuchsobjekt.* Munich: Ernst Reinhardt, 1906.

————. *Handbuch der Friedensbewegung.* Vienna: Oesterrichischen Friedensgesellschaft, 1905.

————. *Die moderne Friedensbewegung.* Leipzig: B. G. Teubner, 1907.

————. *Die Grundlagen des revolutionaren Pacifismus.* Tübingen: Mohr, 1908.

————. *Der kranke Krieg.* Leipzig: Alfred Kröner, 1909.

Fritsch, Theodor. *Handbuch der Judenfrage.* 27th ed. Hamburg: Hanseatische Druck- und Verlags–Anstalt, 1910.

Genoud, Francois, ed. *The Testament of Adolf Hitler: The Hitler–Bormann Documents, February–April 1945.* London: Cassell, 1961.

Gimpl, Georg, ed. *Unter uns gesagt: Friedrich Jodls Briefe an Wilhelm Bolin.* Vienna: Löcker Verlag, 1990.

Gizycki, Georg V. *Philosophische Consequenzen der Lamarck–Darwin'schen Entwicklungstheorie.* Leipzig: C. F. Winter'sche Verlagshandlung, 1876.

————. *Grundzüge der Moral.* Leipzig: Wilhelm Friedrich, 1883.

————. *Moralphilosophie gemeinverständlich dargestellt.* 2nd ed. Leipzig: Hermann Haacke, 1895.

Goldscheid, Rudolf. *Entwicklungstheorie, Entwicklungsökonomie, Menschenökonomie. Eine Programmschrift.* Leipzig: Werner Klinkhardt, 1908.

————. *Darwin als Lebenselement unserer modernen Kultur.* Vienna: Hugo Heller, 1909.

————. *Höherentwicklung und Menschenökonomie. Grundlegung der Sozialbiologie.* Leipzig: Klinkhardt, 1911.

————. *Friedensbewegung und Menschenökonomie.* Berlin: Friedens–Warte, 1912.

Goldschmidt, Richard. *Portraits from Memory: Recollections of a Zoologist.* Seattle: University of Washington Press, 1956.

Goldstein, Kurt. *Über Rassenhygiene.* Berlin: Julius Springer, 1913.

Gross, Hans. *Kriminal-Psychologie.* 2nd ed. Leipzig: F. C. W. Vogel, 1905.

Grotjahn, Alfred. *Geburtenrückgang und Geburtenregelung.* 2nd ed. Berlin: Oscar Coblentz, 1921.

————. *Erlebtes und Erstrebtes:* Erinnerungen eines Sozialistischen Arztes. Berlin: F.A. Herbig, 1932.

Gruber, Max v. *Ursachen und Bekämpfung des Geburtenrückgangs im Deutschen Reich.* 3rd ed. Munich: Lehmann, 1914.

————. *Kreig, Frieden und Biologie.* Berlin: Carl Heymanns, 1915.

Gruber, Max von and Ernst Rüdin. *Fortpflanzung. Vererbung. Rassenhygiene. Illustrierter Führer durch die Gruppe Rassenhygiene der Internationalen Hygieneausstellung 1911 in Dresden.* 2nd ed. Munich: J. F. Lehmanns Verlag, 1911.

Gumplowicz, Ludwig. *Der Rassenkampf. Sociologische Untersuchungen.* Innsbruck: Wanger'schen Univ.-Buchhandlung, 1883.

————. *Letters of Ludwig Gumplowicz to Lester F. Ward,* ed. Bernhard J. Stern. Leipzig: C. L. Hirschfeld, 1933.

Günther, Hans F. K. *Rassenkunde des deutschen Volkes.* 3rd ed. Munich: J. F. Lehmanns Verlag, 1923.

―――. *Mein Eindruck von Adolf Hitler.* Pähl: F. von Bebenburg, 1969.

Heeckel, Ernst. "Ueber die Entwickelungstheorie Darwins." In *Amtlicher Bericht über die 38. Versammlung Deutscher Naturforscher und Ärzte in Stettin im September 1863.* Stettin: F. Hessenalnd, 1864. Pp. 17–30.

―――. *Generelle Morphologie.* 2 vols. Berlin: Georg Reimer, 1866.

―――. *Natürliche Schöpfungsgeschichte.* Berlin: Georg Reimer, 1868.

―――. *Natürliche Schöpfungsgeschichte.* 2nd ed. Berlin: Georg Reimen, 1870.

―――. *Natürliche Schöpfungsgeschichte.* 4th ed. Berlin: Georg Reimer, 1873.

―――. *Freie Wissenschaft und freie Lehre. Eine Entgegnung auf Rudolf Virchow's Münchener Rede über "Die Freiheit der Wissenschaft im modernen Staat."* Stuttgart: E. Schweizerbart'sche Verlagshandlung, 1878.

―――. *Indische Reisebriefe.* Berlin: Gebrüder Paetel, 1883.

―――. *Der Monismus als Band zwischen Religion und Wissenschaft: Glaubensbekenntniss eines Naturforschers.* Bonn: Emil Strauss, 1892.

―――. *Die Welträthsel: Gemeinverständliche Studien über Monistische Philosophie.* Volksausgabe. Bonn: Emil Strauss, 1903.

―――. *Die Lebenswunder: Gemeirverständliche Studien über Biologische Philosophie.* Stuttgart: Alfred Kröner, 1904.

―――. *Ewigkeit. Weltkriegsgedanken über Leben und Tod, Religion und Entwicklungslehre.* Berlin: Georg Reimer, 1917.

Harpf, Adolf. *Zur Lösung der brennendsten Rassenfrage der heutigen europäischen Menschheit.* Vienna: Breitenstein, 1898.

―――. *Darwin in der Ethik: Festschrift zum 80. Geburtstage Carneri's.* n.p.: Verlag der Neuen Leobener Buchdruckerei J. Hans Prosl, 1901.

Haushofer, Karl. "Einleitung." In Friedrich Ratzel, *Erdenmacht und Völkerschicksal: Eine Auswahl aus seinen Werken.* Stuttgart: Alfred Kröner, 1940. Pp. ix–xxvii.

Hegar, Alfred. *Der Geschlechtstrieb: Eine Social-Medicinische Studie.* Stuttgart: Ferdinand Enke, 1894.

Hellwald, Friedrich. *Culturgeschichte in ihrer natürlichen Entwicklung bis zur Gegenwart.* Augsburg: Lampart, 1875.

―――. *Die Erde und Ihre Völker.* 2 vols. 2nd ed. Stuttgart: W. Spemann, 1877–78.

―――. *Naturgeschichte des Menschen.* 2 vols. Stuttgart: W. Spemann, 1880.

―――. *Kulturgeschichte in ihrer natürlichen Entwickelung bis zur Gegenwart.* 4th ed. 4 vols. Leipzig: P. Friesenhahn, 1896.

Hentig, Hans v. *Strafrecht und Auslese: Eine Anwendung des Kausalgesetzes auf den rechtbrechenden Menschen.* Berlin: Julius Springer, 1914.

Hentschel, Willibald. *Mittgart: Ein Weg zur Erneuerung der germanischen Rasse.* 2nd ed. Leipzig: Hammer–Verlag, 1906.

―――. *Varuna: Das Gesetz des aufsteigenden und sinkenden Lebens in der Geschichte.* Leipzig: Theodor Fritsch, 1907.

―――. *Vom aufsteigenden Leben—Ziele der Rassenhygiene.* 3rd ed. Leipzig: Erich Matthes, 1922.

Hesse, Albert. *Natur und Gessellschaft: Eine kritische Untersuchung der Bedeutung der Deszendenztheorie für das soziale Leben*. Jena: Gustav Fischer, 1904.

Hirsch, Max. *Fruchtabtreibung und Präventivverkehr im Zusammenhang mit dem Geburtenrückgang*. Würzburg: Curt Kabitzsch, 1914.

Hirschfeld, Magnus. *Naturgesetze der Liebe: Eine gemeinverständliche Untersuchung über den Liebes-Eindruck, Liebes-Drang und Liebes-Ausdruck*. Berlin: Alfred Pulvermacher, 1912.

Hitler, Adolf. *Mein Kampf.* 2 vols. in 1. Munich: NSDAP, 1943.

Hitlers Zweites Buch: Ein Dokument aus dem Jahr 1928, ed. Gerhard L. Weinberg. Stuttgart: Deutsche Verlags–Anstalt, 1961.

Hoche, Alfred. *Jahresringe: Innenansicht eines Menschenlebens*. Munich: Lehmann, 1935.

Hoffmann, Geza von. *Die Rassenhygiene in den Vereinigten Staaten von Nordamerika*. Munich: J. F. Lehmann, 1913.

———. *Krieg und Rassenhygiene: Die bevölkerungspolitischen Aufgaben nach dem Kriege*. Munich: J. F. Lehmann, 1916.

Hötzendorf, Conrad von. *Aus meiner Dienstzeit*. 5 vols. Vienna: Rikola Verlag, 1921–25.

———. *Private Aufzeichnungen*, ed. Kurt Peball. Vienna: Amalthea, 1977.

Jäckel, Eberhard, ed. *Hitler. Sämtliche Aufzeichnungen, 1905–1924*. Stuttgart: Deutsche Verlags–Anstalt, 1980.

Jaeger, Gustav. *Die Darwinische Theorie und ihre Stellung zu Moral und Religion*. Stuttgart: Julius Hoffmann, 1869.

Jodl, Friedrich. *Die Culturgeschichtsschreibung, ihre Entwickelung und ihr Problem*. Halle: C. E. M. Pfeffer, 1878.

———. *Die Geschichte der Ethik in der neueren Philosophie*. 2 vols. Stuttgart: J. G. Cotta, 1882.

———. *Wesen und Ziele der ethischen Bewegung in Deutschland*. 4th ed. Frankfurt: Neuer Frankfurter Verlag, 1908.

———. *Vom Lebenswege: Gesammelte Vorträge und Aufsätze*, ed. Wilhelm Börner. 2 vols. Stuttgart: J. G. Cotta, 1916–17.

———. *Allgemeine Ethik*, ed. Wilhelm Börner. Stuttgart: J. G. Cotta, 1918.

Jodl, Margarete. *Friedrich Jodl, sein Leben und Wirken nach Tagebüchern und Briefen*. Stuttgart: J. G. Cotta, 1920.

Jost, Adolf. *Das Recht auf den Tod. Sociale Studie*. Göttingen: Dieterich'sche Verlagsbuchhandlung, 1895.

Kautsky, Karl. *Vermehrung und Entwicklung in Natur und Gesellschaft*. Stuttgart: Dietz, 1910.

Kellogg, Vernon. *Headquarter Nights: A Record of conversations and Experiences at the Headquarters of the German Army in France and Belgium*. Boston: Atlantic Monthly Press, 1917.

Kirchhoff, Alfred. *Darwinismus angewandt auf Völker und Staaten*. Frankfurt: Heinrich Keller, 1910.

Klaatsch, Hermann. *Grundzüge der Lehre Darwins*. 4th ed. Mannheim: Bensheimer, 1919.

Kossmann, Robby. *Züchtungspolitik*. Schmargendorf: Verlag Renaissance, 1905.

Kraepelin, Emil. *Psychiatrie: Ein kurzes Lehrbuch*. 4th ed. Leipzig: Ambr. Abel (Arthur Meiner), 1893.

———. *Die psychiatrische Aufgaben des Staates*. Jena: Gustav Fischer, 1900.

———. *Lebenserinnerungen*. Berlin: Springer, 1983.

Krause, Ernst (written under pseudonym Carus Sterne). *Die Krone der Schöpfung. Vierzehn Essays über die Stellung des Menschen in der Natur*. Vienna: Karl Prochaska, 1884.

———. *Werden und Vergehen. Eine Entwicklungsgeschichte des Naturganzen*. 2 vols. 6th ed. Ed. by Wilhelm Bölsche. Berlin: Gebrüder Borntraeger, 1905.

Kubizek, August. *The Young Hitler I Knew*. Trans. E. V. Anderson. Boston: Houghton Mifflin, 1955.

Kurella, Hans. *Cesare Lombroso: A Modern Man of Science*. Trans. M. Eden Paul. New York: Rebman, n.d.

———. *Naturgeschichte des Verbrechers: Grundzüge der criminellen Anthropologie und Criminalpsychologie*. Stuttgart: Ferdinand Enke, 1893.

———. *Die Grenzen der Zurechnungsfähigkeit und die Kriminal-Anthropologie*. Halle: Gebauer–Schwetschke, 1903.

Lange, Friedrich Albert. *Die Arbeiterfrage in ihrer Bedeutung für Gegenwart und Zukunft*. Duisburg: W. Falk und Volmer, 1865. Rprt. Duisburg, 1975.

Lehmann, Melanie. *Verleger J. F. Lehmann. Ein Leben im Kampf um Deutschland. Lebenslauf und Briefe*. Munich: J. F. Lehmann, 1935.

Lenz, Fritz. *Über die krankhaften Erbanlagen des Mannes und die Bestimmung des Geschlechtes beim Menschen*. Jena: Gustav Fischer, 1912.

———. *Die Rasse als Wertprinzip, Zur Erneuerung der Ethik*. Munich: J. F. Lehmann, 1933.

Luschan, Felix von. "Die gegenwärtigen Aufgaben der Anthropologie," *Verhandlungen der Gesellschaft deutscher Naturforscher und Ärzte. 81. Versammlung zu Salzburg 19.–25. September 1909*. Leipzig: F. C. W. Vogel, 1910. 2 parts. 2: 201–8.

———. "Anthropological View of Race," in *Papers on Inter-Racial Problems, Communicated to the First Universal Races Congress*, ed. G. Spiller. London: P. S. King and Son, 1911. Pp. 13–24.

———. *Rassen und Völker*. Berlin: Carl Heymanns, 1915.

Marr, Wilhelm. *Der Sieg des Judentums über das Germanenthum*. 12th ed. Bern: Rudolph Costenoble, 1879.

Matzat, Heinrich. *Philosophie der Anpassung mit besondern Berücksichtigung des Rechtes und des Staates*. Jena: Gustav Fischer, 1903.

Michaelis, Curt. *Prinzipien der natürlichen und sozialen Entwicklungsgeschichte des Menschen*. Jena: Gustav Fischer, 1904.

Mitchell, P. Chalmers. *Evolution and the War*. New York: E. P. Dutton, 1915.

Nicolai, G. F. *Biologie des Krieges*. Zurich: Art. Institut Orell Füssli, 1917.

Nietzsche, Friedrich. *Werke in Drei Bänden*. 3 vols. Ed. Karl Schlechta. Munich: Carl Hanser, 1966.

Nordau, Max. *Der Sinn der Geschichte*. Berlin: Carl Duncker, 1909.

Nordau, Max. *Die konventionellen Lügen der Kulturmenschheit.* Leipzig: B. Elischer Nachfolger, 1909.

———. *Morals and the Evolution of Man* (trans. of *Biologie und Ethik*). London: Cassell, 1922.

———. *Degeneration.* New York: Howard Fertig, 1968.

Ostwald, Wilhelm. *Die Philosophie der Werte.* Leipzig: Alfred Kröner, 1913.

Peschel, Oscar. *Völkerkunde.* 2nd ed. Leipzig: Duncker und Humblot, 1875.

Peters, Karl. *Deutsch-national: Kolonialpolitische Aufsätze.* 2nd ed. Berlin: Walther und Apolant, 1887.

Ploetz, Alfred. *Die Tüchtigkeit unsrer Rasse und der Schutz der Schwachen.* Berlin: S. Fischer, 1895.

Preyer, Wilhelm. *Der Kampf um das Dasein.* Bonn: Weber, 1869.

———. *Die Concurrenz in der Natur.* Breslau: S. Schottlaender, 1882.

Problems in Eugenics: Report of Proceedings of the First International Eugenics Congress. Vol. 2. London: Eugenics Education Society, 1913.

Ratzel, Friedrich. *Sein und Werden der organischen Welt.* Leipzig: Fuess Verlag, 1869.

———. *Anthropogeographie,* 2 vols. Stuttgart: J. Engelhorn, 1882–91.

———. *The History of Mankind.* 3 vols. Trans. A. J. Butler. London: Macmillan, 1896.

———. *Der Lebensraum: Eine biogeographische Studie.* Tübingen: H. Laupp, 1901.

———. *Politische Geographie oder die Geographie der Staaten, des Verkehres und des Krieges.* 2nd ed. Munich: R. Oldenbourg, 1903.

———. *Kleine Schriften.* Ed. Hans Helmolt. 2 vols. Munich: R. Oldenbourg, 1906.

Ratzenhofer, Gustav. *Wesen und Zweck der Politik.* 2 vols. Leipzig: Brockhaus, 1893.

———. *Der positive Monismus und das einheitliche Princip aller Erscheinungen.* Leipzig: Brockhaus, 1899.

———. *Positive Ethik: Die Verwirklichung des Sittlich–Seinsollenden.* Leipzig: F. A. Brockhaus, 1901.

———. *Soziologie: Positive Lehre von den menschlichen Wechselbeziehungen.* Leipzig: Brockhaus, 1907.

Rée, Paul. *Der Ursprung der moralischen Empfindungen.* Chemnitz: Ernst Schmeitzner, 1877.

Reimer, Josef Ludwig. *Ein Pangermanisches Deutschland.* Berlin: Friedrich Luckhardt, 1905.

Ribbert, Hugo. *Rassenhygiene: Eine gemeinverständliche Darstellung.* Bonn: Friedrich Cohen, 1910.

———. *Heredity, Disease and Human Evolution.* Trans. Eden and Cedar Paul. New York: Critic and Guide, 1918.

Rolle, Friedrich. *Der Mensch, seine Abstammung, und Gesittung im Lichte der Darwin'schen Lehre.* Frankfurt: Joh. Christ. Hermann, 1866.

Ruppin, Arthur. *Darwinismus und Sozialwissenschaft.* Jena: Gustav Fischer, 1903.

Schäffle, Albert E. F. *Bau und Leben des Socialen Korpers.* 4 vols. Tübigen: H. Laupp'schen Buchhandlung, 1881.

Schallmayer, Wilhelm. *Die drohende physische Entartung der Culturvölker.* 2nd ed. [of *Ueber die drohende körperliche Entartung der Culturmenschheit,* 1891] Berlin: Heuser's Verlag, 1891.

———. *Vererbung und Auslese im Lebenslauf der Völker. Eine Staatswissenschaftliche Studie auf Grund der neueren Biologie.* Jena: Gustav Fischer, 1903.

———. *Beiträge zu einer Nationalbiologie.* Jena: Hermann Costenoble, 1905.

Schemann, Ludwig. *Lebensfahrten eines Deutschen.* Leipzig: Erich Matthes, 1925.

———. *Die Rasse in den Geisteswissenschaften: Studien zur Geschichte des Rassengedankens.* 3 vols. 2nd ed. Munich: J. F. Lehmann, 1928–43.

Schmidt, Heinrich. *Ernst Haeckel. Leben und Werke.* Berlin: Deutsche Buch–Gemeinschaft, 1926.

———. ed. *Was wir Ernst Haeckel verdanken.* 2 vols. Leipzig: Unesma, 1914.

Schmid, Rudolf. *Die Darwinischen Theorien und ihre Stellung zur Philosophie, Religion und Moral.* Stuttgart: Paul Moser, 1876.

Schmidt, Oscar. "Darwinismus und Socialdemokratie." *Deutsche Rundschau* 17 (1878): 278–92.

Schönerer, Georg. *Zehn Reden Georg Schönerers aus den Jahren 1882 bis 1888.* Vienna: Alldeutsche Verein für die Ostmark, 1914.

Schreiber, Adele, ed. *Mutterschaft: Ein Sammelwerrk für die Probleme des Weibes als Mutter.* Munich: Albert Langen, 1912.

Seeck, Otto. *Geschichte des Untergangs der antiken Welt.* Vol. 1. 4th ed. Stuttgart: J. B. Metzlersche, 1921. Rprt. 1966.

Sommer, Robert. *Kriminalpsychologie und strafrechtliche Psychopathologie auf naturwissenschaftliche Grundlage.* Leipzig: Johann Ambrosius Barth, 1904.

Steinmetz, Sebald Rudolf. *Die Philosophie des Krieges.* Leipzig: Johann Ambrosius Barth, 1907.

Stengel, Karl. *Weltstaat und Friedensproblem.* Berlin: Reichl, 1909.

Stöcker, Helene. *Die Liebe und die Frauen.* 2nd ed. Minden: J. C. C. Bruns, 1906.

Strauss, David Friedrich. *Der alte und der neue Glaube: Ein Bekenntnis.* Leipzig: S. Hirzel, 1872.

———. *Ausgewählte Briefe von David Friedrich Strauss.* Ed. Eduard Zeller. Bonn: Emil Strauss, 1895.

Suttner, Bertha. *Inventarium einer Seele.* Leipzig: Wilhelm Friedrich, 1883.

———. *Die Waffen nieder! Eine Lebensgeschichte.* Volksausgabe. Dresden: E. Pierson, 1900.

———. *Daniela Dormes.* In *Gesammelte Schriften,* Vol. 7. Dresden: E. Pierson, 1906.

———. *Lebenserinnerungen.* Ed. Fritz Böttger. 4th ed. Berlin: Verlag der Nation, 1972.

———. *Maschinenzeitalter.* 3rd ed. 1898. Rprt. Düsseldorf: Zwiebelzwerg, 1983.

Tille, Alexander. *Volksdienst.* Berlin: Wiener'sche Verlagsbuchhandlung, 1893.

———. *Von Darwin bis Nietzsche. Ein Buch Entwicklungsethik.* Leipzig: C. G. Naumann, 1895.

Unold, Johannes. *Aufgaben und Ziele des Menschenlebens.* Leipzig: B. G. Teubner, 1899.

Unold, Johannes. *Organische und soziale Lebensgesetze.* Leipzig: Theod. Thomas, 1906.

———. *Der Monismus und seine Ideale.* Leipzig: Theod. Thomas, 1908.

———. *Politik im Lichte der Entwicklungslehre.* Munich: Ernst Reinhardt, 1912.

Uschmann, Georg, ed. *Ernst Haeckel: Biographie in Briefen.* Gütersloh: Prisma, 1984.

Virchow, Rudolf. *Die Freiheit der Wissenschaft im modernen Staat.* Berlin: Verlag von Wiegandt, Hempel and Parey, 1877.

Vogt, Carl. *Vorlesungen über den Menschen, seine Stellung in der Schöpfung und in der Geschichte der Erde.* 2 vols. Giessen: J. Ricker'sche Buchhandlung, 1863.

Wagner, Klaus. *Krieg: Eine politisch-entwicklungsgeschichtliche Untersuchung.* Jena: Hermann Costenoble, 1906.

Wagner, Moritz. *Die Entstehung der Arten durch räumliche Sonderung, gesammelte Aufsätze.* Basel: B. Schwabe, 1889.

Waldeyer, W. *Die im Weltkriege stehenden Völker in anthropologischer Betrachtung.* Berlin: Carl Heymanns, 1915.

Waldeyer-Hartz, Wilhelm von. *Lebenserinnerungen.* 2nd ed. Bonn: Friedrich Cohen, 1921.

Walser, Hans H., ed. *August Forel: Briefe, Correspondance, 1864–1927.* Bern: Hans Huber, 1968.

Weismann, August. *Aufsätze über Vererbung und Verwandte Biologische Fragen.* Jena: Gustav Fischer, 1892.

Werner, Johannes, ed. *Franziska von Altenhausen. Ein Roman aus dem Leben eines berühmten Mannes in Briefen aus den Jahren 1898/1903* [the famous man is none other than Ernst Haeckel]. Biberach: Koehler und Voigtländer Verlag, 1927.

Wilser, Ludwig. *Die Ueberlegenheit der germanischen Rasse.* Stuttgart: Strecker und Schröder, 1915.

Woltmann, Ludwig. *Kritische und genetische Begründung der Ethik.* Freiburg: Hch. Epstein, 1896.

———. *System der moralischen Bewusstseins.* Düsseldorf: Hermann Michels, 1898.

———. *Die Darwinische Theorie und der Sozialismus.* Dusseldorf: Hermann Michels Verlag, 1899.

———. *Der historische Materialismus: Darstellung und Kritik.* Düsseldorf: Hermann Michels, 1900.

———. *Politische Anthropologie: Eine Untersuchung über den Einfluss der Deszendenztheorie auf die Lehre von der politischen Entwicklung der Völker.* Jena: Eugen Diederichs, 1903.

———. *Die Germanen und die Renaissance in Italien.* Leipzig: Thüringische Verlagsanstalt, 1905.

———. *Die Germanen in Frankreich: Eine Untersuchung über den Einfluss der germanischen Rasse auf die Geschichte und Kultur Frankreichs.* Jena: E. Diederichs, 1907.

Ziegler, Heinrich Ernst. *Die Naturwissenschaft und die Socialdemokratische Theorie, ihr Verhältnis dargelegt auf Grund der Werke von Darwin und Bebel.* Stuttgart: Ferdinand Enke, 1893.

———. "Einleitung zu dem Sammelwerke Natur und Staat," Jena: Gustav Fischer, 1903. In *Natur und Staat*, Vol. 1 (bound with Heinrich Matzat, *Philosophie der Anpassung*).

SECONDARY SOURCES

Allen, Ann, Taylor. "German Radical Feminism and Eugenics, 1900–1918," *German Studies Review* 11 (1988): 31–56.

———. "Feminism and Eugenics in Germany and Britain, 1900–48: A Comparative Perspective." *German Studies Review* 23 (2000): 477–505.

Altner, Günter. *Weltanschauliche Hintergründe der Rassenlehre des Dritten Reiches.* Zürich: E V Z Verlag, 1968.

Arendt, Hannah. *The Origins of Totalitarianism.* New York: Harcourt Brace Jovanovich, 1973.

Ash, Mitchell G. *Gestalt Psychology in German Culture, 1880–1967: Holism and the Quest for Objectivity.* Cambridge: Cambridge University Press, 1995.

Aschheim, Steven E. *The Nietzsche Legacy in Germany, 1890–1990.* Berkeley: University of California Press, 1992.

———. *In Times of Crisis: Essays on European Culture, Germans, and Jews* Madison: University of Wisconsin Press, 2001.

Auerbach, Hellmuth. "Hitlers politische Lehrjahre und die Münchener Gesellschaft 1919–1923," *Vierteljahrshefte für Zeitgeschichte* 25 (1977): 1–45.

Baader, Gerhard. "Sozialdarwinismus—Vernichtungsstrategien im Vorfeld des Nationalsozialismus." In *"Bis endlich der langersehnte Umschwung kam ..." Von der Verantwortung der Medizin unter dem Nationalsozialismus.* Marburg: Schüren, 1991. Pp. 21–35.

Bannister, Robert C. *Social Darwinism: Science and Myth in Anglo-American Social Thought.* Philadelphia: Temple University Press, 1979.

Bassin, Mark "Imperialism and the Nation State in Friedrich Ratzel's Political Geography," *Progress in Human Geography* 11 (1987): 473–95.

———. "Race *contra* Space: The Conflict between German *Geopolitik* and National Socialism," *Political Geography Quarterly* 6 (1987): 115–34.

Bäumer, Änne. *NS–Biologie.* Stuttgart: S. Hirzel, 1990.

Bayertz, Kurt. "Naturwissenschaft und Sozialismus. Tendenzen der Naturwissenschaft-Rezeption in der deutschen Arbeiterbewegung des 19. Jahrhundert." *Social Studies of Science* 13 (1983): 355–94.

———. "Darwinismus als Politik: Zur Genese des Sozialdarwinismus in Deutschland 1860–1900," in *Welträtsel und Lebenswunder: Ernst Haeckel—Werk, Wirkung und Folgen.* Ed. Erna Aescht et al. Linz: Oberösterreichischen Landesmuseums, 1998.

Becker, Peter Emil. Vol 1: *Zur Geschichte der Rassenhygiene. Wege ins Dritte Reich.* Vol 2: *Sozialdarwinismus, Rassismus, Antisemitismus und volkischer Gedanke: Wege ins Dritte Reich.* Stuttgart: Georg Thieme, 1988–90.

Benton, Ted. "Social Darwinism and Socialist Darwinism in Germany: 1860 to 1900," *Rivista di filosofia* 73 (1982): 79–121.

Benzenhöfer, Udo. *Der gute Tod? Euthanasie und Sterbehilfe in Geschichte und Gegenwart.* Munich: C. H. Beck, 1999.

Beyl, Werner. *Arnold Dodel (1843–1908) und die Popularisierung des Darwinismus.* Frankfurt: Verlag Peter Lang, 1984.

Blasius, Dirk. *Umgang mit Unheilbarem: Studien zur Sozialgeschichte der Psychiatrie.* Bonn: Psychiatrie–Verlag, 1986.

Bley, Helmut. *South-West Africa under German Rule, 1894–1914.* Trans. Hugh Ridley. Evanston: Northwestern University Press, 1971.

Bock, Gisela. "Sterilization and 'Medical' Massacres in National Socialist Germany: Ethics, Politics, and the Law," in *Medicine and Modernity: Public Health and medical Care in Nineteenth- and Twentieth-Century Germany.* Ed. Manfred Berg and Geoffrey Cocks. Washington: Cambridge University Press, 1997. Pp. 149–72.

———. *Zwangssterilisation im Nationalsozialismus: Studien zur Rassenpolitik und Frauenpolitik.* Opladen: Westdeutscher Verlag, 1986.

Bowler, Peter. *The Non-Darwinian Revolution: Reinterpreting a Historical Myth.* Baltimore: Johns Hopkins University Press, 1988.

Bridgman, Jon. *The Revolt of the Hereros.* Berkeley: University of California Press, 1981.

Bridgman, Jon and Leslie J. Worley. "Genocide of the Hereros." In *Genocide in the Twentieth Century: Critical Essays and Eyewitness Accounts.* Ed. Samuel Totten, William S. Parsons, and Israel W. Charny. NY: Garland, 1995. Pp. 3–48.

Broberg, Gunnar, and Nils Roll-Hansen, eds. *Eugenics and the Welfare State: Sterilization Policy in Denmark, Sweden, Norway, and Finland.* East Lansing: Michigan State University Press, 1996.

Brobjer, Thomas. *Nietzsche's Ethics of Character: A Study of Nietzsche's Ethics and Its Place in the History of Moral Thinking.* Uppsala: Department of History of Science and Ideas of Uppsala University, 1995.

———. "Nietzsche's Reading and Private Library, 1885–89." *Journal of the History of Ideas* 58 (1997): 663–93.

Brooke, John Hedley. *Science and Religion: Some Historical Perspectives.* Cambridge: Cambridge University Press, 1991.

Burleigh, Michael. *Death and Deliverance: Euthanasia in Germany, 1900–1945.* Cambridge: Cambridge University Press, 1994.

Burleigh, Michael and Wolfgang Wippermann. *The Racial State: Germany, 1933–1945.* Cambridge: Cambridge University Press, 1991.

Butcher, Barry W. "Darwinism, Social Darwinism and the Australian Aborigines: A Reevaluation." In *Darwin's Laboratory: Evolutionary Theory and Natural History in the Pacific.* Ed. Roy MacLeod and Philip F. Rehbock. Honolulu: University of Hawaii Press, 1994. Pp. 371–94.

Buttmann, Gunther. *Friedrich Ratzel: Leben und Werk eines deutschen Geographen.* Stuttgart: Wissenschaftliche Verlagsgesellschaft, 1977.

Byer, Doris. *Rassenhygiene und Wohlfahrtspflege: Zur Entstehung eines sozialdemokratischen Machtdispositivs in Österreich bis 1934.* Frankfurt: Campus, 1988.

Call, Lewis. "Anti-Darwin, Anti-Spencer: Nietzsche's Critique of Darwin and 'Darwinism,' " *History of Science* 36 (1998): 1–22.

Chickering, Roger. *Imperial Germany and a World without War: The Peace Movement and German Society, 1892–1914.* Princeton: Princeton University Press, 1975.

Crook, Paul. *Darwinism, War and History: The Debate over the Biology of War from the 'Origin of Species' to the First World War*. Cambridge: Cambridge University Press, 1994.

Daim, Wilfried. *Der Mann, der Hitler die Ideen Gab: Jörg Lanz von Liebenfels*. 3rd ed. Wien: Ueberreuter, 1994.

Daum, Andreas W. *Wissenschaftspopularisierung im 19. Jahrhundert: Bürgerliche Kultur, naturwissenschaftliche Bildung und die deutsche Öffentlichkeit, 1848–1914*. Munich: R. Oldenbourg, 1998.

Deak, Istvan. *Beyond Nationalism: A Social and Political History of the Habsburg Officer Corps, 1848–1918*. New York: Oxford University Press, 1990.

Dedering, Tilman. "The German–Herero War of 1904: Revisionism of Genocide or Imaginary Historiography?" *Journal of Southern African Studies* 19 (1993), 80–88.

Deichmann, Ute. *Biologists under Hitler*. Cambridge: Harvard University Press, 1996.

Desmond, Adrian. *Huxley: From Devil's Disciple to Evolution's High Priest*. Reading, MA: Addison–Wesley, 1997.

Desmond, Adrian, and James Moore. *Darwin*. London: Michael Joseph, 1991.

Dickinson, Edward Ross. "Reflections on Feminism and Monism in the Kaiserreich, 1900–1913." *Central European History* 34 (2001): 191–230.

Dikötter, Frank. *Imperfect Conceptions: Medical Knowledge, Birth Defects and Eugenics in China*. New York: Columbia University Press, 1998.

Doeleke, W. *Alfred Ploetz (1860–1940): Sozialdarwinist und Gesellschaftsbiologe*. Dissertation: University of Frankfurt, 1975.

Donnellan, B. "Friedrich Nietzsche and Paul Rée: Cooperation and Conflict," *Journal of the History of Ideas* 43 (1982): 595–612.

Dowbiggin, Ian. *A Merciful End: The Euthanasia Movement in Modern America*. Oxford: Oxford University Press, 2003.

Drechsler, Horst. *Let Us Die Fighting: The Struggle of the Herero and Nama against German Imperialism (1884–1915)*. Trans. Bernd Zöllner. London: Zed Press, 1980.

Eckart, Wolfgang. *Medizin und Kolonialimperialismus: Deutschland 1884–1945*. Paderborn: Ferdinand Schöningh, 1997.

Efron, John. *Defenders of the Race: Jewish Doctors and Race Science in Fin-de-Siécle Europe*. New Haven: Yale University Press, 1994.

Eley, Geoff, ed. *Society, Culture, and the State in Germany, 1870–1930*. Ann Arbor: University of Michigan Press, 1996.

Evans, Richard J. *The Feminist Movement in Germany, 1894–1933*. London: Sage Publications, 1976.

———. "In Search of German Social Darwinism: The History and Historiography of a Concept." In *Medicine and Modernity: Public Health and Medical Care in 19th- and 20th-Century Germany*. Ed. Manfred Berg et al. Washington, DC: Cambridge University Press, 1997. Pp. 55–79.

Fabian, Reinhard, ed. *Christian von Ehrenfels: Leben und Werk*. Amsterdam: Rodopi, 1986.

Fangerau, Heiner. *Etablierung eines rassenhygienischen Standardwerkes 1921–1941: Der Baur–Fischer–Lenz im Spiegel der zeitgenössischen Rezensionsliteratur.* Frankfurt: Peter Lang, 2001.

Farber, Paul L. *The Temptations of Evolutionary Ethics.* Berkeley: University of California Press, 1994.

Fest, Joachim. *Hitler.* Trans. Richard and Clara Winston. New York: Harcourt Brace Jovanovich, 1974.

Field, Geoffrey. *Evangelist of Race: The Germanic Vision of Houston Stewart Chamberlain.* New York: Columbia University Press, 1981.

Fischer, Klaus P. *The History of an Obsession.* New York: Continuum, 1998.

Frei, Norbert. "Wie modern war der Nationalsozialismus?" *Geschichte und Gesellschaft* 19 (1993): 367–87.

Frewer, Andreas and Clemens Eickhoff, eds. *"Euthanasie" und die aktuelle Sterbehilfe-Debatte: Die historischen Hintergründe medizinischer Ethik.* Frankfurt: Campus, 2000.

Friedlander, Henry. *The Origins of Nazi Genocide: From Euthanasia to the Final Solution.* Chapel Hill: University of North Carolina Press, 1995.

Fye, Bruce. "Active Euthanasia: An Historical Survey of Its Conceptual Origins and Introduction into Medical Thought." *Bulletin of the History of Medicine* 52 (1978): 492–502.

Gadebusch Bondio, Mariacarla. *Die Rezeption der kriminalanthropologischen Theorien von Cesare Lombroso in Deutschland von 1880–1914.* Husum: Matthiesen Verlag, 1995.

Gasman, Daniel. *The Scientific Origins of National Socialism: Social Darwinism in Ernst Haeckel and the German Monist League.* London: MacDonald, 1971.

———. *Haeckel's Monism and the Birth of Fascist Ideology.* New York: Peter Lang, 1998.

Gassert, Phillip, and Daniel S. Mattern, eds. *The Hitler Library: A Bibliography.* Westport, CT: Greenwood Press, 2001.

Gayon, Jean. "Nietzsche and Darwin." In *Biology and the Foundation of Ethics.* Eds. Jane Maienschein and Michael Ruse. Cambridge: Cambridge University Press, 1999. Pp. 154–97.

Gellately, Robert. *Backing Hitler: Consent and Coercion in Nazi Germany.* Oxford: Oxford University Press, 2001.

Goldhagen, Daniel. *Hitler's Willing Executioners: Ordinary Germans and the Holocaust.* New York: Knopf, 1996.

Gollwitzer, Heinz. *Die gelbe Gefahr: Geschichte eines Schlagwortes: Studien zum imperialistischen Denken.* Göttingen: Vandenhoek und Ruprecht, 1962.

Goodrick-Clark, Nicholas. *The Occult Roots of Nazism: Secret Aryan Cults and Their Influence on Nazi Ideology.* New York: New York University Press, 1992.

Greene, John. *Science, Ideology and World View: Essays in the History of Evolutionary Ideas.* Berkeley: University of California Press, 1981.

Gregorio, Mario di. "Entre Méphistophélès et Luther: Ernst Haeckel et la réforme de l'univers." In *Darwinisme et Société.* Ed. Patrick Tort. Paris: Presses universitaires de France, 1992. Pp. 237–83.

————. "Reflections of a Nonpolitical Naturalist: Ernst Haeckel, Wilhelm Bleek, Friedrich Müller and the Meaning of Language," *Journal of the History of Biology* 35 (2002): 79–109.

Gregory, Frederick. *Nature Lost? Natural Science and the German Theological Traditions of the Nineteenth Century.* Cambridge: Harvard University Press, 1992.

————. *Scientific Materialism in Nineteenth Century Germany.* Dordrecht: D. Reidel, 1977.

Greven-Aschoff, Barbara. *Die bürgerliche Frauenbewegung in Deutschland 1894–1933.* Göttingen: Vandenhoeck & Ruprecht, 1981.

Groschopp, Horst. *Dissidenten: Freidenkerei und Kultur in Deutschland.* Berlin: Dietz Verlag, 1997.

Grosse, Pascal. *Kolonialismus, Eugenik und Bürgerliche Gesellschaft in Deutschland, 1850–1918.* Frankfurt: Campus Verlag, 2000.

Grossman, Atina. *Reforming Sex: The German Movement for Birth Control and Abortion Reform, 1920–1950.* Oxford: Oxford University Press, 1995.

Haas, Peter J. *Morality after Auschwitz: The Radical Challenge of the Nazi Ethic.* Philadelphia: Fortress Press, 1988.

Hackett, Amy. "Helene Stöcker: Left-Wing Intellectual and Sex Reformer." In *When Biology became Destiny: Women in Weimar and Nazi Germany.* Ed. Renate Bridenthal et al., New York: Monthly Review Press, 1984. Pp. 109–30.

Hahn, Susanne. " 'Minderwertige, widerstandlose Individuen . . .': Der erste Weltkrieg und das Selbstmordproblem in Deutschland." In *Die Medizin und der Ersten Weltkrieg.* Ed. Wolfgang Eckert et al. Pfaffenweiler: Centaurus, 1996. Pp. 273–97.

Hamann, Brigitte. *Bertha von Suttner: Ein Leben für den Frieden.* Munich: Piper, 1986.

————. *Hitler's Vienna: A Dictator's Apprenticeship.* Trans. Thomas Thornton. New York: Oxford University Press, 1999.

Hamelmann, Gudrun. *Helene Stöcker, der "Bund für Mutterschutz" und die "Neue Generation."* Frankfurt: Haag und Herchen, 1992.

Harrington, Anne. *Reenchanted Science: Holism in German Culture from Wilhelm II to Hitler.* Princeton: Princeton University Press, 1996.

Hawkins, Mike. *Social Darwinism in European and American Thought, 1860–1945: Nature as Model and Nature as Threat.* Cambridge: Cambridge University Press, 1997.

Hecht, Jennifer Michael. "The Solvency of Metaphysics: The Debate over Racial Science and Moral Philosophy in France, 1890–1919." *Isis* 90 (1999): 1–24.

————. "Vacher de Lapouge and the Rise of Nazi Science." *Journal of the History of Ideas* 61 (2000): 285–304.

Henning, Max. *Handbuch der freigeistigen Bewegung Deutschlands, Österreichs, und der Schweiz.* Frankfurt: Neuer Frankfurter Verlag, 1914.

Herf, Jeffrey. *Reactionary Modernism: Technology, Culture, and Politics in Weimar and the Third Reich.* Cambridge: Cambridge University Press, 1984.

Herlitzius, Anette. *Frauenbefreiung und Rassenideologie: Rassenhygiene und Eugenik im politischen Programm der "Radikalen Frauenbewegung" (1900–1933).* Wiesbaden: Deutscher Universitäts–Verlag, 1995.

Holt, Niles. "The Social and Political Ideas of the German Monist Movement, 1871–1914." Dissertation: Yale University, 1967.

Hübinger, Gangolf. "Die monistische Bewegung: Sozialingenieure und Kulturprediger." In *Kultur und Kulturwissenschaften um 1900*. Vol. 2: *Idealismus und Positivismus*. ed. Gangolf Hübinger et al. Stuttgart: Franz Steiner, 1997. pp. 246–59.

Hunter, James M. *Perspective on Ratzel's Political Geography*. Lanham, MD: University Press of America, 1983.

Iggers, Georg. *The German Conception of History: The National Tradition of Historical Thought from Herder to the Present*. Middleton, CT: Wesleyan University Press, 1983.

———. "Historicism: The History and Meaning of the Term." *Journal of the History of Idea* 56 (1995): 129–52.

Jäckel, Eberhard. *Hitler's Weltanschauung: A Blueprint for Power*. Trans. Herbert Arnold. Middleton, CN: Wesleyan University Press, 1972.

Johnston, William M. *The Austrian Mind: An Intellectual and Social History, 1848–1938*. Berkeley: University of California Press, 1972.

Jones, Greta. *Social Darwinism and English Thought: The Interaction between Biological and Social Theory*. Sussex: Harvester Press, 1980.

Katz, Jacob. *From Prejudice to Destruction: Anti-Semitism, 1700–1933*. Cambridge: Harvard University Press, 1980.

Kaufmann, Walter. *Nietzsche: Philosopher, Psychologist, Antichrist*. 4th ed. Princeton: Princeton University Press, 1974.

Kelly, Alfred. *The Descent of Darwin: The Popularization of Darwinism in Germany, 1860–1914*. Chapel Hill: University of North Carolina Press, 1981.

Kemp, N. D. A. *"Merciful Release": The History of the British Euthanasia Movement*. Manchester: Manchester University Press, 2002.

Kempf, Beatrix. *Woman for Peace: The Life of Bertha von Suttner*. Trans. R. W. Last. Park Ridge, NJ: Noyes Press, 1973.

Kershaw, Ian. *Hitler*. 2 vols. New York: Norton, 1998–2000.

Kevles, Daniel J. *In the Name of Eugenics: Genetics and the Uses of Human Heredity*. Berkeley: University of California Press, 1985.

Klee, Ernst. *"Euthanasie" im NS-Staat: Die "Vernichtung lebensunwerten Lebens."* Frankfurt: Fischer Taschenbuch, 1985.

———. *Deutsche Medizin im Dritten Reich: Karrieren vor und nach 1945*. Frankfurt: S. Fischer, 2001.

Kloppenberg, James T. *Uncertain Victory: Social Democracy and Progressivism in European and American Thought, 1870–1920*. Oxford: Oxford University Press, 1986.

Koch, H. W. "Social Darwinism as a Factor in the 'New Imperialism.'" In *The Origins of the First World War: Great Power Rivalry and German War Aims*, ed. H. W. Koch. New York: Taplinger, 1972.

Koch, Hannsjoachim W. *Der Sozialdarwinismus: Seine Genese und sein Einfluss auf das imperialistische Denken*. Munich: D. H. Beck, 1973.

Köhnke, Klaus Christian. *The Rise of Neo-Kantianism: German Academic Philosophy between Idealism and Positivism*. Trans. R. J. Hollingdale. Cambridge: Cambridge University Press, 1991.

Krausse, Erika. *Ernst Haeckel*. 2nd ed. Leipzig: BSB Teubner, 1984.

Kudlien, Fridolf. "Max v. Gruber und die frühe Hitlerbewegung." *Medizinhistorisches Journal* 17 (1982): 373–89.

Kühl, Stefan. *The Nazi Connection: Eugenics, American Racism, and German National Socialism*. Oxford: Oxford University Press, 1994.

Larson, Edward J. and Darrel W. Amundsen. *A Different Death: Euthanasia and the Christian Tradition*. Downers Grove, IL: Intervarsity Press, 1998.

La Vergata, Antonello, "Evolution and War, 1871–1918," *Nuncius* 9 (1994): 143–63.

Lindberg, David and Ronald Numbers, eds. *God and Nature*. Berkeley: University of California Press, 1986.

Lifton, Robert. *The Nazi Doctors: Medical Killing and the Psychology of Genocide*. New York: Basic Books, 1986.

Livingstone, David N. *Darwin's Forgotten Defenders: The Encounter between Evangelical Theology and Evolutionary Thought*. Grand Rapids: Eerdmanns, 1987.

Livingstone, David N., D. G. Hart, and Mark A. Noll, eds. *Evangelicals and Science in Historical Perspective*. New York: Oxford University Press, 1999.

Lösch, Niels. *Rasse als Konstrukt: Leben und Werk Eugen Fischers*. Frankfurt: Lang, 1997.

Löwenberg, Dieter. *Willibald Hentschel (1858–1947): Seine Pläne zur Menschenzüchtung, sein Biologismus und Antisemitismus*. Dissertation: University of Mainz, 1978.

Mandelbaum, Maurice. *History, Man, and Reason: A Study in Nineteenth-Century Thought*. Baltimore: Johns Hopkins University Press, 1974.

Mann, Gunther. *Biologismus im 19. Jahrhundert*. Stuttgart: Ferdinand Enke, 1973.

Marten, Heinz-Georg. *Sozialbiologismus. Biologische Grundpositionen der politischen Ideengeschichte*. Frankfurt: Campus, 1983.

———. "Racism, Social Darwinism, Anti-Semitism, and Aryan Supremacy." *International Journal of the History of Sport* 16, 2 (1999): 23–41.

Massin, Benoit. "From Virchow to Fischer: Physical Anthropology and 'Modern Race Theories' in Wilhelmine Germany." In *Volksgeist as Method and Ethic*. Ed. George W. Stocking. Madison: University of Wisconsin Press, 1996. Pp. 79–154.

Massing, Paul. *Rehearsal for Destruction: A Study of Political Antisemitism in Imperial Germany*. New York: Harper and Brothers, 1949.

Mehnert, Ute. *Deutschland, Amerika und die "gelbe Gefahr": Zur Karriere eines Schlagworts in der grossen Politik, 1905–1917*. Stuttgart: Steiner, 1995.

Meyer, Alfred. *The Feminism and Socialism of Lily Braun*. Bloomington: Indiana University Press, 1985.

Moore, Gregory. *Nietzsche: Biology and Metaphor*. Cambridge: Cambridge University Press, 2002.

Moore, James R. *The Post-Darwinian Controversies: A Study of the Protestant Struggle to Come to Terms with Darwin in Great Britain and America, 1870–1900.* Cambridge: Cambridge University Press, 1979.

Moreland, J. P., and Scott Rae, *Body and Soul: Human Nature and the Crisis in Ethics.* Downers Grove, IL: Intervarsity Press, 2000.

Mosse, George Lachmann. *The Crisis of German Ideology: Intellectual Origins of the Third Reich.* New York: Grosset and Dunlap. 1964.

———. *Towards the Final Solution: A History of European Racism.* New York: Howard Fertig, 1978.

Mühlen, Patrik von zur. *Rassenideologien. Geschichte und Hintergründe.* 2nd ed. Berlin: Dietz, 1979.

Müller, Gerhard H. *Friedrich Ratzel (1844–1904): Naturwissenschaftler, Geograph, Gelehrter.* Stuttgart: Verlag für Geschichte der Naturwissenschaften und der Technik, 1996.

Müller, Joachim. *Sterilisation und Gesetzgebung bis 1933.* Husum: Matthiesen, 1985.

Müller-Hill, Benno. *Murderous Science: Elimination by Scientific Selection of Jews, Gypsies, and Others, Germany, 1933–1945.* Trans. George R. Fraser. Oxford: Oxford University Press, 1988.

Munro, Robert. *From Darwinism to Kaiserism.* Glasgow: James Maclehose & Sons, 1919.

Nasmyth, George. *Social Progress and the Darwinian Theory: A Study of Force as a Factor in Human Relations.* New York: Putnam's Sons, 1916.

Nehamas, Alexander. *Nietzsche: Life as Literature.* Cambridge: Harvard University Press, 1985.

Nicosia, Francis R. and Jonathan Huener, eds. *Medicine and Medical Ethics in Nazi Germany: Origins, Practices, Legacies.* New York: Berghahn Books, 2002.

Nowak, Kurt. *"Euthanasie" und Sterilisierung im "Dritten Reich": Die Konfrontation der evangelischen und katholischen Kirche mit dem "Gesetz zur Verhütung erbkranken Nachwuchese" und der "Euthanasie"-Aktion.* Göttingen: Vandenhoek und Ruprecht, 1977.

Nyhart, Lynn K. *Biology Takes Form: Animal Morphology and the German Universities, 1800–1900.* Chicago: University of Chicago Press, 1995.

Paul, Diane B. *Controlling Human Heredity, 1865 to the Present.* Atlantic Highlands, NJ: Humanities Press, 1995.

Pauley, Bruce F. *From Prejudice to Persecution: A History of Austrian Anti-Semitism.* Chapel Hill: University of North Carolina Press, 1992.

Peukert, Detlev. "The Genesis of the 'Final Solution' from the Spirit of Science." In *Reevaluating the Third Reich.* Ed. Thomas Childers and Jane Caplan. New York: Holmes and Meier, 1993. Pp. 234–52.

Pick, Daniel. *Faces of Degeneration: A European Disorder, c.1848–c.1918.* Cambridge: Cambridge University Press, 1989.

———. *War Machine: The Rationalisation of Slaughter in the Modern Age.* New Haven: Yale University Press, 1993.

Pickhardt, Thomas. "Sozialdarwinismus: Ein panoramabild deutscher bevölkerungskundlicher Fachzeitschriften vor dem Ersten Weltkrieg." *Historische Mitteilungen* 10 (1997): 14–55.

Poliakov, Leon. *The Aryan Myth: A History of Racist and Nationalist Ideas in Europe.* London: Sussex University Press, 1974.

Prinz, Michael and Rainer Zitelmann, eds. *Nationalsozialismus und Modernisierung.* Darmstadt: Wissenschaftliche Buchgesellschaft, 1991.

Proctor, Robert. *Racial Hygiene: Medicine under the Nazis.* Cambridge: Harvard University Press, 1988.

Pross, Christian and Götz Aly, eds. *Der Wert des Menschen: Medizin in Deutschland, 1918–1945.* Berlin: Edition Hentrich, 1989.

Pulzer, Peter. *The Rise of Political Antisemitism in Germany and Austria.* New York: John Wiley and Sons, 1964.

Puschner, Uwe, Walter Schmitz, and Justus H. Ulbricht, eds. *Handbuch zur "Völkischen Bewegung" 1871–1918.* Munich: K. G. Saur, 1996.

Repp, Kevin. " 'More Corporeal, More Concrete': Liberal Humanism, Eugenics, and German Progressives at the Last Fin de Siècle." *Journal of Modern History* 72 (2000): 683–730.

———. *Reformers, Critics, and the Paths of German Modernity, 1890–1914.* Cambridge: Harvard Univesity Press, 2000.

Richards, Robert J. *Darwin and the Emergence of Evolutionary Theories of Mind and Behavior.* Chicago: University of Chicago Press, 1987.

Rissom, Renate. *Fritz Lenz und die Rassenhygiene.* Husum: Matthiesen, 1983.

Ritvo, Lucille B. *Darwin's Influence on Freud: A Tale of Two Sciences.* New Haven: Yale University Press, 1990.

Rolston, Holmes III, *Science and Religion: A Critical Survey.* Philadelphia: Temple University Press, 1987.

Rooy, Piet de. "In Search of Perfection: The Creation of a Missing Link." *In Ape, Man, Apeman: Changing Views since 1600.* Ed. Raymond Corbey and Bert Theunissen. Leiden: Leiden University, 1995. Pp. 195–207.

Rothfels, Nigel. *Savages and Beasts: The Birth of the Modern Zoo.* Baltimore: Johns Hopkins Univeristy Press, 2002.

Rubenstein, Richard. "Modernization and the Politics of Extermination." In *A Mosaic of Victims.* Ed. Michael Berenbaum. New York: New York University Press, 1990. Pp. 3–19.

Rupp-Eisenreich, Britta. "Le darwinisme social en Allemagne." In *Darwinisme et Société.* Ed. Patrick Tort. Paris: Presses universitaires de France, 1992.

Sablik, Karl. *Julius Tandler. Mediziner und Sozialreformer. Eine Biographie.* Vienna: A. Schendl, 1983.

Safranski, Rüdiger. *Nietzsche: A Philosophical Biography.* Trans. Shelley Frisch. New York: Norton, 2002.

Sandmann, Jürgen. *Der Bruch mit der humanitären Tradition. Die Biologisierung der Ethik bei Ernst Haeckel und anderen Darwinisten seiner Zeit.* Stuttgart: Gustav Fischer Verlag, 1990.

Sandmann, Jürgen. "Ernst Haeckels Entwicklungslehre als Teil seiner biologistischen Weltanschauung." In *Die Rezeption von Evolutionstheorien im 19. Jahrhundert.* Ed. Eve-Marie Engels. Frankfurt: Suhrkamp, 1995. Pp. 326–46.

———. "Ansätze einer biologistischen Ethik bei Ernst Haeckel und ihre Auswirkungen auf die Ideologie des Nationalsozialismus." In *Heilen–Verwahren–Vernichten.* Ed. G. Wahl and W. Schmitt. Reichenbach: Kommunikative Medien und Medizin, 1997. Pp. 83–92.

Schmidt, Günter. *Die Literarische Rezeption des Darwinismus: Das Problem der Verebung bei Émile Zola und im Drama des deutschen Naturalismus.* Berlin: Akademie Verlag, 1974.

Schmitt-Egner, Peter. *Kolonialismus und Faschismus. Eine Studie zur historischen und begrifflichen Genesis faschistischer Bewusstseinsformen am deutschen Beispiel.* Giessen: Andreas Achenbach, 1975.

Schmuhl, Hans Walter. *Rassenhygiene, Nationalsozialismus, Euthanasie. Von der Verhütung zur Vernichtung "lebensunwerten Lebens" 1890–1945.* Göttingen: Vandenhoek und Ruprecht, 1987.

Schneider, William H. *Quality and Quantity: The Quest for Biological Regeneration in Twentieth-Century France.* Cambridge: Cambridge University Press, 1990.

Schorske, Carl E. *Fin-de-Siecle Vienna: Politics and Culture.* New York: Vintage, 1981.

Schungel, Wilfried. *Alexander Tille (1866–1912): Leben und Ideen eines Sozialdorwinisten.* Husum: Matthiesen, 1980.

Schwabe, Klaus. *Wissenschaft und Kriegsmoral. Die deutschen Hochschullehrer und die politischen Grundfragen des Ersten Weltkrieges.* Göttingen: Musterschmidt Verlag, 1969.

Schwartz, Michael. *Sozialistische Eugenik: Eugenische Sozialtechnologien in Debatten und Politik der deutschen Sozialdemokratie, 1890–1933.* Bonn: Dietz Nachf., 1995.

———. " 'Euthanasie'—Debatten in Deutschland (1895–1945)." *Vierteljahrshefte für Zeitgeschichte* 46 (1998): 617–65.

Simon-Ritz, Frank. *Die Organisation einer Weltanschauung: Die freigeistige Bewegung im Wilhelminischen Deutschland.* Gütersloh: Christian Kaiser, 1997.

Sluis, I. van der. "The Movement for Euthanasia, 1875–1975." *Janus* 66 (1979): 131–72.

Smith, Helmut Walser. "The Talk of Genocide, the Rhetoric of Miscegenation: Notes on Debates in the German Reichstag concerning Southwest Africa, 1904–14." In *The Imperialist Imagination: German Colonialism and Its Legacy.* Ed. Sara Friedrichsmeyer, Sara Lennox, and Susanne Zantop. Ann Arbor: Univeristy of Michigan Press, 1998. Pp. 107–23.

Smith, Woodruff D. *The German Colonial Empire.* Chapel Hill: University of North Carolina Press, 1978.

———. *The Ideological Origins of Nazi Imperialism.* New York: Oxford University Press, 1986.

Smolka, Wolfgang J. *Völkerkunde in München: Voraussetzungen, Möglichkeiten und Entwicklungslinien ihrer Institutionalisierung, ca. 1850–1933.* Berlin: Duncker & Humboldt, 1994.

Sondhaus, Lawrence. *Franz Conrad von Hötzendorf: Architect of the Apocalypse.* Boston: Humanities Press, 2000.

Spraul, Gunter. "Der 'Völkermord' an den Herero. Untersuchungen zu einer neuen Kontinuitätsthese." *Geschichte in Wissenschaft und Unterricht* 39 (1988): 713–39.

Stackelberg, Roderick. *Idealism Debased: From Völkisch Ideology to National Socialism.* Kent, OH: Kent State University Press, 1981.

Stark, Gary. *Entrepreneurs of Ideology: Neoconservative Publishers in Germany, 1890–1933.* Chapel Hill: University of North Carolina Press, 1981.

Steigmann-Gall, Richard. *The Holy Reich: Nazi Conceptions of Christianity, 1919–1945.* Cambridge: Cambridge University Press, 2003.

Steinberg, Hans-Josef. *Sozialismus und deutsche Sozialdemokratie. Zur Ideologie der Partei vor dem I. Weltkrieg.* Hanover: Verlag für Literatur und Zeitgeschehen, 1967.

Stepan, Nancy. *The Idea of Race in Science: Great Britain, 1800–1960.* Hamden, CN: Archon Books, 1982.

———. " 'Nature's Pruning Hook': War, Race and Evolution, 1914–1918." In *The Political Culture of Modern Britain.* Ed. J. M. W. Bean. London: Hamish Hamilton, 1987. Pp. 129–48.

———. *"The Hour of Eugenics": Race, Gender and Nation in Latin America.* Ithaca: Cornell University Press, 1991.

Stern, Fritz. *The Politics of Cultural Despair: A Study in the Rise of the Germanic Ideology.* Berkeley: University of California Press, 1963.

Stoecker, Helmuth, ed. *German Imperialism in Africa: From the Beginnings until the Second World War.* Trans. Bernd Zöller. London: C. Hurst, 1986.

Stromberg, Roland. *Redemption by War: The Intellectuals and 1914.* Lawrence: Regents Press of Kansas, 1982.

Sulloway, Frank. *Freud: Biologist of the Mind.* New York: Norton, 1979.

Swan, Jon. "The Final Solution in Southwest Africa." *MHQ: The Quarterly Journal of Military History* 3 (1991): 36–55.

Thomann, Klaus-Dieter, and Werner Friedrich Kümmel, "Naturwissenschaft, Kapital und Weltanschauung: Das Kruppsche Preisausschreiben und der Sozialdarwinismus." *Medizinhistorisches Journal* 30 (1995): 99–143, 205–43.

Vogel, C. "Rassenhygiene—Rassenideologie—Sozialdarwinismus: die Wurzeln des Holocaust." In *Dienstbare Medizin. Äerzte betrachten ihr Fach im Nationalsozialismus.* Ed. Hannes Friedrich and Wolfgang Matzow. Göttingen. Vandenhoek und Ruprecht, 1992. Pp. 11–31.

Weber, Matthias. *Ernst Rüdin: Eine kritische Biographie.* Berlin: Springer, 1993.

Wehler, Hans-Ulrich. *Das deutsche Kaiserreich 1871–1918.* 2nd ed. Göttingen: Vandenhoek und Ruprecht, 1975.

Weikart, Richard. "The Origins of Social Darwinism in Germany, 1859–1895." *Journal of the History of Ideas* 54 (1993): 469–88.

―――. "A Recently Discovered Darwin Letter on Social Darwinism." *Isis* 86 (1995): 609–11.

―――. *Socialist Darwinism: Evolution in German Socialist Thought from Marx to Bernstein.* San Francisco: International Scholars Publications, 1999.

―――. "Darwinism and Death: Devaluing Human Life in Germany, 1860–1920." *Journal of the History of Ideas* 63 (2002): 323–44.

―――. " 'Evolutionäre Aufklärung'? Zur Geschichte des Monistenbundes," in *Wissenschaft, Politik, und Öffentlichkeit: Von der Wiener Moderne bis zur Gegenwart.* Ed. Mitchell G. Ash and Christian H. Stifter. Vienna: WUV Universitätsverlag, 2002. Pp. 131–48.

―――. "Progress through Racial Extermination: Social Darwinism, Eugenics, and Pacifism in Germany, 1860–1918." *German Studies Review* 26 (2003): 273–94.

Weindling, Paul. *Health, Race and German Politics between National Unification and Nazism, 1870–1945.* Cambridge: Cambridge University Press, 1989.

Weingart, Peter. "Biology as Social Theory: The Bifurcation of Social Biology and Sociology in Germany, circa 1900." In *Modernist Impulses in the Human Sciences, 1870–1930.* Ed. Dorothy Ross. Baltimore: Johns Hopkins University Press, 1994.

Weingart, Peter, Jürgen Kroll, and Kurt Bayertz. *Rasse, Blut, und Gene. Geschichte der Eugenik und Rassenhygiene in Deutschland.* Frankfurt: Suhrkamp, 1988.

Weiss, Sheila Faith. *Race Hygiene and National Efficiency: The Eugenics of Wilhelm Schallmayer.* Berkeley: University of California Press, 1987.

―――. "The Race Hygiene Movement in Germany." *Osiris,* 2nd series 3 (1987): 193–236.

Welträtsel und Lebenswunder: Ernst Haeckel—Werk, Wirkung, und Folgen. Ed. Erna Aescht et al. Linz: Oberösterreichischen Landesmuseums, 1998.

Wetzell, Richard F. *Inventing the Criminal: A History of German Criminology, 1880–1945.* Chapel Hill: University of North Carolina Press, 2000.

Whiteside, Andrew G. *The Socialism of Fools: Goerg Ritter von Schönerer and Austrian Pan-Germanism.* Berkeley: University of California Press, 1975.

Wickert, Christl. *Helene Stöcker, 1869–1943: Frauenrechtlerin, Sexualreformerin und Pazifistin: Eine Biographie.* Bonn: Dietz, 1991.

Williams, C. M. *A Review of the Systems of Ethics founded on the Theory of Evolution.* London: Macmillan, 1893.

Winau, Rolf. "Erst Haeckels Vorstellungen von Wert und Werden menschlicher Rassen und Kulturen." *Medizinhistorisches Journal* 16 (1981): 270–9.

Wittkau-Horgby, Annette. *Materialismus: Entstehung und Wirkung in den Wissenschaften des 19. Jahrhunderts.* Göttingen: Vandenhoek und Ruprecht, 1998.

Young, Robert. "Darwinism *Is* Social." In *The Darwinian Heritage.* Ed. David Kohn. Princeton: Princeton University Press, 1985. Pp. 609–38.

Zimmerman, Andrew. *Anthropology and Antihumanism in Imperial Germany.* Chicago: University of Chicago Press, 2001.

Zitelmann, Rainer. *Hitler: Selbstverständnis eines Revolutionärs*. Hamburg: Berg, 1987.

Zmarzlik, Hans-Günter. "Social Darwinism in Germany, Seen as a Historical Problem." In *Republic to Reich: The Making of the Nazi Revolution*. Ed. Hajo Holborn. Trans. Ralph Manheim. New York: Pantheon, 1972. Pp. 435–74.

Index

abortion   2, 6, 75, 98, 145–49, 152,
    153, 156–60, 210, 213
Adler, Felix   59
adultery   131
alcohol (see temperance)
Amira, Carl von   33
Ammon, Otto   79, 93, 120–22, 197
anthropology   5–6, 12, 17, 41, 52,
    54, 63, 79, 95, 104, 110,
    113–16, 120–21, 126, 168, 171,
    190, 196, 197, 223, 225, 226
anti-Semitism   3, 4, 6, 55, 69, 111,
    112, 117, 122–24, 195, 204–05,
    217–25 (see also Jews)
Archery Club   200, 221
Ariosophy (see Lanz von Liebenfels
    and List)
Aschheim, Steven   4, 46
atheism   12
Auerbach, Hellmuth   221
Auschwitz   4–5, 6, 227

Bastian, Adolf   104, 113–15
Bäumer, Gertrud   97
Baur, Erwin   96, 121, 122,
    221–23
Bavarian Academy of Science   64
Bebel, August   171
Becker, Peter Emil   219
Bentham, Jeremy   16, 30
Benzenhöfer, Udo   75, 146
Bernhardi, Friedrich von   174
bestiality   131
Binding, Karl   155–56
Binswanger, Otto   40

birth control   4, 97, 133–36,
    138, 140
Bismarck, Otto von   165, 167, 216
Blaschko, Alfred   93
Bleuler, Eugen   40, 138
Bluhm, Agnes   86, 155, 158, 159
Bock, Gisela   209
Bolin, Wilhelm   33
Bölsche, Wilhelm   86, 178
Börner, Wilhelm   151
Braun, Lily   62, 93–94, 157
Braune, A.   151
Bré, Ruth   97
Britain   10, 23–24, 135, 146
Bryan, William Jennings   1,
    163–64
Büchner, Ludwig   12–14, 36, 37,
    80–81, 91–92, 94, 111, 117,
    178, 191
Buckle, Thomas Henry   178
Bülow, Bernhard von   206
Bunge, Gustav von   56, 132
Burleigh, Michael   6–7, 224
Butler, Joseph   23
Byr, Robert   189

capital punishment   146, 149, 155,
    159–60
Carneri, Bartholomäus von   26–28,
    30, 31, 33–35, 43, 45, 63,
    110–11, 149, 178, 204
Center Party   206
Chamberlain, Houston Stewart   121,
    122, 124, 125, 195, 219–22
Christaller, Erdmann Gottreich   48

Christianity   12, 16, 21, 24–25, 33,
    51, 52, 61, 62, 66–68, 75, 76,
    81, 92, 94, 103, 104, 145, 148,
    183, 185, 203, 206, 219, 231,
    232 (*see also* religion and Judeo-
    Christian ethics)
Class, Heinrich   122
Conrad von Hötzendorf, Franz
    173–74
criminal anthropology   37–41, 115,
    137, 159
Cronau, Rudolf   183

Dahn, Felix   117
Daim, Wilfried   217
Dante   216
Darwin, Charles   1, 3, 4, 6–11,
    15–17, 21–24, 26–28, 31,
    33–35, 38, 44, 55, 61, 73, 74,
    76–78, 81, 89, 95, 104–05, 108,
    110, 117, 124, 166, 174, 178,
    179, 185, 188, 189, 191, 212
Darwinism, passim
David, Eduard   83, 94, 130, 140, 157
Deak, Istvan   173
death   3, 9, 17, 73–75, 80–83, 86,
    92, 100, 124, 129, 145, 147,
    154, 156, 160, 164, 230
degeneration   3, 48–49, 52–56,
    68–69, 83–86, 98, 120, 139,
    141, 142, 157, 158, 164,
    174–78, 181, 196, 205, 211, 217
Dennett, Daniel   2
Desmond, Adrian   22, 74
determinism   13–14, 16, 23–26, 33,
    35–36, 41, 46, 47, 50, 51, 68,
    76–77, 89, 114, 115, 117, 123,
    171, 172, 220
Deutschlands Erneuerung   121,
    221–24
devaluing human life (*see* sanctity
    of life)
Dikötter, Frank   232
Dilthey, Wilhelm   13, 42, 53

disabled   3, 10, 17, 48, 49, 89,
    95–103, 126, 136, 137, 140,
    145–61, 230–32
divorce   144
Dodel, Arnold   28, 43, 78, 83, 93,
    151–52, 178, 181
Dowbiggin, Ian   10
Drexler, Anton   221
Driesmans, Heinrich   87, 119,
    122, 199
dualism   12, 13, 16, 25, 26, 33, 35,
    66, 76–77, 103, 147, 148
duel   160
Durkheim, Emile   59

Eckart, Dietrich   221
Ecker, Alexander   120, 168
Eckstein, Anna   176
Egidy, Moritz von   62
Ehinger, Otto   157
Ehrenfels, Christian von   43, 53–54,
    93, 132, 141–44, 150, 202
Engels, Friedrich   4, 93
Enlightenment   21, 23, 32, 33, 41,
    59, 103–04, 123
equality   4, 7, 10, 17, 34, 47, 48,
    89–133, 171, 172, 178, 183,
    189–91, 199, 203, 223, 225, 230
Ethical Culture Society (*see* Society for
    Ethical Culture)
ethics   6–9, 12–70, 98, 101, 115, 121,
    130, 132, 154, 179, 189, 209–27,
    229 (*see also* evolutionary ethics,
    morality, and sexual morality)
eugenics, passim
euthanasia   2, 3, 9, 10, 49, 70,
    75–76, 145–56, 160, 214, 217,
    225–27, 232
Evans, Richard J.   4, 232–33
evil, problem of   73
evolutionary ethics   1–6, 8, 11,
    14–16, 19–70, 116, 130, 146,
    210, 225, 230 (*see also* ethics)
evolutionary psychology   129, 231

Fatherland Party   174, 176
Feder, Gottfried   221
feminism   4–6, 46, 70, 93, 97,
    133–36, 138, 158, 201
Feuerbach, Ludwig   33
Fick, Adolf   122
Fick, Heinrich   176
Fischer, Eugen   120–22, 197–99,
    221–23, 226
Fischer, Klaus   232
Flechsig, Paul   63
Foerster, Friedrich Wilhelm   61
Foerster, Wilhelm   60–62
Forel, August   35–37, 40, 56, 67, 77,
    82, 86, 93, 101, 130–32, 136,
    138, 140, 143, 153–54, 201
France   56
Franco-Prussian War   168
Frank, Hans   220–21
Frederick the Great   165
free will (*see* determinism)
Freud, Sigmund   129
Fried, Alfred   175, 178, 180–81
Fritsch, Gustav   115
Fritsch, Theodor   55–56, 69, 123,
    142–43, 204–05, 224–25
Fürth, Henriette   97, 135, 157

Galton, Francis   15, 36, 48, 98
Gasman, Daniel   70, 216–17
Gayon, Jean   46
genocide   6, 74, 110, 215, 232–33
    (*see also* racial extermination and
    Holocaust)
Gerhard, Ute   5
Gerkan, Roland   150–51
German Anthropological Society   110
German Peace Society   199
German Renewal Community   69
Gizycki, Georg von   30–32, 43,
    60–63, 81, 96
Gobineau, Artur de   108, 119, 121,
    122, 192, 200, 219
Gobineau Society   119, 122

Goethe   108
Golden Rule   21, 24
Goldhagen, Daniel   204
Goldscheid, Rudolf   101–02
Goldschmidt, Richard   11
Goldstein, Kurt   96
Gollwitzer, Heinz   185, 205
Goodricke-Clarke, Nicholas   219
Gottstein, A.   99
Graham, W.   186
Grant, Madison   10
Greven-Aschoff, Barbara   5
Grossmann, Atina   4
Grotjahn, Alfred   12, 81, 93, 94,
    96–97, 119, 135, 136, 149, 180
Gruber, Max von   74, 81, 93, 99,
    130, 134–36, 144, 174, 199,
    221, 222
Gumplowicz, Ludwig   149, 192,
    194, 195
Günther, Hans F. K.   122, 221–24

Haeckel, Ernst   9, 11–13, 15, 24–29,
    33, 34, 36, 37, 43, 50, 52,
    60–63, 65–67, 70, 76–78, 80,
    81, 83–84, 86, 89–93, 105–15,
    117, 122–23, 132, 139, 146–49,
    151, 153, 156–57, 159–61,
    164–65, 168, 169, 175–78,
    186–87, 189, 191, 193, 215–17,
    219, 231
Hagen, Adolf (*see* Adolf Harpf)
Hague Peace Conference   180, 198
Hamann, Brigitte   8, 219
Harpf, Adolf (pseudonym of Adolf
    Hagen)   204–5, 218
Harrer, Karl   221
Hauptmann, Carl   117–18
Hauptmann, Gerhard   86, 117,
    144, 200
Haushofer, Karl   225
Hecht, Jennifer Michael   197
Hegar, Alfred   132, 140, 155
Hegelianism   26–27, 42

Hellwald, Friedrich   34, 81, 91,
    111–13, 151, 168–70, 183,
    188–89
Hentig, Hans von   79, 150
Hentschel, Willibald   43, 122–23,
    141–43, 160, 199
Herder, Johann Gottfried von
    104, 216
Hereditary Health Courts   225–26
Herero Revolt   184, 205–06
Hess, Rudolf   222
Hesse, Albert   65
Hirsch, Max   138, 144, 159
historicism   42, 59, 229
Hitler, Adolf   3–4, 6–10, 17, 70, 122,
    184, 206, 209–27, 232–33
Hoche, Alfred   155–57
Hoffmann, Geza von   138
Holocaust   3–4, 5, 6, 232
homosexuality   4, 70, 130, 131, 210,
    213, 217
Hull, David   21
human rights   26, 33–35, 75, 116,
    145, 146, 152, 153, 160, 164,
    173, 176, 177, 179, 189–90,
    211, 214
Hume, David   23, 33
Hutcheson, Francis   23
Huxley, Thomas H.   24, 74, 76

Ice Age   119
inequality (*see* equality)
infanticide   2, 3, 49, 75, 95,
    145–55, 157, 159, 214–15,
    223, 232
instincts   21–22, 24–25, 27, 30,
    35–36, 48, 53, 55, 65, 75, 119,
    124, 129, 141, 152–53, 166,
    172, 178, 186, 189, 201, 229
International Order for Ethics and
    Culture   67

Jäckel, Eberhard   209, 220
Jaeger, Gustav   167–68

Jens, Ludwig   99
Jews   6, 8, 9, 111, 112, 117, 118,
    124, 184, 204, 213, 220,
    222–24, 226 (*see also* anti-
    Semitism)
Jodl, Friedrich   28, 30, 32–34, 43,
    54, 62, 63, 66–67
Jordan, David Starr   119, 180
Jost, Adolf   153
Judeo-Christian ethics   7, 21–22,
    24–25, 27, 32, 41, 43, 45, 46,
    48, 49, 51–55, 59, 64, 66, 73,
    75, 77, 103–4, 130, 132, 144,
    145, 149, 154, 158, 210–11,
    213–15, 218, 229, 230 (*see also*
    Christianity)

Kaiser Wilhelm Institute   225–26
Kammerer, Paul von   102
Kant, Immanuel   16, 21, 31,
    108, 119
Kantian ethics   27, 30, 41, 43, 61,
    66, 77, 210–11, 224, 229
Kaup, Ignaz   99, 137
Kautsky, Karl   93–94, 137
Kautsky, Karl jr.   140
Kellogg, Vernon   163–64
Kemp, Nick   10
Kershaw, Ian   221
Kirchhoff, Alfred   171, 184
Klaatsch, Hermann   115
Kloppenberg, James   42
Koch, Julius   98
Koehler, Wolfgang   64
Kossmann, Robby   2, 78, 81
Kraepelin, Emil   40, 77, 85, 137
Krause, Ernst (pseudonym, Carus
    Sterne)   111, 178, 191–92
Krupp, Friedrich   15
Krupp Prize   15, 35, 65, 119–20,
    136, 154
Kubizek, August   216
Kühl, Stefan   10
Kurella, Hans   38–40, 85, 99–100

Lagarde, Paul 221
laissez faire 3
Lamarck, Jean-Baptiste 108
Lamarckism 9, 31, 36–37, 47, 111
La Mettrie, Julien de 231
Langbehn, Julius 123
Lange, Friedrich Albert 11
Lanz von Liebenfels, Jörg 199, 204, 217–19
Lapouge, Georges Vacher de 89, 120–22, 196–97, 219
League for the Protection of Mothers 68, 70, 97–98, 132, 134, 135, 139–40, 157, 201
League of German Women's Organizations 158
Lebensraum 192
Lehmann, Julius Friedrich 122, 221–24
Lenz, Fritz 7, 8, 121–22, 143, 200, 221–23, 226
liberalism 9, 26, 37, 75, 82, 89, 91, 100, 104, 111, 113, 114, 145, 216
List, Guido von 199, 217, 219
Locke, John 75, 104, 145
Lombroso, Cesare 37–40, 85, 159
Luschan, Felix von 41, 54, 95, 116, 135, 137, 138, 144, 171, 198, 200
Lyell, Charles 31

Mackintosh, James 23
Malthus, Thomas Robert 17, 74, 185
Marcuse, Max 139
marriage 41, 48, 93, 106, 130–33, 137–45, 155, 213, 224, 230
Marx, Karl 4, 42, 119
Marxism 4, 42, 119, 222–23
Massin, Benoit 116
materialism 12–14, 16, 33, 41, 46, 62, 76, 90, 114, 116, 120, 123, 124, 145, 157, 178, 203, 231

Max, Gabriel 108
Mehnert, Ute 205
Mengele, Josef 227
militarism 1, 3, 6, 9, 22, 28, 126, 149, 161, 163–81, 189, 198–200
Moens, Bernelot 110
Moleschott, Jakob 14
monism 12–14, 24–26, 44, 51, 60, 62, 63, 66, 77, 114, 132, 145, 146, 179
Monist League 65–68, 70, 98, 100, 101, 132, 139, 140, 150–51, 175–76
Moore, G. E. 13
Moore, Gregory 46
Moore, James 22
morality 6, 13, 17, 105, 115, 124, 132, 143, 149, 169, 171–73, 178, 179, 183–84, 186, 189–90, 218, 222, 224, 227, 229, 231, 233 (*see also* ethics)
moral insanity 40
moral relativism (*see* relativism, ethical)
More, Sir Thomas 75
Morel, Augustin 85
Much, Matthäus 197
Müller-Lyer, F. 139, 144
Munk, Hermann 63–64

Näcke, Paul 137–138
Nägeli, Karl 47
National Socialism (*see* Nazism)
nationalism 46
natural selection 9, 11, 16, 17, 22, 25, 27, 29, 31, 44, 51, 65, 73, 75, 79, 81, 84–86, 90, 101, 145, 163, 172, 185, 193, 195, 218
naturalism 12–15
naturalistic fallacy 13
Nazism 4–7, 9, 10, 70, 105, 121, 144, 209, 216, 221–22, 226, 227
Neanderthal 12, 115–116
Nicolai, G. F. 202

Nietzsche, Friedrich   7, 45–49, 53, 59, 68, 82, 201, 211, 216
Nietzscheanism   4, 45, 55, 87, 94, 132, 133, 176, 211, 222
Nordau, Max   29–30, 92, 141
Nordenholz, Anastasius   52
Nordic Ring   118
Nuremberg Laws   226

Olberg, Oda   153, 157
Ostwald, Wilhelm   67, 70, 151, 176

pacifism   1, 4, 9, 70, 163–65, 170, 172–81, 185, 187, 191, 199–203, 216–17
Pan-German League   122, 187, 193, 197, 221
pantheism   12, 26, 62
Pappritz, Anna   158–59
Penzig, Rudolph   61
Peschel, Oscar   8, 111–13, 167, 187–89
Peters, Karl   195
Plate, Ludwig   15
Ploetz, Alfred   15, 40, 50–52, 56, 68, 70, 82, 85, 86, 92, 93, 97–98, 100, 117–21, 132–36, 140, 144, 152–53, 158, 177–78, 180, 200, 221, 226
Poland   114
polygamy   54, 131, 138, 140–44, 202, 230
population policy   5, 81, 135–36, 157, 170, 176, 185–86, 199, 215, 226
positivism   12, 16, 23, 26, 33, 46, 47, 123, 203, 231
Potthoff, Heinz   100–01
Preyer, Wilhelm   10, 28, 80–81, 91
Prussian Academy of Science   64
psychiatry   37–41, 52, 56, 69, 77, 85–86, 96, 98, 131, 137, 150, 155–56, 225

Rachels, James   2, 232
race hygiene   5, 51, 118, 121, 134
racial extermination   6, 8, 110, 126, 165, 172, 175, 177, 179–81, 183–206, 215, 216, 218, 232
racism   3–10, 17, 70, 82, 87, 95, 97, 102–26, 135, 136, 161, 165, 168, 173, 181, 183–206, 217–25, 230–31
Ranke, Johannes   104, 114–15
Ratzel, Friedrich   112–14, 192–94, 225
Ratzenhofer, Gustav   44–45, 66, 173, 194–95
recapitulation   11, 147, 152, 157
Rée, Paul   47
Reich, Emil   93
Reimer, Josef   219
relativism, ethical   16, 21–44, 54, 64, 67, 95, 170, 211, 224, 229
religion   11, 12–14, 16, 17, 21–23, 26, 28, 31, 33, 34, 51, 60–62, 66, 76–78, 90, 101, 138, 144, 153, 168, 205, 210, 218, 219 (*see also* Christianity and Judeo-Christian ethic)
Repp, Kevin   5
Ribbert, Hugo   54–55, 96, 135–38, 140, 149
right to life (*see* sanctity of life)
rights (*see* human rights)
Rohleder, H.   110
Rolle, Friedrich   165, 189–91, 193
Rolph, William   47–48
Roosevelt, Theodore   205
Rosenberg, Alfred   222
Roux, Wilhelm   47
Royer, Clemence   89
Rüdin, Ernst   86, 120, 134, 135, 138, 159, 221, 225–26
Ruppin, Arthur   65, 136–37

Safranski, Rüdiger   46, 49
Samson, Albert   63–65

sanctity of life   1–3, 6, 8–10, 16–17,
    49, 75–83, 86, 98–102, 108–09,
    130, 145–47, 151, 153, 155, 158,
    160, 164, 175, 181, 205, 230–32
Sanger, Margaret   135
Scandinavia   10, 114
Schaffhausen, Hermann   12
Schäffle, Albert E. F.   28–29, 170
Schallmayer, Wilhelm   6, 15, 35, 40,
    50–51, 65, 82–83, 85–86, 93,
    101, 118–19, 121, 132, 140,
    143, 150, 154, 177, 200–1
Schemann, Ludwig   119, 122,
    200, 221
Schiller, Friedrich   216
Schlieffen, General von   206
Schmid, Rudolf   1–2
Schmidt, Oscar   28, 91, 190–91
Schmidt-Gibichenfels, Otto   173
Schmuhl, Hans-Walter   146
Schneider, Georg Heinrich   48
Schönerer, Georg von   219–20
Schopenhauer, Arthur   216
Schreiber, Adele   97–98, 135, 157
Schwalbe, Gustav   115
Schwartz, Michael   146
Sedgwick, Adam   1, 2, 8
Seeck, Otto   195
Semon, Richard   37, 147
sexual morality   54, 56, 79, 129–44
sexual reform   4, 41, 56, 129–45, 158
sexual selection   129, 131, 139, 142
Shaftesbury, Earl of   23
Shirer, William   220
Siebert, Friedrich   151
Singer, Peter   2, 232
Slavs   111, 119
social Darwinism, passim
Social Democrats   4, 62, 83, 94, 97,
    135, 164, 171, 206
socialism   9, 32, 35, 46, 52, 62, 63,
    80, 90–94, 101, 133, 153, 171,
    178, 201, 203, 217 (*see also* Social
    Democrats)

Society for Ethical Culture (Germany)
    32, 59–63, 67
Society for Race Hygiene   15, 56, 68,
    70, 82, 92–93, 102, 118, 120,
    121, 132, 134, 139, 140, 144,
    221
Society of German Scientists and
    Physicians   54, 90, 91, 95, 171
sociobiology   129, 231
Sombart, Werner   68
Sommer, Robert   38–40
soul (*see* dualism)
Southwest Africa   121
Spencer, Herbert   23, 38, 48, 178
Stackelberg, Roderick   220
Stark, Gary   221
Steinmetz, Sebald   171–72, 198–99
Stephen, Leslie   23
sterilization   10, 137–38, 159, 160,
    225–26, 232
Sterne, Carus (*see* Krause, Ernst)
Stöcker, Helene   68, 97–98, 132–35,
    138, 144, 156, 157, 158, 201–02
Stoddard, Lothrop   222
Stopes, Marie   135
Strauss, David Friedrich   10–11, 33,
    46, 61, 168
struggle for existence, passim (*see also*
    natural selection)
suicide   49, 75, 145–46, 148, 149,
    150, 160, 172
Sumner, William Graham   165–66
Suttner, Bertha von   175, 176, 178–81

Tandler, Julius   99, 102, 140, 159
teleology   5, 26, 31, 73
temperance   56, 68–69
Ten Commandments   33
Thayer, William Roscoe   163
Thule Society   221–22
Tille, Alexander   34–35, 45–46, 49,
    51, 68, 82, 94, 132, 176–77
Tönnies, Ferdinand   59
Trotha, General   205–06

Umfrid, Otto   199
United States   10, 24, 75, 114, 135, 138, 146, 163, 222, 232
Universal Race Congress   198, 200
Unold, Johannes   66, 82
utilitarianism   16, 27, 30–32, 50, 51, 53, 66, 96

value of human life (*see* sanctity of life)
Verschuer, Otmar von   226, 227
Verworn, Max   11
Vienna Sociological Society   102
Virchow, Rudolf   90–91, 104, 113–15
Vogt, Karl   14, 95, 110, 115, 117, 124
Voltaire   104

Wagner, Klaus   171–72, 198–99
Wagner, Moritz   119, 193
Wagnerism   4
Waldeyer, Wilhelm   63–65, 116

war (*see* militarism or World War I)
Weinberg, Siegfried   156–57
Weismann, August   15, 36–37, 61, 82–85
Weiss, Sheila Faith   6
Wells, H. G.   185
Wilhelm II   184, 205–06
Wilser, Ludwig   120–22, 219
Wilson, Edward O.   129
Wippermann, Wolfgang   6, 224
Wolfsdorf, Eugen   151
Woltmann, Ludwig   11, 93, 97, 105, 119–22, 132, 196–99, 219
World War I   1, 150, 163–65, 173–74, 176, 177, 180, 187, 201, 202

Ziegler, Heinrich Ernst   15, 63, 65, 67, 91, 170–71
Zimmerman, Andrew   5–6
Zmarzlik, Hans-Günther   4